# of OREGON

Second Edition

MARLI B. MILLER

PHOTOGRAPHS BY THE AUTHOR

2014
Mountain Press Publishing Company
Missoula, Montana

Seventh Printing, February 2023

Geological maps for the road guides are based on the *Geologic Map of Oregon* by G. W. Walker and N. S. MacLeod, published in 1991

Library of Congress Cataloging-in-Publication Data

Miller, Marli Bryant, 1960-
Roadside geology of Oregon / Marli B. Miller ; photographs by Marli B. Miller. — Second edition.
pages cm. — (Roadside geology series)
Previous edition by David D. Alt.
ISBN 978-0-87842-631-7 (pbk. : alk. paper)
1. Geology—Oregon—Guidebooks. I. Title.
QE155.A47 2014
557.95—dc23

2014025206

Printed in Hong Kong by Mantec Production Company

P.O. Box 2399 • Missoula, MT 59806 • 406-728-1900
800-234-5308 • info@mtnpress.com
www.mountain-press.com

For my daughters, Lindsay and Megan.

You make the world a better place.

And to their grandparents, Katie, Lloyd, Audrey, and Alan.

Thank you.

Acknowledgments

Writing this book has felt like coming home. I've always loved Oregon, but this process of pulling together its geology into a single handbook taught me more about this state than I could ever imagine. For one thing, I discovered the incredible wealth of published research that details the geology of Oregon. I benefited especially from the excellent work of geologists at the Oregon Department of Geology and Mineral Industries (DOGAMI) and the US Geological Survey (USGS). I used the USGS geologic map of Oregon by Walker and MacLeod (1991) as the base for drafting the many geologic road maps in this book. I used many other more detailed maps and reports as supplements where needed.

But no matter how much I enjoyed becoming more and more intimate with Oregon's geology, the best part by far were my interactions with the many people who helped me complete this project.

First and foremost, I want to thank my editor, Jennifer Carey, who was open-minded and supportive from the start of this project, was flexible to my need for additional time, and applied her editing and geologic expertise to my all-too-long manuscript. Every step along the way (and there were plenty!), she greatly improved the project. Being from Oregon herself, Jenn lent a perspective and asked innumerable questions that broadened this book's scope and made it far more interesting and informative.

I'm blessed with wonderful colleagues at the University of Oregon, *all* of whom expressed enthusiasm and support for this project. Most notably, Greg Retallack and Ray Weldon spent hours with me, helping me understand certain complexities of the geology or describing localities. I also benefited immensely from discussions with Ilya Bindeman, Dave Blackwell, Kathy Cashman, Edward Davis, Natalia Deligne, Becky Dorsey, Jon Erlandson, Ted Fremd, Samantha Hopkins, Gene Humphreys, Allan Kays, Win McLaughlin, Mark Reed, Josh Roering, Craig Tozer, Paul Wallace, Jim Watkins, and Lili Weldon. Colleagues at other institutions greatly helped me in discussions as well. These folks include Ellen Bishop (Whitman College), Jody Bourgeois (University of Washington), Roger Brandt (formerly of the National Park Service), Darrel Cowan (University of Washington), Todd LaMaskin (University of North Carolina, Wilmington), Danielle McKay (Oregon State University, Bend), Barb Nash (University of Utah), and Ray Wells (USGS).

I benefited greatly from thoughtful reviews of different parts of this book by Charlie Bacon (USGS), Jody Bourgeois, Roger Brandt, Scott Burns (Portland State University), Darrel Cowan, Ted Fremd, Anita Grunder (Oregon State University), Allan Kays, Todd LaMaskin, Danielle McKay, Andrew Meigs

(Oregon State University), Pat Pringle (Centralia College), Willie Scott (USGS), Martin Streck (Portland State University), Bob Walter, and Lili Weldon. Of course, the mistakes in this book are my own.

My friends Lorna Baldwin, Sammy Castonguay, Steve Downey, Jeffrey Freeman, Birgitta Jansen, Donna Rose, and Craig Tozer—and my daughters Lindsay and Megan—accompanied me on numerous road trips. Birgitta and I logged more than 5,000 miles together—and Lorna and Jeff each put in well over 1,000 with me. Jeff's expertise in geology helped immensely when we were together on the road. In addition, Lindsay drafted several of the road maps for southern Oregon. I should also apologize to my fellow Oregon drivers, who sometimes had to put up with my sudden stops on the shoulders of busy highways! Chas Rogers and Dan Tyler took me for some photo trips in their single engine airplanes. Joe, Alyse, Tyler, and Emily Gass; Craig Tozer and Faye Ameredes; and Doug Norseth and Bruce Hegna opened their homes to me while I was traveling.

And a special thanks goes to Robert Thomas of University of Montana, Western, who helped talk Mountain Press into giving me this contract, and who helped talk me into taking it.

And to the many other people who helped in myriad other ways, thank you all!

Simplified geologic map of Oregon and legend. –Modified from Walker and MaLeod, 1991

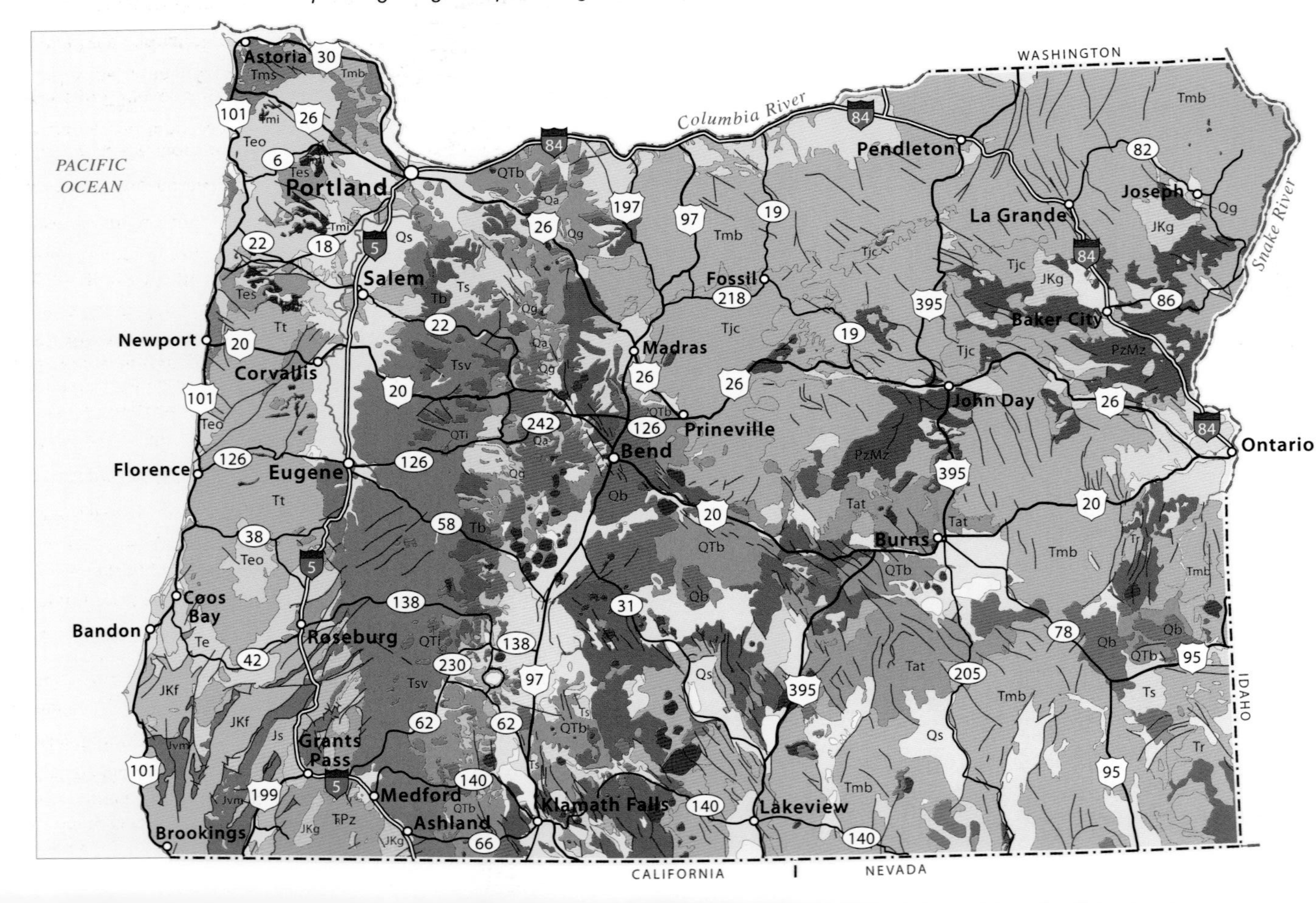

QUATERNARY

- Qs — sediment; includes alluvium, colluvium, landslide, coastal dune, and terrace deposits
- Qm — Holocene Mazama Ash; includes ash-fall as well as ash-flow deposits
- Qg — glacial deposits
- Qa — andesite
- Qb — basalt and basaltic andesite; forms some stratovolcanoes in the Cascades and large basalt flows in eastern and central Oregon

TERTIARY

- QTb — Quaternary, Pliocene, and some Late Miocene basalt flows
- Ts — Miocene, Pliocene, and some Quaternary sedimentary rocks, dominantly tuffaceous and contains tuff and volcanic breccia; grades into Tb in places
- QTsv — Quaternary and Late Tertiary silicic vent complexes (erupted rhyolite)
- QTmv — Quaternary and Late Tertiary mafic vent complexes (erupted basalt)
- Tb — Middle to Late Miocene and some Pliocene basalt
- Tmb — Middle Miocene basalt; predominantly the Columbia River Basalt Group, including the Saddle Mountains, Wanapum, Grande Ronde, Picture Gorge, and Imnaha Basalts, as well as the basalt of Steens Mountain and other undifferentiated Early Miocene basalts; also includes minor intervening tuffaceous sedimentary rocks, lakebed deposits, and rhyolitic flows
- Tms — Early to Middle Miocene marine sedimentary rock of the Coast Range
- Tsv — Eocene through Early Miocene sedimentary and volcanic rock of the Western Cascades; includes the marine Eugene Formation, and the nonmarine Fisher Formation of the southern Willamette Valley
- Tat — Miocene ash-flow tuff
- Tr — Eocene to Miocene rhyolite and rhyolitic tuff
- Tjc — Eocene to Early Miocene sedimentary and volcanic rocks of the Clarno and John Day Formations and related rocks
- Teo — Eocene to Oligocene marine volcanic-rich siltstone and sandstone and volcanic rocks; includes the Yamhill, Cowlitz, and Alsea Formations as well as the Tillamook Volcanics
- Tt — Eocene Tyee Formation of the Coast Range
- Te — Eocene marine sandstone and siltstone that predates Tyee Formation

ROCKS OF ACCRETED TERRANE

- Tes — Paleocene to Eocene basaltic and related rocks of the Siletz River Volcanics (Siletzia)
- JKf — Jurassic and Cretaceous sedimentary rock of the Franciscan Complex, found in the Klamath Mountains
- Js — Jurassic sedimentary rock of the Klamath Mountains that does not include the Franciscan Complex
- Jvm — Jurassic volcanic, ultramafic, and metamorphic rock of the Klamath Mountains
- TRPz — Paleozoic and Triassic rock of the Klamath Mountains
- PzMz — Paleozoic and Mesozoic rock of eastern Oregon

INTRUSIVE ROCK

- QTi — Quaternary and Tertiary silicic and mafic intrusions of the Cascades
- Tmi — Tertiary, mostly Miocene mafic intrusions of the Coast Range
- JKg — Jurassic and Cretaceous granitic intrusions of the Klamath and Blue Mountains

faults

N

0 25 50 100 miles

0 50 100 kilometers

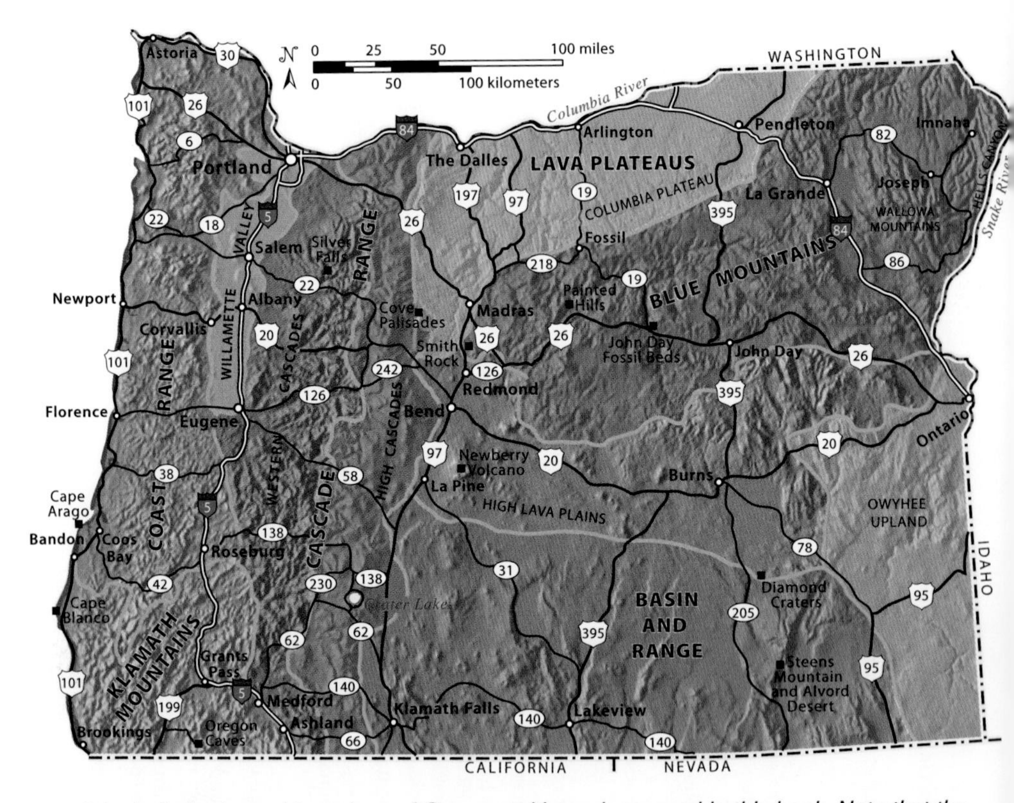

Principal physiographic regions of Oregon, with roads covered in this book. Note that the Coast Range, Cascade Range, and Lava Plateaus are subdivided into subregions. The map also shows some sites highlighted in the text. –Physiography from Thelin and Pike, 1991

Contents

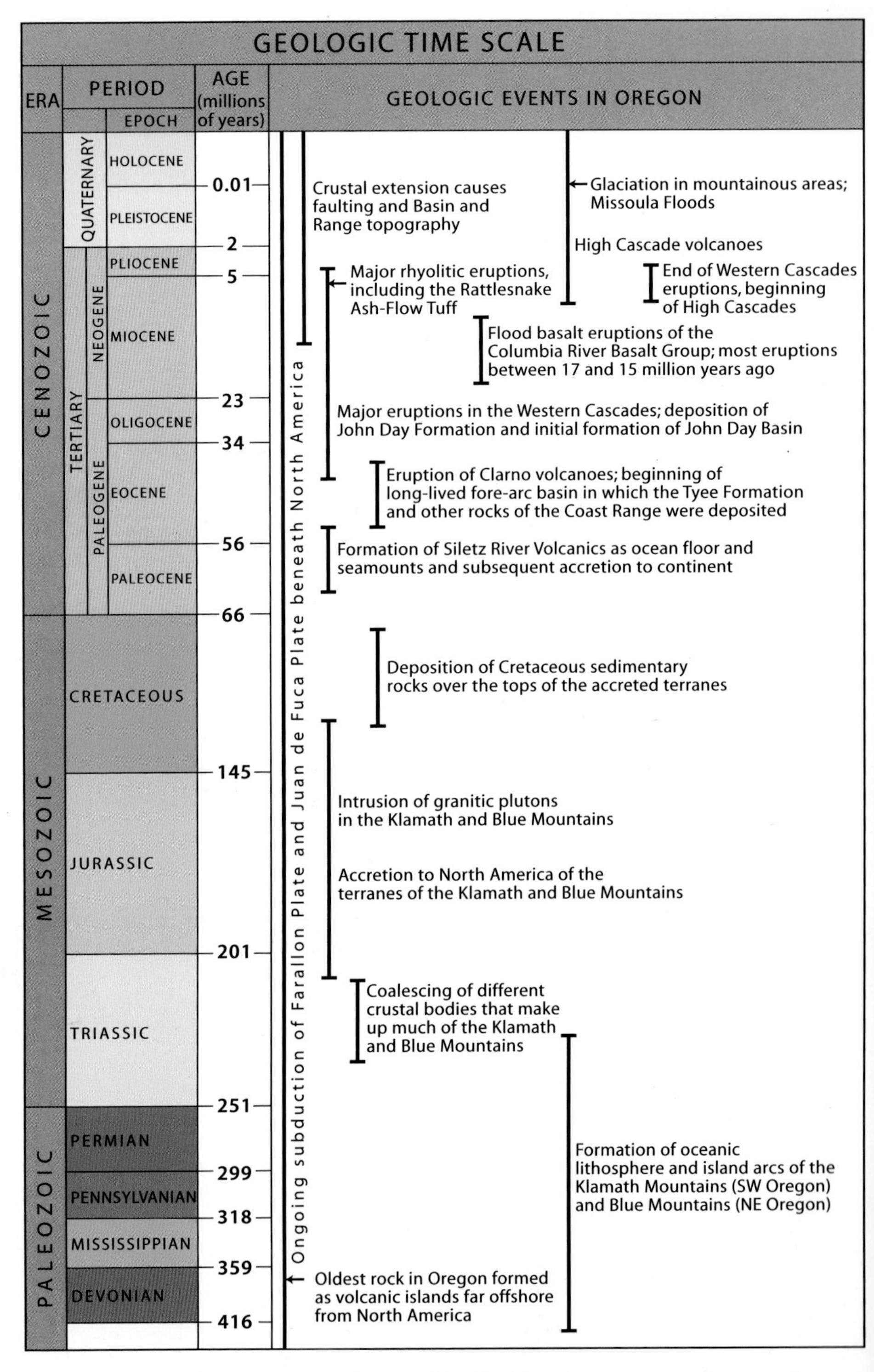

Simplified geologic timescale of Oregon. The black brackets represent the approximate duration of the corresponding event; arrows represent narrow ranges of time. Note that subduction has been occurring along the west coast since the Paleozoic Era.

Oregon's Geologic History

Oregon's geologic history is relatively short but unusually eventful when compared to that of most other parts of the continent. Oregon's oldest exposed rocks are from the Devonian Period, some 400 million years ago, and found only in central Oregon. Many other states, including every state from the Rocky Mountains westward except Oregon and Hawaii, contain at least some rocks older than 1 billion years. Despite its relatively young age, Oregon's geologic history showcases an unusually wide range of geologic processes, and today the state is one of the most geologically active places in the United States.

Oregon encompasses six physiographic regions. The Klamath Mountains in the southwestern corner of the state consist of a variety of smaller ranges, including the Siskiyous. The Coast Range, which runs from northwestern Oregon south along the coast to the Klamaths, includes the Willamette Valley, arguably a region unto itself. Immediately to the east lies the Cascade Range, with its snow-capped, active volcanoes of the High Cascades and the deeply eroded, older volcanoes of the Western Cascades. The other three regions—the Lava Plateaus, Blue Mountains, and Basin and Range—cover the arid, eastern two-thirds of the state. The Blue Mountains contain many of Oregon's best-known natural landmarks, including the John Day Fossil Beds and the Wallowa Mountains. The Lava Plateaus contain three subareas, the Columbia Plateau, the High Lava Plains, and the Owyhee Upland.

This chapter introduces some important geologic concepts as they relate to Oregon, including plate tectonics, earthquakes, and volcanic activity. It also outlines Oregon's geologic history, beginning with its oldest rocks and ending with some features that we see forming today. While Oregon's regions illustrate different parts of this history, together they paint a complete picture dictated by plate tectonics.

Plate Tectonics and the Pacific Northwest

The theory of plate tectonics holds that Earth's rigid outer shell, the lithosphere, is broken into a dozen or so fragments called plates, which move gradually over the softer asthenosphere below. Most geologic activity, in terms of earthquakes, mountain building activity, and volcanism, occurs at the margins of plates, where two different plates come together. Oregon's history reflects the story of the plate margins along the western edge of North America. Today, the western edge of the North American Plate lies some 50 miles (80 km) offshore, where it meets the small Juan de Fuca Plate. West of that lies the Pacific Plate.

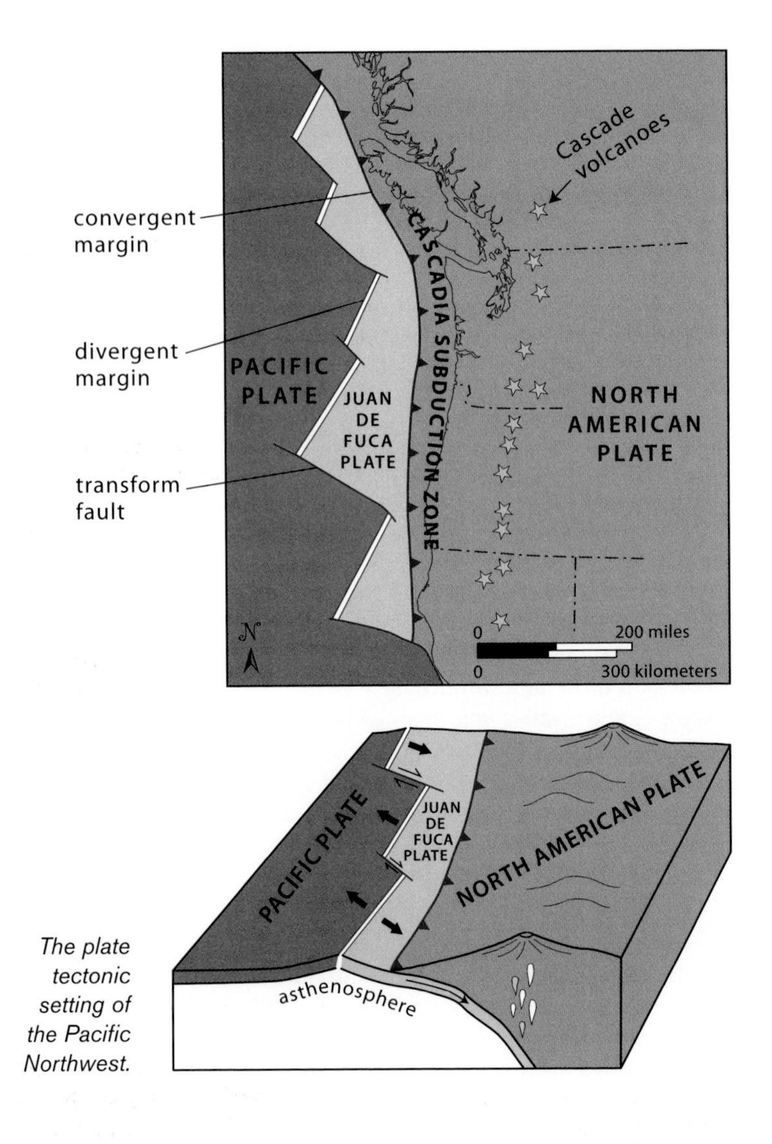

The plate tectonic setting of the Pacific Northwest.

Adjoining plates can move in only three different ways with respect to each other: toward each other, away from each other, or sideways past each other. These different types of relative motion define the type of plate boundary. Where plates move toward each other, their margins are convergent; where plates move apart, their margins are divergent; and where plates slide past each other, their margins are transform faults. The Pacific Northwest hosts each type of plate boundary. The Juan de Fuca and North American Plates move toward each other to form a convergent margin, while 200 miles (320 km) or so to the west, the Juan de Fuca Plate and the Pacific Plate move apart from each other. Their boundary, a divergent margin, is broken in several places by fracture zones where the Pacific and Juan de Fuca Plates slide past each other along transform faults.

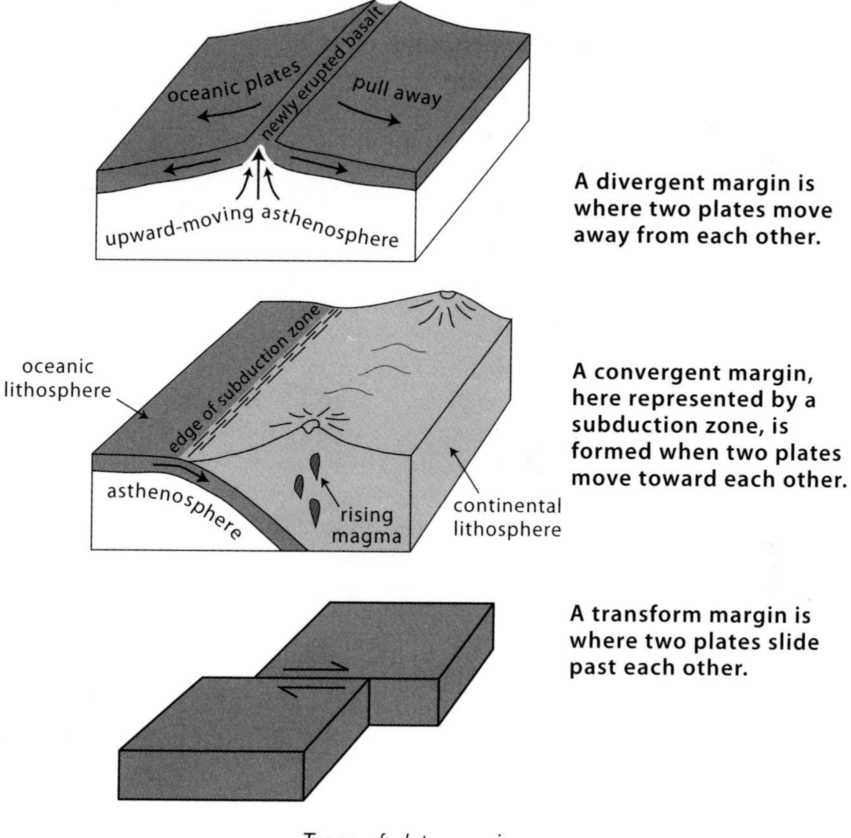

Types of plate margins.

The divergent margin off the coast of Oregon coincides with a mid-ocean ridge, a topographically elevated part of the seafloor. There, new oceanic lithosphere is formed by the eruption of seafloor basalt. As more basalt erupts along the divergent margin, the two plates continually move apart, carrying the seafloor with them. On the Juan de Fuca Plate, this seafloor moves toward the convergent margin and North America. On the Pacific Plate, the seafloor moves northwestward toward Asia.

At the convergent margin, the oceanic lithosphere of the Juan de Fuca Plate encounters the continental lithosphere of the North American Plate. Because oceanic lithosphere is denser than continental lithosphere, the Juan de Fuca Plate sinks, or subducts, beneath North America. The line of convergence of these two plates is therefore a line of sinking, the edge of which is called the Cascadia subduction zone. Today, the Juan de Fuca Plate is subducting at a rate of about 1.6 inches per year (4 cm/year). Compared to what we can achieve on an interstate highway, this rate might seem slow, but given enough time, it can accomplish a great deal. In 10 million years, for example, 250 miles (400 km) of ocean floor would be subducted at this rate. In fact, the subduction zone has dictated much of Oregon's past and present geology. It causes major earthquakes and volcanic activity and, through time, has been responsible for the westward growth of North America, including the foundation of Oregon.

Earthquakes occur along subduction zones because the converging plates create high stresses that cause rock to break and move along large fractures in the crust called faults. Movement of rock along a fault releases a great deal of energy—an earthquake—that travels through the crust. Larger fault movements release more energy and form bigger earthquakes. Where two plates converge, such as at a subduction zone, some of the world's largest earthquakes can occur. Large and small faults extend hundreds of miles inland from the subduction zone.

Subduction zones also create magmatic arcs, which are narrow, broadly arcuate zones of intensive magmatic activity, such as igneous intrusions and volcanic eruptions. Modern magmatic arcs consist of chains of volcanoes on the plate overriding the subduction zone, in a line parallel to the zone. Ancient magmatic arcs appear as the deeply eroded roots of the volcanoes, manifest as numerous bodies of intrusive igneous rock, also parallel to the subduction zone. Volcanoes in the Cascades of Oregon are part of the Cascades magmatic arc, a line of volcanoes extending northward as far as southern British Columbia and southward as far as Lassen Peak in northern California. The line of volcanoes parallels the Cascadia subduction zone. Through a complex process, the Juan de Fuca Plate heats and releases water as it sinks beneath the North American Plate, melting some of the overriding North American lithosphere. The melted rock, called magma, rises to fuel the volcanoes.

Subduction causes continents to grow through a process called accretion. Parts of the subducting oceanic floor get scraped off and added, or accreted, to the edge of the continent as subduction proceeds. The parts of the ocean floor that get accreted typically protrude above the rest of the seafloor, such as island arcs, seamounts, or oceanic sediment that is lower in density than the seafloor basalt. Less frequently, entire sections of the oceanic lithosphere, called ophiolites, may be accreted to the continent. Ophiolites are preserved in Oregon's Klamath and Blue Mountains.

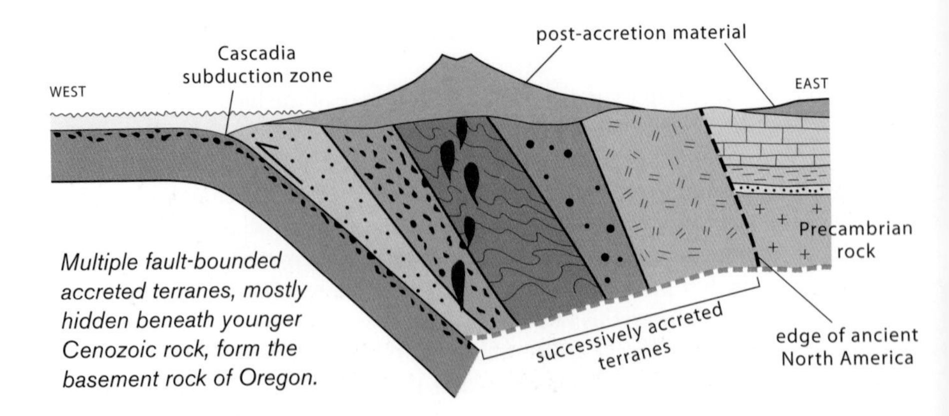

Multiple fault-bounded accreted terranes, mostly hidden beneath younger Cenozoic rock, form the basement rock of Oregon.

Accretion is an especially relevant process for Oregon, because nearly all of Oregon's deepest-level, or basement, rock has been accreted since Jurassic time, some 200 million years ago. This rock is visible at the surface in only those places where erosion has removed the younger overlying rock. Oregon's accreted basement rock ranges in age from Devonian (about 400 million years old) through Early Eocene (about 55 million years old). The only place in Oregon where the basement might not have been accreted during this time lies beneath thousands of feet of Cenozoic volcanic rock in the very southeastern corner of the state.

Oregon's basement rock consists of igneous, metamorphic, and sedimentary rock, most of which originated in oceanic settings and was subsequently fragmented during the accretion process. Individual fragments, separated from other seemingly unrelated ones by fault zones, are called terranes. Over most of the state, these terranes are buried, but in places where they are exposed, they reveal evidence of Oregon's past. These places lie in the Klamath Mountains, Coast Range, and Blue Mountains, including the Wallowa Mountains and Hells Canyon.

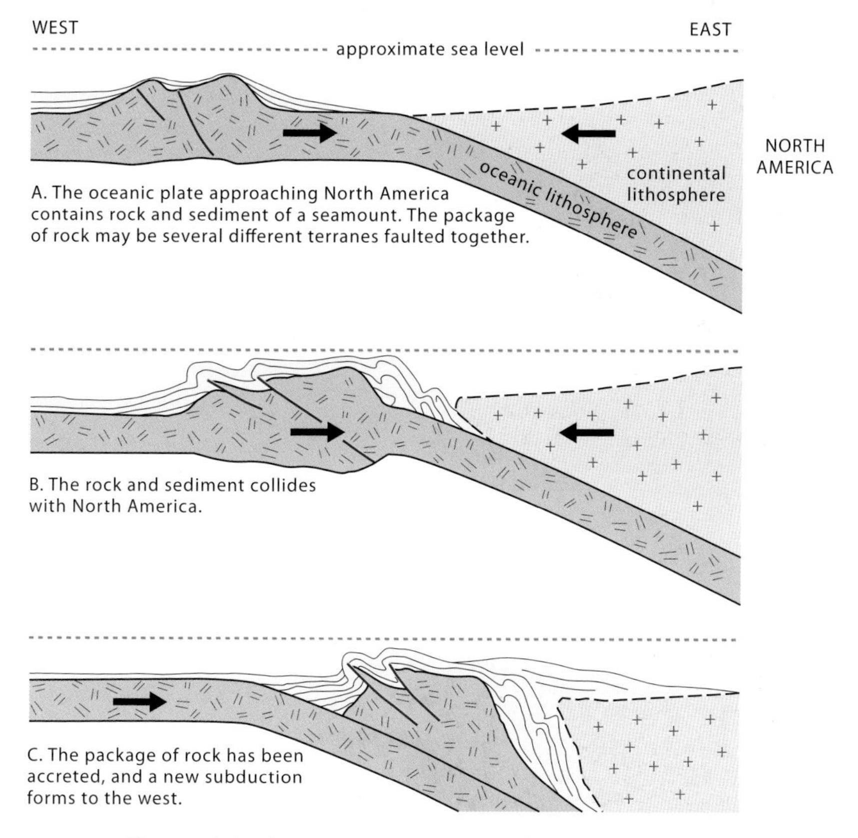

Along subduction zones, seamounts and oceanic sediment can be accreted to the edge of the continent.

Earthquakes and Faults

The standard way of describing the size of an earthquake is with its moment magnitude. This number is derived primarily from modeling of the seismic record or from estimations of the size of the fault's area that actually slipped during the earthquake, the amount it slipped, and the amount of shaking. The number rates the energy released during an earthquake on an exponential scale from 0 to 10. Each number in the scale reflects an increase in seismic energy of just over thirty times. For example, a magnitude 5 earthquake releases about thirty-two times the energy of a magnitude 4 earthquake, and about one thousand times the energy of a magnitude 3. The largest known earthquake on Earth was the Chilean earthquake of 1960, which measured a 9.5; the second largest known earthquake was the Alaskan earthquake of 1964, which measured 9.2. For comparison, the 1989 Loma Prieta earthquake in San Francisco registered a magnitude of 6.7, more than one thousand times smaller than a magnitude 9.

The largest fault zones are capable of generating the largest earthquakes, and most of Earth's largest fault zones lie right at the interface of a subducting plate and the overriding plate. These faults generate what are known as subduction zone earthquakes, many of which exceed magnitudes of 9. Both the 1960 Chilean and 1964 Alaskan earthquakes, for example, were subduction zone earthquakes. In Oregon, the last subduction zone earthquake probably also exceeded a magnitude 9. What's more, all the available evidence, reviewed in

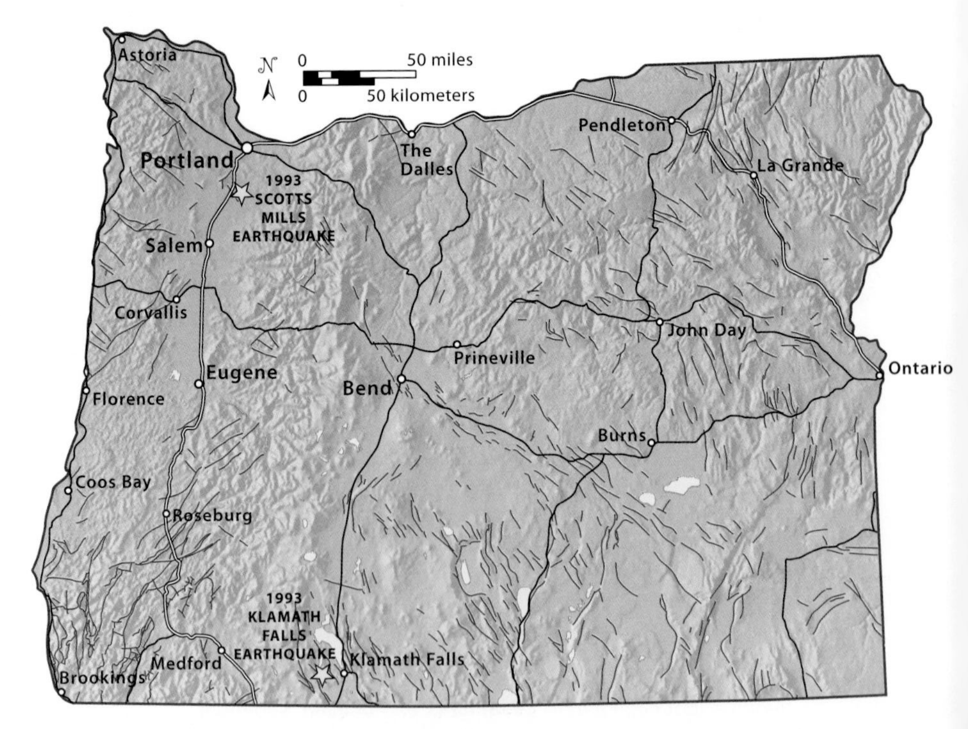

Oregon fault map, showing the locations of the 1993 Scotts Mills and Klamath Falls earthquake

this book in the chapter on the Coast Range, indicates that an earthquake such as this occurs about every three hundred to five hundred years. The most recent of these gigantic earthquakes in Oregon occurred in 1700, more than three hundred years ago.

Away from the subduction zone, smaller but still damaging earthquakes can occur on other faults. Oregon is broken by thousands of ancient and modern fault zones. In the last two hundred years, these faults have generated more than a dozen moderate to large earthquakes. In 1993 two well-known earthquakes occurred: the magnitude 5.7 Scotts Mills earthquake south of Portland and the magnitude 6 Klamath Falls earthquake. These two earthquakes originated on two different types of faults, a reverse fault and a normal fault.

Analogous to plate boundaries, faults are classified according to the direction of movement of the rock on either side of the fault. Faults that exhibit up-and-down movement, parallel to the direction of a fault's inclination, or the

strike-slip faults (parallel to strike)

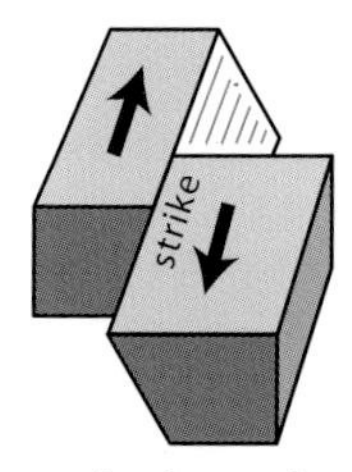

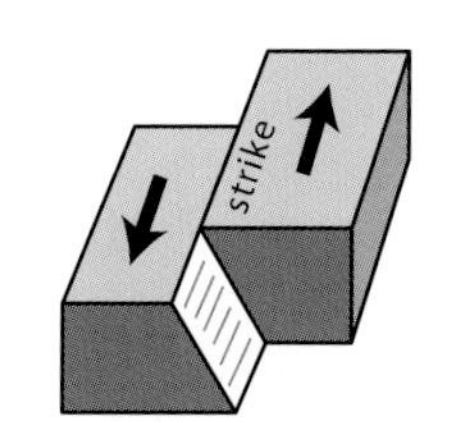

dip-slip faults (parallel to dip)

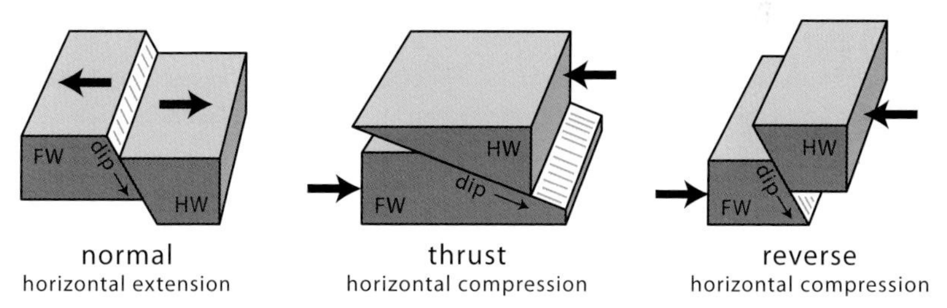

oblique-slip fault

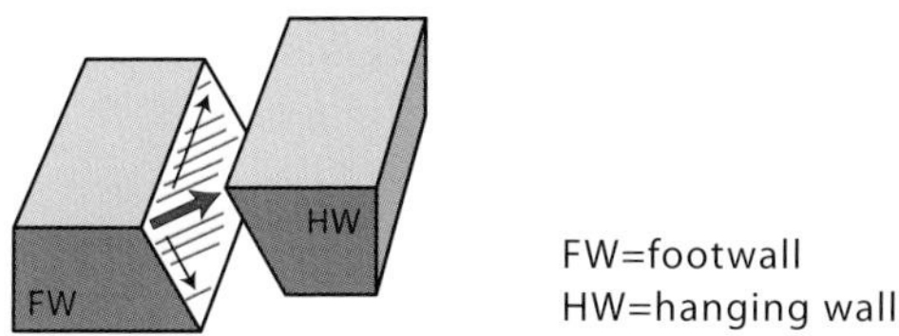

Block diagrams that illustrate the major types of faults.

dip, are called dip-slip faults; those that show side-by-side movement, parallel to a horizontal line on the fault plane, or the strike, are called strike-slip faults.

Dip-slip faults, because they are inclined, contain a hanging wall and footwall, the blocks of rock above and below the fault surface, respectively. Those dip-slip faults in which the hanging wall moves down relative to the footwall are called normal faults; those in which the hanging wall moves up relative to the footwall are called thrust faults and reverse faults. Reverse faults are those in which the fault plane dips more steeply than 45 degrees; thrust faults dip more gently. By contrast, strike-slip faults are classified as right-lateral or left-lateral, depending on the relative movement in a horizontal sense. To distinguish between the two, one needs to look across the fault. If the opposing side moves right, it's a right-lateral fault; if the opposing side moves left, it's a left-lateral fault. Oblique-slip faults are those in which the direction of movement has components of both strike-slip and dip-slip.

Different types of faults form as the result of different types of crustal stresses. Horizontal compressive stresses, for example, cause thrust and reverse faults to form, and they can also fold the rock. Horizontal extensional stresses cause normal faults to form. A horizontal shear stress, in which the two sides of the fault are pushed in opposing directions horizontally, is required for strike-slip faulting. The Scotts Mills earthquake occurred on a reverse fault, indicating crustal compression in the area near Portland. The Klamath Falls earthquake occurred on a normal fault, indicating crustal extension in that area.

Volcanic Activity

A brief glance at the geologic map of Oregon shows that volcanism dominates Oregon's modern geology as well as its geologic history. Young as well as ancient volcanic rock covers most of the state! The High Cascades, which run down the western third of Oregon, contain more than a dozen large volcanoes, at least five of which have erupted in the last 10,000 years. These volcanoes receive a relatively continuous supply of magma derived from the subduction of the Juan de Fuca Plate. Much of eastern Oregon is covered by lava flows of the Columbia River Basalt Group, most of which erupted about 15 million years ago and is most likely related to the formation of the Yellowstone hot spot. Many of the rocky headlands of the Oregon coast consist of ancient lava flows of varying ages. Even much of the basement rock of the Coast Range and Blue Mountains is volcanic, formed in oceanic volcanic environments, including seamounts and island arcs, and accreted to North America by subduction.

Oregon hosts each major type of volcano. Shield volcanoes, like Newberry Volcano south of Bend, exhibit relatively gentle slopes, resembling shields in profile. Cinder cones, which may form on shield volcanoes or as separate entities, are relatively small and steep-sided and consist mostly of cinders. Hundreds of cinder cones, including many that erupted in the last 10,000 years, decorate Oregon's Cascades and High Lava Plains. Stratovolcanoes, also called composite volcanoes, are the most prominent volcanoes in the Cascades, including Mt. Hood and Mt. Jefferson. Stratovolcanoes exhibit steep slopes and, in Oregon, reach elevations greater than 9,000 feet (2,700 m). Calderas, such as Crater

Lake, form where a volcano has collapsed in on itself after an eruption emptied much of its magma chamber. Domes are steep-sided and relatively small. Some examples of domes include Hayrick Butte near Santiam Pass on US 20 and the dacite domes of Mt. Mazama at today's Crater Lake.

In general, basaltic lavas, which are relatively fluid and can flow over gentle gradients away from their eruption locations, or vents, form shield volcanoes. More viscous (less fluid) lavas, such as andesite, tend to form the steeper stratovolcanoes. Dacite and rhyolite, which are even more viscous, frequently plug up their volcanic conduits to form domes. The concentration of silica in the lava controls the lava type and its viscosity. Basalts are low in silica and viscosity, andesites are intermediate, and dacites and then rhyolites are high in silica and viscosity.

Many lavas have compositions that are intermediate between the three main types described above. One especially important intermediate rock type is dacite, which is between andesite and rhyolite in composition. Another important intermediate rock type is basaltic andesite. In Oregon, much of Crater Lake consists of dacite and another intermediate rock type called rhyodacite, while many of the stratovolcanoes are basaltic andesites. The road guides in this book describe the rocks as they appear in the field. Therefore, in most cases, they avoid the intermediate terms, which are largely reliant on chemical analyses.

Explosive eruptions tend to occur when the erupting magma is rich in dissolved gases, such as water or carbon dioxide. The decrease in pressure experienced by magmas as they rise through the crust allows these gases to bubble out in a process similar to the release of gases when one opens a carbonated beverage. If the magma is also highly viscous because of a high silica content, then the eruptions can be extraordinarily violent, such as the 1980 eruption of Mt. St. Helens, just across the border in Washington State. These explosive eruptions produce huge amounts of ash, pumice, lava bombs, or fragmental wall rocks, collectively called pyroclastic material. Pyroclastic flows can form when ash and other pyroclastics become concentrated and flow away from the vent, at speeds exceeding 100 miles per hour (160 km/hr). The resulting deposit, called an ash-flow tuff, can give important clues to the eruptive history of the volcano. For example, Crater Lake formed through the explosive eruption and collapse of Mt. Mazama about 7,700 years ago. One reason we know the eruption was so explosive is because the area is surrounded by hundreds of feet of ash-flow tuff that extend more than 30 miles (48 km) away, a volume about fifty times as great as the material exploded from Mt. St. Helens. And the Rattlesnake Ash-Flow Tuff, which erupted 7 million years ago near Burns, Oregon, covers nearly one-tenth of the state!

Volcanoes also produce lahars, or volcanic mudflows, which form when loose rock debris or pyroclastic flows mix with water or ice to become a slurry that flows in a manner similar to wet concrete. Lahars are the greatest hazard of the Cascade Range, largely because of the preponderance of glacial ice on the mountains. In addition to the lahars that form during eruptions, some lahars form from rock avalanches off steep hillsides when the rock contains groundwater. The rock avalanches can be triggered by earthquakes, which can physically shake the rock loose, or by far more subtle things, such as the ongoing chemical alteration and resultant weakening of the rock. Not only are lahar deposits an important process in the modern High Cascades, they are also a major component of the rock record in the Western Cascades and in the Clarno Formation of eastern Oregon.

basalt

andesite

dacite-rhyolite

shield volcano

Gentle slopes and large; consists of basaltic lava flows that were relatively fluid; may have cinder cones on flanks

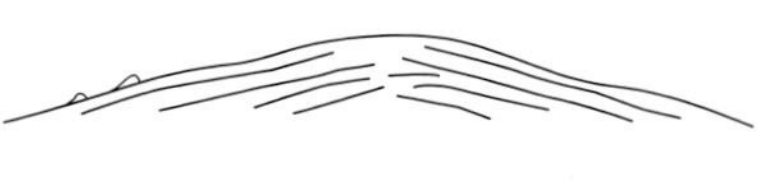

cinder cone

Steep, relatively small conical hills that consist of cinders erupted from a central vent

stratovolcano/ composite volcano

Steep and large; consists mostly of andesitic flows and pyroclastics, which are explosive products like ash and pumice; may consist of two or more coalesced volcanoes

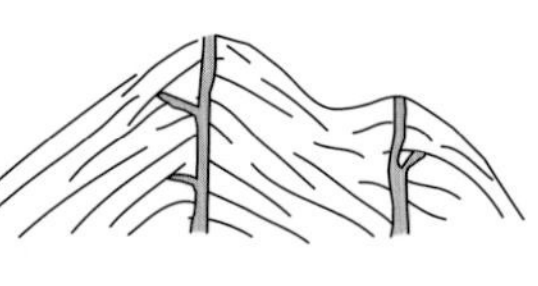

caldera

Steep-sided depression, typically hundreds to thousands of feet deep and more than 1 mile (1.6 km) across; formed by collapse of a volcano into its emptied magma chamber after a major eruption

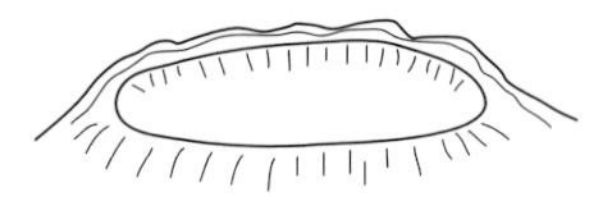

dome volcano

Very steep slopes, but relatively small features; made out of dacitic or rhyolitic lava, which tends to be very viscous (sticky) and so cannot travel far from the vent

Different types of volcanoes and their main types of lava. Note the wide range of volcanoes that can consist of basaltic lavas, which is partly because of the prevalence of basaltic andesite.

PALEOZOIC ERA

543 to about 250 million years ago

Oregon's Paleozoic history is one of mostly marine conditions in and around a chain of volcanic islands. Rocks from this era, which are exposed in the Blue and Klamath Mountains, were mostly deposited in deep to shallow oceans, but also include volcanic rock, some terrestrial, or land-based, deposits, and even a slice of oceanic lithosphere preserved in the mountains immediately south of John Day. Some of the fossils preserved in these rocks resemble animals that lived in the western Pacific, far from North America.

Oregon's geologic history begins with its oldest known rocks, two relatively small outcrops of limestone in the Suplee area in the central part of the state. These rocks contain fossilized creatures, such as corals and brachiopods, that lived in shallow seas during the Devonian Period. The presence of nearby sandstone with abundant volcanic particles indicates that the limestone was deposited in shallow water near active volcanoes. Based partly on these lines of evidence, most researchers agree that these rocks probably formed as reefs that fringed an oceanic volcanic arc.

Devonian coral in limestone from the Suplee area of Oregon.

Quite unlike the Grand Canyon in Arizona, where you can pick out a rock layer and follow it for a seemingly endless distance, the limestone outcrops near Suplee appear as isolated blocks, surrounded by outcrops of other rock units and of other ages. In turn, these other rocks also exist as discontinuous blocks, creating a mélange, or body of mixed rock. They include chert, mudstone, sandstone, shale, and other limestone that ranges in age from Mississippian through Permian time.

The wide variety of rock types and ages represented in these mélange exposures give a fragmentary history of Oregon for the remainder of Paleozoic time.

The characteristics of most individual blocks indicate they mostly originated in oceanic environments, at times in deep marine environments, at other times in shallower reef settings near volcanoes. Rocks of similar ages elsewhere in the Blue Mountains and the Klamath Mountains, although not abundant, corroborate and lend detail to this story.

The most abundant Paleozoic rocks in Oregon are those that formed at the end of the Paleozoic, during the Permian Period. These mostly sedimentary rocks, many of which contain volcanic particles, are in scattered localities of the Blue Mountains and the northeastern Klamath Mountains. Some Permian volcanic rocks occur near Baker City, and a slice of Permian oceanic lithosphere occurs immediately south of John Day. Together, these rocks paint a complicated picture of deep ocean floor environments, shallow marine settings, and a volcanic arc, mixed together in a fashion analogous to the mélange at Suplee.

Some of the Permian-age limestone contains fossils of certain types of corals and single-celled organisms called foraminifera, or forams. While corals and forams are relatively common fossils, these particular ones do not appear to be from the ocean that existed west of Laurasia, the ancestor of the North American and Asian continents. Instead they resemble those that lived in the ancient Tethys Sea, on the other side of Laurasia. By Permian time, all the landmasses on Earth had coalesced into a single supercontinent called Pangaea. The Tethys Sea formed in equatorial regions at the end of Paleozoic time and widened into a continuous seaway between Laurasia and Gondwana as the supercontinent Pangaea began to break up in early Mesozoic time. As the continents continued to rearrange themselves, moving toward their present configuration, the Tethys Sea became progressively smaller. Today, its last remnant is the Mediterranean Sea.

Permian coral from the Tethys Sea.

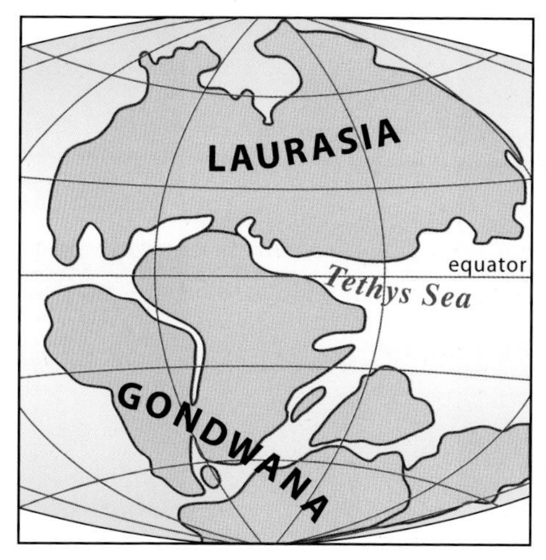

The Tethys Sea.

These exotic Tethyan fossils exist only in limestone bodies that are surrounded by fault zones. In North America, they can be found only in scattered locations along the western margin, including parts of Oregon. Because corals are stationary creatures, their discovery indicated the rocks had moved great distances. This was one of the main lines of evidence used to develop the ideas behind accretionary tectonics. The Tethyan fossils belonged to various terranes that became rafted together with other terranes to become larger composite terranes. Eventually, during Mesozoic time, the first composite terranes accreted onto the edge of North America.

MESOZOIC ERA

About 250 to 66 million years ago

The first half of the Mesozoic Era, until the middle of the Jurassic Period, was primarily a time of terrane building in Oregon, when individual fragments of oceanic regions came together through plate motions. The second half was one of terrane accretion, when plate motions added the composite terranes onto North America. Accretion, which continued into the Cenozoic Era, also caused mountain building along much of North America's western edge. The sedimentary record, ancient fault zones, and some of the metamorphic rocks show evidence of this mountain building. Igneous intrusions called stitching plutons intruded some of the terranes during Late Jurassic and Cretaceous time.

Although Mesozoic rocks are much more abundant in Oregon than Paleozoic ones, they still crop out in a comparatively small area of Oregon, in the Blue and Klamath Mountains. Oregon's Mesozoic igneous, metamorphic, and sedimentary rocks mostly belong to the basement complex that formed elsewhere and was eventually accreted to the continent. Some of these rocks formed a

A mélange is a mixture of crunched rock that has been caught between colliding plates.

Radiolarian chert, also called ribbon chert, is a fine-bedded rock of siliceous skeletons of radiolaria that settle to the quiet seafloor.

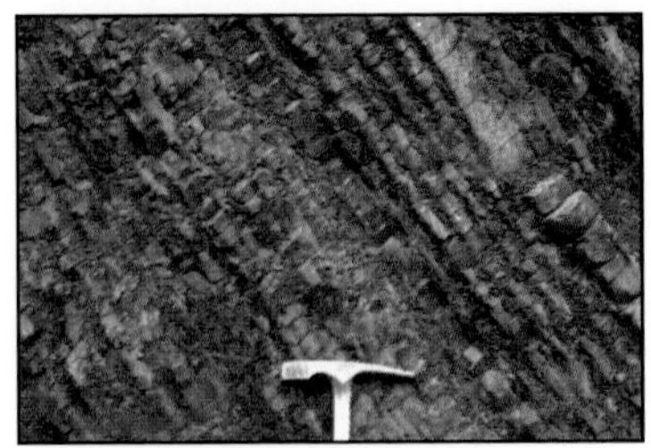

Turbidites, which contain thin graded beds, are deposited on the seafloor of fore-arc basins.

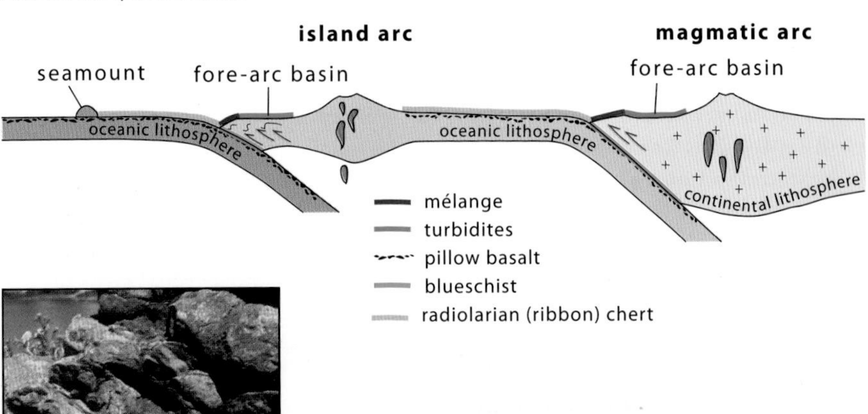

Pillow basalt forms when magma erupts underwater, usually when new ocean crust forms at spreading ridges.

Blueschist, a metamorphic rock, only forms in subduction zones.

Important rock types form in the various marine and on-land environments that exist in different plate tectonic settings. The presence of each rock type provides evidence for the existence of that geologic environment in the geologic past.

great distance away, as in the case of limestone that contains Tethyan fossils, but most of them probably formed much closer to North America in oceanic settings, including volcanic arcs, shallow and deep marine environments, and an ancient subduction zone. In addition, we know these settings were mostly tropical because they contain fossilized warm-water organisms. Like the older Paleozoic rocks, these rocks exist as fault-bounded slices, or terranes, that do not necessarily relate to each other in any predictable way. Individual terranes of the Blue and Klamath Mountains are described in more detail in those respective chapters.

Many individual terranes merged into larger composite terranes before they accreted onto the North American continent. This merging can take place through a variety of mechanisms, all involving large-scale movements of crustal blocks. Merging can occur, for example, when subduction takes place between two oceanic plates, and fragments of marine basins, seamounts, or the underlying lithosphere break off from the subducting plate and become added to the volcanic arc. Another cause might be large-scale strike-slip faults that move different slices of crust together laterally.

The evidence for merging of terranes comes largely from field relations between the terranes and intrusive or younger sedimentary rocks. Where an intrusive body cuts across the boundary between two terranes, we can infer that the terranes were joined prior to the intrusion. These intrusions, called stitching plutons, can be found throughout the Klamath and Blue Mountains and are mostly Late Jurassic and Cretaceous in age.

Where sedimentary rocks overlie the boundary between two terranes, we can infer that the terranes merged before deposition of the sedimentary rocks. Such younger assemblages of sedimentary rocks exist in the Klamath and Blue Mountains and are largely Mesozoic in age. If we can also determine that the sediments in the sedimentary rocks did not erode from North American

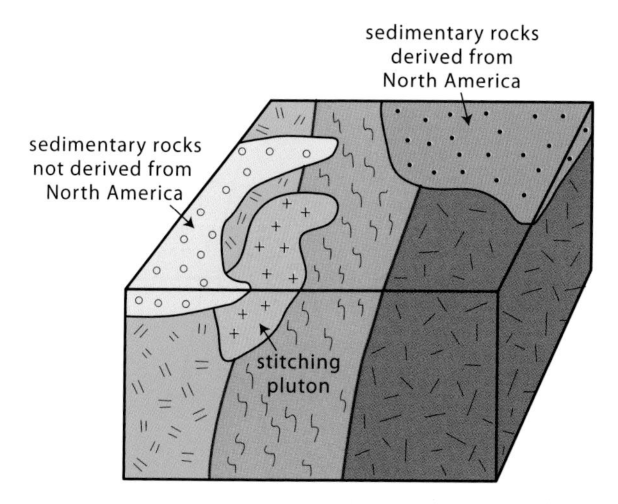

Where a stitching pluton intrudes two or more terranes, it indicates the terranes had come together prior to the intrusion. Where a sedimentary rock unit lies over the top of a boundary between terranes, it indicates the terranes had come together before deposition of the sedimentary rock.

rocks, then we can infer that merging and deposition of the younger sedimentary assemblage occurred before the terrane was accreted. If the sedimentary rocks appear to be derived from North America, then we infer that they were deposited after accretion. In Oregon, pre-accretionary sedimentary rocks accumulated during Triassic and Jurassic time, whereas post-accretionary ones first accumulated during Cretaceous time.

The Cretaceous sedimentary assemblages were mostly deposited in a geologic setting called a fore-arc basin, essentially a depositional basin between the subduction zone and the magmatic arc. The Willamette Valley is a modern fore-arc basin that lies between the High Cascades (the magmatic arc) and the subduction zone. During Cretaceous time, the magmatic arc lay at the site of the Idaho Batholith,

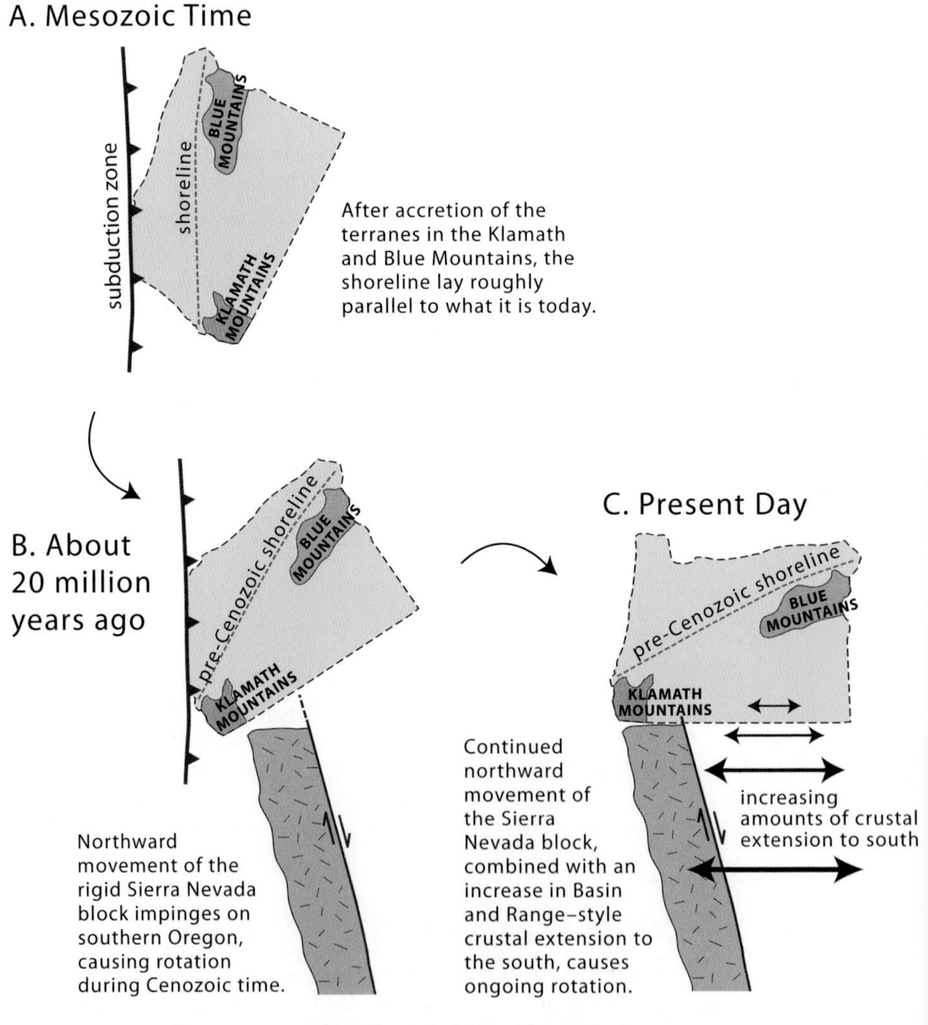

Terranes moved northward during Mesozoic time and then rotated clockwise during Cenozoic time.

a large area of Cretaceous-age granitic intrusions in central Idaho. The fore-arc basin lay between the arc and the subduction zone. Scraps of the Cretaceous fore-arc basin are preserved as the Hornbrook Formation in the Klamaths, and as a variety of Cretaceous sedimentary rocks near Mitchell, Oregon.

Remnants of the adjacent subduction zone are also preserved in Oregon's southwestern corner. There, one can see a mélange of marine sandstone and shale with a variety of other rock types, including scattered blocks of blueschist, a metamorphic rock that seems to only form deep in subduction zones. Collectively, this body of rock is called the Franciscan Complex and extends as a nearly continuous belt southward into California. The fore-arc basin and magmatic arc, which are discontinuous in Oregon, also extend southward into California, where they form continuous belts.

The many similarities between terranes of the Blue Mountains and those of the Klamath Mountains suggest that they reflect similar events but at different locations along the same subduction zone. The presence of pieces of the Cretaceous fore-arc and subduction zone in these provinces further substantiates this idea. A line drawn approximately parallel to the ancient shoreline and subduction zone runs in an east-northeast direction from about Roseburg to La Grande.

It is likely that the Mesozoic shoreline originally ran approximately parallel to the modern shoreline. Studies of ancient magnetism in Oregon's rocks indicate that much of Oregon's bedrock rotated some 70 degrees clockwise during Cenozoic time. The principal difference was therefore one of location: during early Mesozoic time, the shoreline lay across what is now eastern Oregon, but as accretion continued, the continent grew westward and the subduction zone migrated with it. Today, western Oregon rotates about 1 degree clockwise every million years, but this rate decreases eastward, away from the subduction zone. Most of the rotation seems to be caused by two effects: the northward impingment of the Sierra Nevada on southwestern Oregon and crustal extension in south-central and southeastern Oregon.

CENOZOIC ERA

66 million years ago to present

Unlike the Mesozoic and Paleozoic Eras, the rock record of the Cenozoic Era is anything but fragmentary. Cenozoic rock covers about 90 percent of the state. Cenozoic time was an incredibly diverse and active geological era that created the Oregon landscape we see today. During Cenozoic time, accretion continued along the continental margin, different types of volcanoes erupted, a whole new set of faults formed, the climate changed, and the modern landscape developed.

Siletz River Volcanics and Tyee Formation

The oldest Cenozoic rocks in Oregon are the 62- to 56-million-year-old Siletz River Volcanics, mostly basalt flows with a minor fraction of sedimentary rock. Many of the basalt flows contain rounded, bulbous features called pillows,

Pillow basalt of the Siletz River Volcanics exposed in a cliff face near Roseburg.

which form when basalt is erupted underwater. Individual bits of lava congeal into bulbous shapes during underwater eruptions and cool quickly on their outside edges but remain hot and malleable on their insides. When the pillow settles to the seafloor, it conforms to the shapes of other already-formed pillows. These basalts were probably part of a chain of seamounts because they are interbedded with marine sandstone. The seamounts likely stood up in topographic relief with respect to the surrounding seafloor. We know they were accreted to North America between about 55 and 50 million years ago, because the 54- to 50-million-year-old Umpqua Group, a sequence of mostly deep marine sandstone and shale, rests on top the basalt but is highly faulted and folded in places, along with the basalt. On top of these rocks lies the Tyee Formation, which is only mildly deformed. The Siletz River Volcanics, which are also called the Siletz terrane or Siletzia, now form the basement of the Oregon Coast Range.

The Tyee Formation was deposited in a new fore-arc basin that formed on top of the accreted Siletzia terrane between 49 and 48 million years ago. This marine basin received sediments in mostly deep water in a submarine fan complex. Overlying rock units tend to reflect shallower water conditions. Later uplift in the Coast Range displaced the site of the fore-arc basin to the modern-day Willamette Valley.

The lower slopes of Marys Peak (right) near Corvallis and nearby mountains of the Coast Range are made of pillow basalt of Siletzia.

Clarno and John Day Formations and the Western Cascades

During and after the accretion of Siletzia, Oregon experienced its first land-based or continental volcanism, the evidence of which appears in the Clarno Formation, a rock unit exposed over much of central Oregon in the area around Fossil and John Day. It consists of andesite flows, tuff-rich sandstone, andesite, and volcanic mudflows (lahars), as well as volcanic necks and plugs near Mitchell. These types of volcanic deposits and features are usually found in association with stratovolcanoes, so we can assume a similar setting for the Clarno. The Clarno Formation ranges in age from 54 to 37 million years old.

The Clarno Formation reflects a resumption of volcanic activity after a quiet period at the end of Mesozoic time. Some geophysicists argue that this resumption reflects changes in the subducting plate. They suggest that the angle of the subducting plate was shallow toward the end of Mesozoic time, which put much of the western United States under compression. This compressive stress hampered magmatic activity and caused the quiet period. Then, during the accretion of Siletzia, the subduction zone jammed and eventually stepped oceanward, in the process breaking off the edge of the subducting slab. No longer being pushed from behind, the broken slab dropped down at a steeper angle, allowing the overlying lithosphere to extend and prompting renewed volcanism in the Clarno eruptions and the Challis eruptions in Idaho. It likely also caused later caldera eruptions in eastern and western Oregon between about 40 and 30 million years ago.

Oceanward of Siletzia, the new subduction zone also plunged fairly steeply beneath the continent and through time began to cause volcanic activity to the west of the Clarno eruptions. This new volcanism formed the Western Cascades, a true magmatic arc and the precursor to the modern High Cascades. It is not entirely clear just when the Western Cascades formed, but their oldest preserved rocks are between 40 and 35 million years old. Many geologists argue that eruptions from the Western Cascades deposited the tuff and ash that make up much of the John Day Formation, which overlies the Clarno Formation. The John Day Formation was deposited between 37 and 19 million years ago, about the same time as the Western Cascades. However, some researchers now argue that much of the early John Day Formation came from a series of calderas that lie east of the Cascades.

The Clarno and John Day Formations give us much more information about Oregon's past than just the nature of its changing plate margin. Together, these rock units record changing conditions on land from about 54 to 22 million years ago. The overlying Mascall and Rattlesnake Formations extend that record to about 7 million years ago. These sedimentary and volcanic rocks filled the John Day Basin of central Oregon and contain such an abundance of ancient soils and plant and animal fossils that parts of the area have received special designation as a national monument, administered by the National Park Service. The fossils, in turn, give us information about the changing landscape that we might not otherwise have been able to read from only the rocks. The

EAST

magmatism of Idaho Batholith

NORTH AMERICA

A. Subduction of oceanic lithosphere beneath North America (continental lithosphere) in the latter part of Mesozoic time. The subduction takes place at a moderate angle, melting the rock that later became the Idaho Batholith, just east of Oregon.

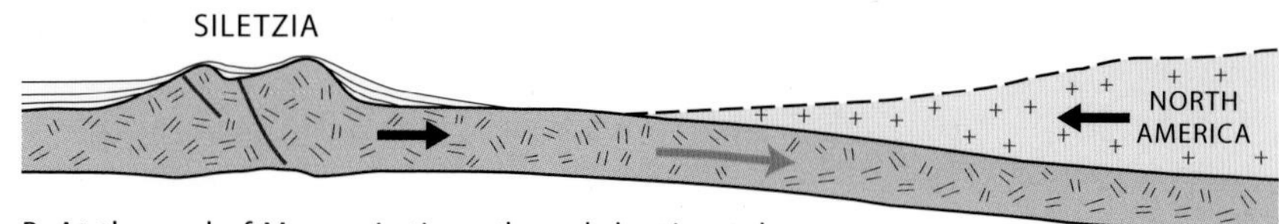

B. At the end of Mesozoic time, the subduction takes place at a low angle, causing compressional stress in western North America and inhibiting magmatism. The oceanic plate contains Siletzia.

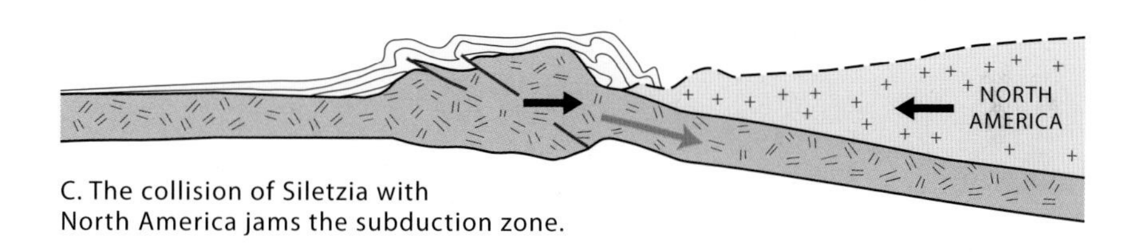

C. The collision of Siletzia with North America jams the subduction zone.

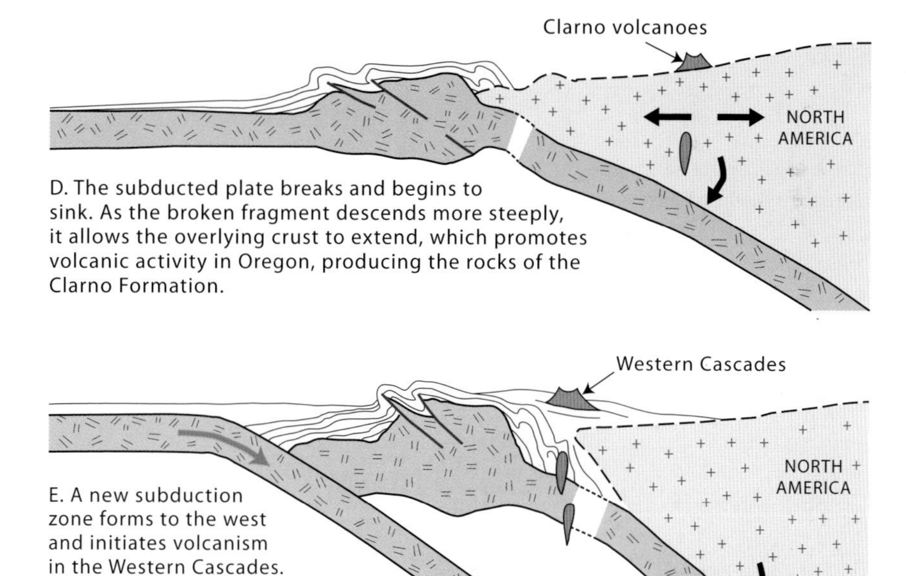

D. The subducted plate breaks and begins to sink. As the broken fragment descends more steeply, it allows the overlying crust to extend, which promotes volcanic activity in Oregon, producing the rocks of the Clarno Formation.

E. A new subduction zone forms to the west and initiates volcanism in the Western Cascades.

Possible sequence of events for the accretion of Siletzia and reconfiguration of the plate margin during early Cenozoic time. The green arrows indicate movement of the subducting plate, the large black arrows indicate if the crust is under compression or extension, and the orange bodies indicate magma and lava.

The Painted Hills Unit of the John Day Fossil Beds National Monument preserves numerous ancient soils, called paleosols. Here, the red color in paleosols near the base of the John Day Formation reflects the subtropical conditions in which they formed. Younger paleosols reflect more temperate conditions.

fossils within the Clarno Formation indicate tropical to subtropical forest environments, whereas those in the John Day Formation indicate more temperate forests that evolved into grasslands. The chapter on the Blue Mountains gives a much more detailed description of these fascinating rocks.

Columbia River Basalt Group and Uplift of the Coast Range

Basalt covers more of Oregon than any other rock type. Much of this basalt belongs to the Columbia River Basalt Group, which began erupting 16.8 million years ago, reached a crescendo around 15 million years ago, and continued erupting intermittently until about 6 million years ago. Some 96 percent of the total volume of lava was erupted at its beginning stages, from its inception until 14.5 million years ago. The basalt flows, which originated mostly from fissures in northeastern Oregon and southeastern Washington are called flood basalts, because they literally flooded the landscape. Some individual flows made it all the way from eastern Oregon to the Pacific Ocean. There, they managed to invade the existing sedimentary rock, forming gigantic sills, some of which even erupted to the surface as small undersea volcanoes!

To gain some perspective on the scale of these eruptions, you need only to consider some numbers. The flows covered more than 77,220 square miles (200,000 km^2) of land in Washington and Oregon with a volume that exceeds 52,800 cubic miles (220,000 km^3). By comparison, the National Park Service estimates the volume of the Grand Canyon in Arizona to be about 1,000 cubic miles (4,200 km^3), less than one-fiftieth the volume of the Columbia River Basalt Group.

Many researchers now hypothesize that the basalt eruptions marked the onset of the Yellowstone hot spot—a poorly understood feature, perhaps a plume of hot mantle material, that caused melting in the overriding continental

Lava flows of the Columbia River Basalt Group in northern Oregon.

lithosphere. As the lithosphere drifted west-southwestward over the hot spot, it created a series of volcanic calderas that became progressively younger toward the present site of the hot spot, in a direction opposite the plate motion. Indeed, a well-defined track of calderas lies beneath the Snake River Plain of Idaho. These calderas become increasingly younger in the direction of the Yellowstone Caldera, the site of the most recent and ongoing eruptions.

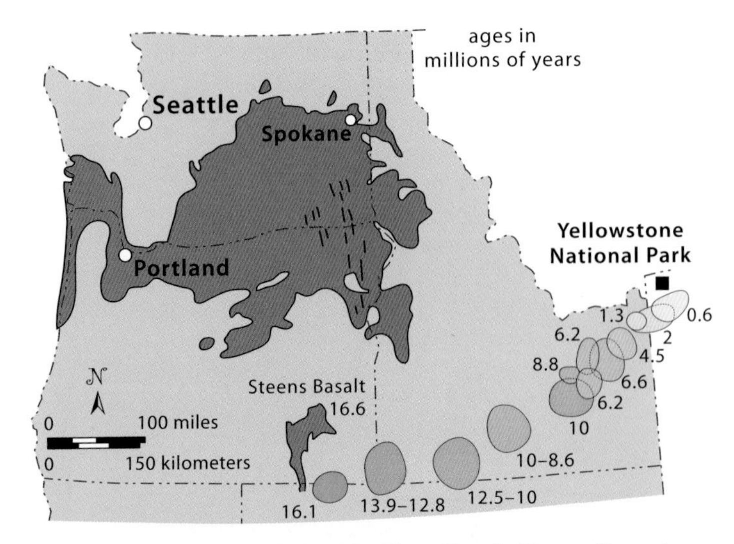

The distribution of Columbia River Basalt Group flows in Oregon and Washington, as well as the distribution and ages of calderas related to the Yellowstone hot spot. The thin purple lines in southeastern Washington and northeastern Oregon show some of the locations for the fissures, now preserved as dikes, through which the magma reached the surface.

Along the line of the known Yellowstone hot spot track, but in the other direction to the southwest, lies Steens Mountain, near the Oregon-Nevada border. The 16.8- to 16.6-million-year-old basalt flows there mark the onset of the Columbia River Basalt Group. From there, flood basalt volcanism migrated northward to the site of today's Wallowa Mountains. This northerly migration, a clear deviation from the narrow northeast-trending path we can see in Idaho, is likely a consequence of the early plume exploiting the edge of Oregon's accreted terranes. There, the relatively thin crust of accreted terranes meets the much thicker crust of the older North American continent.

The Columbia River likely started its history sometime during early eruptions of the Columbia River Basalt Group. At times, it formed a throughgoing river system; at other times it appears to have been an integrated series of lakes and streams. Some deposits of the ancient river system exist east of Portland. Known as the Troutdale Formation, the sediments were deposited between about 15 and 2 million years ago. Its older parts were therefore contemporary with parts of the Columbia River Basalt Group. The distribution of these and other deposits, along with earlier basalt flows that were confined to river channels, preserve a history of the ancient drainage. Lava flows of the Columbia River Basalt Group make the river's history especially interesting because they frequently dammed the channel and forced it to move elsewhere.

Still, the position of the Columbia River did not change drastically during its long history. In western Oregon, it mostly stayed south of modern-day Washington and north of Mt. Hood. This position included the prominent northward bend in the river near Portland, which appears to have been firmly established by 12 million years ago. The bend probably occurred because the ancient river found an easier path in the lowlands of the north-south-trending fore-arc basin as opposed to trying to cross the early Coast Range directly to the west. Farther north, near the vicinity of Kelso, the river found a way through the Coast Range and turned back to the west.

It appears that the Oregon Coast Range formed over a long period of time, beginning before the eruptions of the Columbia River Basalt Group but attaining its present form much more recently. The presence of Columbia River Basalt Group flows along the coast indicates that the Coast Range was not everywhere a significant barrier during the eruption of basalt about 15 million years ago, and that most Coast Range uplift occurred afterward. However, some lava flows appear to be deflected by early topographic features on the west side of Portland, and the rock beneath some flows is tilted, indicating that at least some mountain building took place beforehand. The Willamette Valley is probably much younger, obtaining its clear topographic expression only 2 or 3 million years ago.

Basin and Range

The south-central and southeastern parts of Oregon exhibit the distinctive Basin and Range topography, in which parallel valleys alternate with parallel mountain ranges. In general, the valleys and ranges trend approximately north to northwestward and are separated by normal faults that down-drop the valleys and uplift the ranges. These normal faults are products of crustal extension in

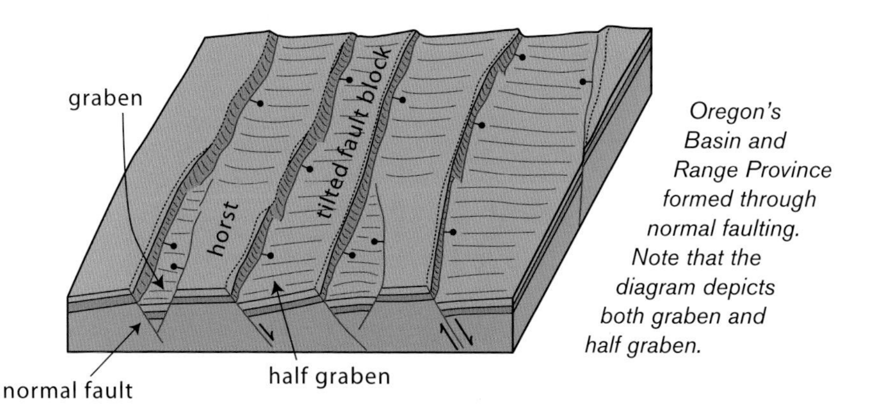

Oregon's Basin and Range Province formed through normal faulting. Note that the diagram depicts both graben and half graben.

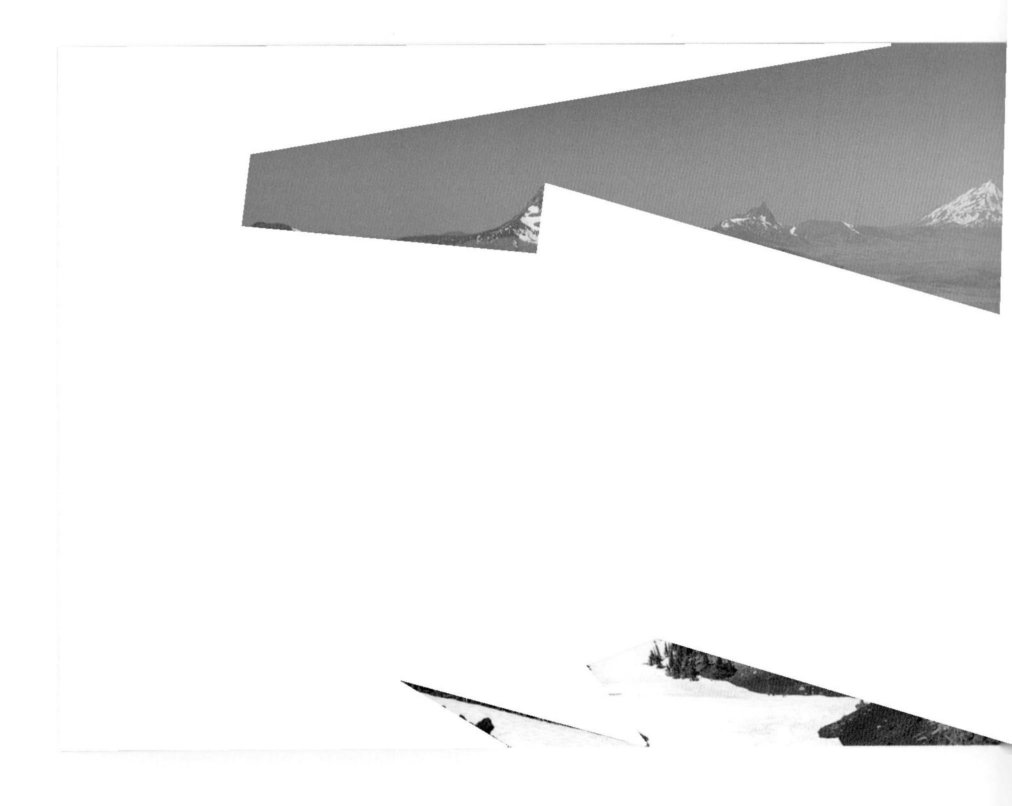

a west-southwest direction, perpendicular to the faults. Crustal extension in southern Oregon began before about 11 to 10 million years ago near Steens Mountain and continues today. Northern Oregon and even Washington are also broken by normal faults and express some of these features but to a much smaller degree.

A distinctive asymmetry to the landscape develops as a range tilts away from its bounding fault. Known as a tilted fault block, this asymmetry is especially common in Nevada, where ranges are uplifted along only one side by faults. Several ranges in southern Oregon, however, are more complicated because they have been uplifted along both sides by faults. A range uplifted on both sides is called a horst, and an intervening valley down-dropped by faults on both sides is called a graben. Some examples of ranges that are part tilted fault block and part horst include Hart Mountain and Steens Mountain, with the Alvord Desert and the southern part of Summer Lake valley forming grabens.

High Cascades

Beginning about 8 million years ago, volcanic eruptions began to shift slightly eastward from the Western Cascades to the site of the High Cascades, where most volcanic activity in Oregon is today. Like so much else in Oregon, this shift likely resulted from changes at the plate margin. A slight decrease in the angle of subduction naturally moved the locus of magmatism eastward. As the site of

View northward of the High Cascades from near McKenzie Pass. From left to right, the peaks are Belknap shield volcano, Mt. Washington, Three Fingered Jack, Mt. Jefferson, and Mt. Hood. Mt. Jefferson looks bigger than Mt. Hood because it's closer.

volcanism shifted away from the Western Cascades, they became deeply incised by river erosion. The active High Cascades were blanketed by abundant basaltic lava that gave them relatively low relief. Just as today, this low relief was punctuated by the occasional andesitic stratovolcano, which rose steeply above the low-relief basalt. However, not all recent volcanic activity has been focused in the High Cascades. A series of lava flows called the Boring Volcanics erupted on the western edge of the Western Cascades near Portland beginning 2.6 million years ago. East of the Cascades, basaltic and rhyolitic eruptions took place at Newberry Volcano on the High Lava Plains southeast of Bend as recently as 1,300 years ago.

Glaciation and the Missoula Floods

Earth's most recent glacial period began about 2 million years ago and ended about 10,000 years ago. During this time, known as the Pleistocene Epoch, alpine glaciers occupied areas of high elevation and continental ice sheets advanced into the northern United States from Canada. In Oregon, stratovolcanoes of the High Cascades and the higher elevations of the Wallowa Mountains hosted glaciers. The continental ice sheets never reached as far south as Oregon, but they still affected Oregon's landscape. Because so much water was tied up as ice, sea level was some 300 feet (90 m) lower. As a result, the coastal rivers cut deeply into the Coast Range. When sea level returned to its original level after the glaciers retreated, the river valleys became estuaries. The Siuslaw and

Umpqua Rivers, for example, are both influenced by tides more than 10 miles (16 km) upstream from their mouths.

Probably the most mind-boggling effects of the glacial period were the Missoula Floods, when a huge ice-dammed lake in Montana burst through its dam and the floodwaters charged across eastern Washington, down the Columbia River, and into the Pacific Ocean. The flood cut deep canyons into basaltic bedrock across thousands of square miles in southeastern Washington. In the Columbia River Gorge, the flood crested near the level of Crown Point, 700 feet (210 m) above the modern river. Water inundated what is now downtown Portland to a depth greater than 350 feet (105 m). Between 18,000 and 15,000 years ago, these floods happened repeatedly. The evidence for these floods comes from a multitude of erosional and depositional features that can be seen along the flood's path, from Missoula, Montana, to Astoria, Oregon.

The Missoula Floods consisted of more than forty floods. At least twenty-five of these floods had discharges greater than 35 million cubic feet (990,000 cubic meters) per second, and at least six of those had discharges greater than 230 million cubic feet (6.5 million cubic meters) per second, nearly four hundred times the average flow of the Mississippi River. Volumes of some floods exceeded 500 cubic miles (2,100 km^3). The larger floods filled the Columbia River Gorge to depths of 800 feet (240 m), overtopped drainage divides, and backed up tributary canyons for tens of miles. In addition, the floodwaters created two sizable

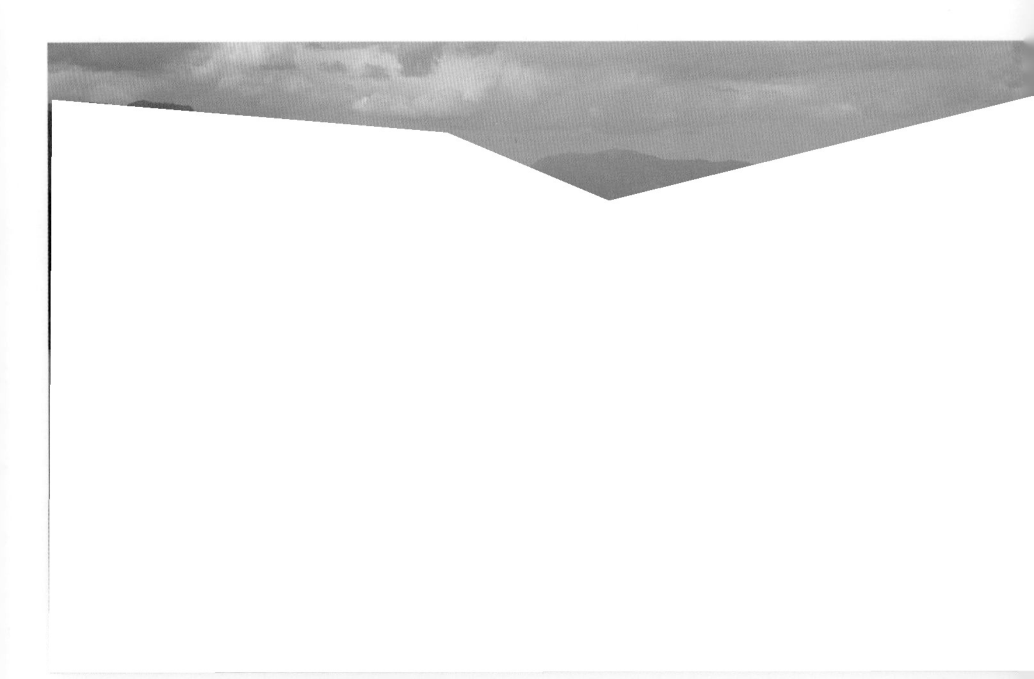

Crown Point and Vista House and the Columbia River Gorge, looking eastward from Portlan Women's Forum State Scenic Viewpoint. Floodwaters of the Missoula Floods crested at the fl surface on which the Visitor Center stands. Interstate 84 cuts through the trees in the lower left.

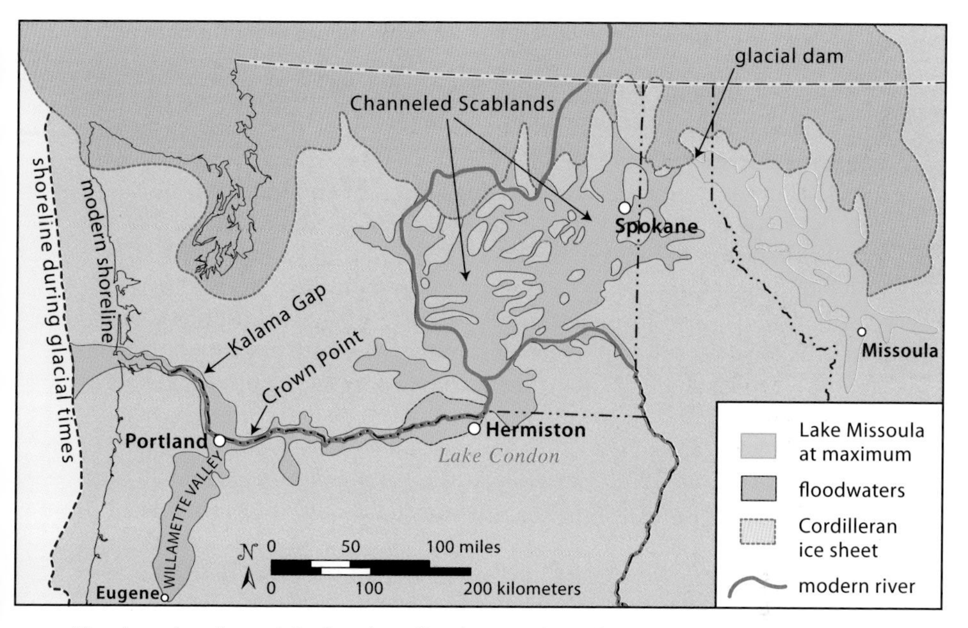

After bursting through its ice dam, floodwaters from Glacial Lake Missoula scoured much of eastern Washington en route to the Columbia River valley and the Pacific.

but short-lived lakes in Oregon where the water backed up behind narrow stretches of the channel. One of these lakes formed behind Kalama Gap downriver of Portland and filled the Willamette Valley all the way south to Eugene. The other, named Lake Condon after Oregon's first state geologist, stretched from its constriction at The Dalles all the way to Hermiston, Oregon.

As their name implies, the Missoula Floods originated near Missoula, Montana. There, prominent shorelines, more than 500 feet (150 m) above the valley floor, indicate the presence of a former lake. Known as Glacial Lake Missoula, it filled the Missoula and Bitterroot Valleys and stretched northwestward to Idaho, covering 3,000 square miles (7,800 km^2), and reaching depths greater than 1,500 feet (460 m) at its greatest extent. The lake formed because a tongue of ice from the Cordilleran ice sheet advanced across and dammed the Clark Fork River. Eventually, the water backed up and deepened enough to exert so much pressure on the ice dam that it failed catastrophically. The actual failure could have occurred through a variety of causes, but most researchers seem to favor the idea that water exploited cracks in the ice, possibly as it overtopped the dam. The suddenly unleashed lake water traveled southwestward to Spokane, then across southeastern Washington and into the Columbia River.

For about 2,000 years, these floods occurred every 50 years or so. After the lake drained, the ice would advance once again across the river and dam it, creating a new lake in the process. This lake would eventually break through the ice, destroy the dam, and cause a new flood. As might be expected, the larger the ice dam, the larger the lake and subsequent flood. Flood sizes tended to decrease through time as the climate warmed and smaller amounts of ice advanced into the area.

The floods stripped soil and sediment, exposing bedrock along nearly their entire route and scoured channels and potholes in places that later became canyons and lakes. Probably the most dramatic erosional features are the Channeled Scablands of southeastern Washington, a series of deep canyons carved into the basaltic bedrock. Grand Coulee and Moses Coulee exceed 500 feet (150 m) in depth. In Oregon, the floods widened the Columbia River Gorge by scouring its sides as well as destabilizing slopes, which resulted in landslides.

Although many of the depositional features are more subtle, they abound along the flood route and provide most of the details about the floods. Grain sizes of individual sedimentary deposits provide information about flow conditions, and wood fragments within them can be dated using carbon-14, to create a detailed chronology. Gravel bars and fans, for example, exist along the Columbia River Gorge, especially at tributary mouths where floodwaters formed huge eddies that swirled back into the tributary. Deltas formed in places where water velocities slowed as the result of a widening of the channel, such as where the floods entered the Portland Basin. Some gravel deposits even exist in canyons separated from the Columbia River by drainage divides, indicating flood stages that overtopped the divides.

In addition to the gravel bars, finer deposits of sand, silt, and clay accumulated in slack-water areas, such as the upper reaches of some of the tributaries, or in the large temporary lakes that formed upstream from channel constrictions. As many as forty flood-deposited beds, each reflecting a different flood event over the 2,000-year period, have been identified in Washington and Oregon. These deposits can be found some 50 miles (80 km) up the Deschutes and John Day Rivers, for example, as well as throughout much of the Willamette Valley and in the area that was submerged beneath Lake Condon. Many of the deposits have been plowed over as farmland, but there are still exposures preserved along streams and rivers.

Some of the more unusual depositional features are glacial erratics, boulders encased in ice that was broken off the advancing ice sheet and carried by the floods like giant ice cubes. A glacial erratic is a rock that originated somewhere else and was carried to its present position. The erratics in Oregon don't resemble the local bedrock. Many, for example, consist of granite derived from intrusions in Idaho. The well-known Sheridan erratic just south of McMinnville consists of slightly metamorphosed shale of the Belt Supergroup, a thick sequence of Precambrian sedimentary rocks found only in northeastern Washington, northern Idaho, western Montana, and southern Alberta. Altogether, more than four hundred erratics have been identified in the Willamette Valley, some of which were found near its southern edge. Most of these erratics lie at elevations of 200 to 400 feet (60–120 m) above the valley floor, indicating the depth of water on which the ice floated.

The most famous erratic in Oregon is the Willamette meteorite, discovered in 1902 on a hill in West Linn, southeastern Portland, by a settler named Ellis Hughes. At more than 15 tons (13,600 kg), this iron-nickel meteorite is the largest meteorite found in North America and the sixth largest in the world. With no trace of a nearby impact crater and the presence of other glacial erratics nearby, it is most likely that the meteorite fell somewhere in southern Canada or Montana and was carried to Oregon along with the other glacial debris. The meteorite is highly fluted, an appearance acquired through a combination of the high temperatures during entry into the atmosphere and subsequent weathering after impact.

The meteorite is on permanent display at the American Museum of Natural History, which acquired it in 1906. However, the meteorite played an important ceremonial role for the Clackamas tribe long before it was discovered by Hughes. After some legal wrangling, the museum and tribe worked out a solution where tribal members could perform an annual private ceremony with the meteorite at the museum. A replica of the meteorite, beautifully restored by sculptor Peter Helzer, is on display in the courtyard of the Museum of Natural and Cultural History in Eugene, Oregon.

Today, small alpine glaciers still inhabit parts of some of the high peaks in the Cascades, such as Mt. Hood, Mt. Jefferson, and North Sister, but they have pretty much disappeared elsewhere. The glaciers did, however, leave behind a great deal of evidence of their more extensive past, from glacially sculpted peaks and valleys to glacial deposits called till. Some of these deposits form large ridges called moraines, such as that surrounding much of Wallowa Lake.

Alpine glacier near the summit of Mt. Jefferson, an eroded volcano of the High Cascades.

Scale replica of the Willamette meteorite in Eugene. The replica measures 10 by 6.5 feet (3 x 2 m).

Oregon's Modern Climate and Waterways

Oregon's climate is largely controlled by its proximity to the Pacific Ocean and the presence of the High Cascades. Moisture-laden air coming off the Pacific Ocean rises and cools as it moves eastward over the Coast Range and then the High Cascades. As it cools, it loses its moisture as precipitation. The part of Oregon west of the Cascade peaks receives abundant rain and snowfall, but east of the Cascades, the air is much drier. Most of eastern Oregon is in the rain shadow of the Cascades.

Still, two extensive wetland areas in Oregon lie east of the High Cascades and are protected as national wildlife refuges. The water in each of these areas originates from deep snowpacks in adjacent mountains. The Klamath Basin National Wildlife Refuge Complex near Klamath Falls obtains most of its water from the nearby High Cascades. The Malheur National Wildlife Refuge near Burns obtains its water from melting snow on Steens Mountain, which reaches an elevation of 9,733 (2,966 m) feet above sea level.

Numerous rivers flow to the sea from the Coast Range, and a few bigger rivers flow west from the Cascades, but only the mighty Columbia River and the smaller Klamath River on the Oregon-California border manage to cut through the Cascade Range in Oregon. Most rivers in eastern Oregon flow north to the Columbia or Snake River and then on to the Pacific. Some rivers in the Basin and Range of eastern Oregon flow into closed basins, valleys that don't drain out. Water that flows into a closed basin forms a lake at the lowest elevation. Much or all of the lake water may evaporate in the dry, hot summers.

In the cooler, wetter climate of the Pliocene and Pleistocene Epochs, large lakes existed in these closed basins, eroding shorelines and depositing sediments. Summer Lake and Alvord Lake are two remnants of these glacial lakes.

Western Oregon, with its abundant moisture and numerous streams, hosts the vast majority of Oregon's waterfalls. There is no way to generalize how all waterfalls form; they simply form wherever water flows over a cliff. Differential erosion, where a rock that resists erosion sits next to one that is more susceptible to erosion, is responsible for many of the cliffs. As erosion proceeds and the resistant rock is especially hard, a cliff can develop. In Oregon, many of the basalt flows form nearly horizontal layers. Some of these layers are resistant to erosion while others are not, giving a stair-stepped appearance to the landscape, with resistant cliffs alternating with less resistant slopes. This type of differential erosion makes Oregon particularly well-suited to the formation of waterfalls. Then, as waterfalls form, their plunge pools erode the bases of the cliffs, making the cliffs even steeper. Eventually, this erosion will undercut the cliff and cause it to fail, and the waterfall will migrate upstream.

The White River pours over lava flows of the Columbia River Basalt Group at White River State Park in Tygh Valley, just east of US 197.

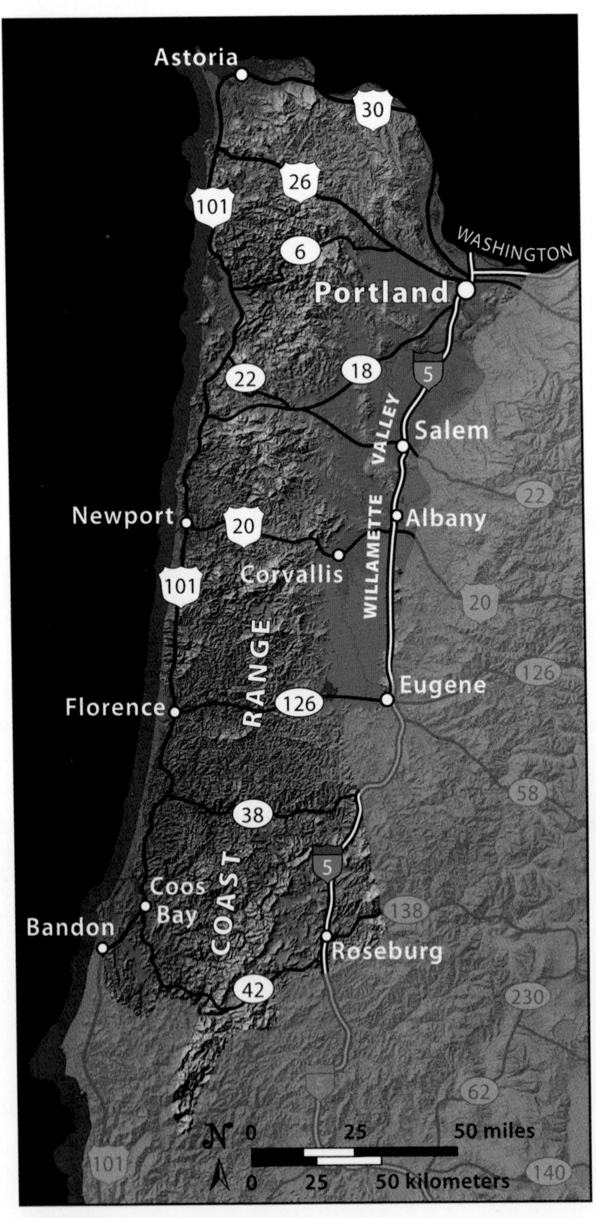

Physiography and principal cities and towns of the Coast Range and Willamette Valley. –Base image from US Geological Survey, National Elevation Data Set Shaded Relief of Oregon

Coast Range

Three geographic areas of Oregon, the coastline north of Bandon, the Coast Range, and the Willamette Valley, display different landscapes but have similar underlying geologic stories. They are fundamentally the same, just modified by different modern processes. The Willamette Valley is a part of the Coast Range that has subsided to form a broad valley, while the Oregon coastline is simply the Coast Range eroded by the Pacific Ocean.

Formation of the Coast Range

The Coast Range, a belt of sandstone- and igneous-capped peaks and ridges, runs from southern Oregon to northern Washington. It exceeds 50 miles (80 km) in width in some places. The range in Oregon is drained by a dozen or so rivers that cut deeply into the bedrock. In many places, the topographic relief from valley floor to ridgetop exceeds 3,000 feet (900 m). The roadside geology of the Coast Range is pretty remarkable, despite the heavy forests, which can make even the most mundane bedrock exposure seem exciting. The roads traversing the range pass enough roadcuts and features to illustrate the geologic story. And on the coastline, you can see unparalleled scenery with outstanding rock exposures.

The Coast Range bedrock consists mostly of Eocene-age sandstone deposited on top of a basement of Eocene oceanic basalt that is only slightly older. The basalt is part of the Siletz terrane of seamounts and ocean floor that was accreted to North America between 55 and 50 million years ago. Much of the sandstone belongs to the Umpqua Group and Tyee Formation. The Umpqua Group was deposited in marine environments during the accretion of Siletzia, whereas the Tyee was deposited after accretion. The Tyee Formation was deposited in a submarine fan complex with shallow delta deposits near the coast and increasingly deeper water deposits farther north. In many places, the Tyee Formation reaches several thousand feet in thickness.

The source of the Tyee Formation sandstones was long thought to be the Klamath Mountains to the south. Ancient current directions in the Tyee pointed northward away from the Klamaths, and the specific mineral grains within the Tyee all had counterparts in the Klamath bedrock. Beginning in the mid-1980s, however, studies of chemical isotopes within potassium feldspar and white mica grains of the Tyee indicated a profound difference from the Klamaths, which ruled them out as the primary source. Instead, researchers now believe most Tyee sands eroded from the granitic rocks of the Idaho Batholith. During

Eocene time, before the clockwise rotation of Oregon, the Idaho Batholith and the Tyee were much closer to each other.

Overlying the Tyee Formation in much of the northern and parts of the southern Coast Range are sedimentary rock and interlayered igneous sills and volcanic flows that formed later in the Eocene Epoch and into the Oligocene and Miocene Epochs. One rock formation of particular note is the Columbia River Basalt Group, which forms many of the headlands of the northern coast. Intrusive features, such as basaltic dikes and sills intruding sedimentary rocks, led many researchers to argue these rocks came from local magma that reached the surface here. In the 1980s, however, a series of geochemical studies showed that these basalts were part of the Columbia River Basalt Group that originated in eastern Oregon or Washington. Having flowed from eastern Oregon, they are some of the longest basalt flows in the world. The intrusive features along the coast formed when lava at the base of the massive basalt flows invaded the sedimentary rock downward along its layers and then worked its way back up to the surface along fractures. This mind-boggling sequence of events could happen because the surface of the flow had cooled upon entering the ocean, but the lava inside remaind hot and continued to flow, exerting pressure on and escaping into the underlying soft sediments.

The entire Coast Range forms a single gigantic fold, with the rock on the east side tilting gently toward the east, and the rock on the west side tilting gently toward the west. Somebody driving across the range encounters the oldest rocks in the core and progressively younger rocks outward in either direction. This type of fold, in which the rocks tilt away from and are oldest in the core, is called an anticline. A fold in which the rocks tilt toward and are youngest in the core is called a syncline.

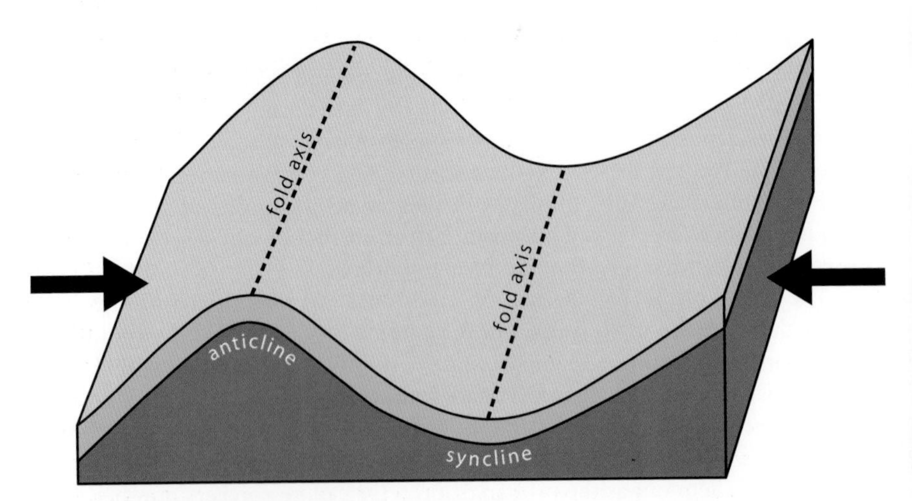

Block diagram of an anticline and syncline. The arrows signify the direction of compression, perpendicular to the long axes of the folds. Note that in the anticline, the rocks in the core are the oldest, and in the syncline, the rocks in the core are the youngest.

Most folds form because rock becomes compressed—and the long direction of the fold (the axis) is perpendicular to the direction of compression. It makes sense that the Coast Range is folded with a north-south trending axis, because the convergence of the North American and Juan de Fuca Plates is causing east-west compression. It is surprising, however, how little deformation there is, given that the Coast Range resides next to this plate boundary. Aside from the large anticline, the rocks appear fairly undisturbed. One explanation for this lack of deformation is that the Siletz terrane, which lies at shallow levels beneath the Tyee Formation, acts as a rigid block that stabilizes the rock above it and prevents it from folding.

The Oregon Coast Range began forming before the first eruptions of the Columbia River Basalt Group but attained its current form much more recently. The presence of Columbia River Basalt Group flows along the coast indicates that the Coast Range was not a significant barrier to the flows during their eruption at about 15 million years ago and that most of the uplift occurred later. However, some lava flows appear to be deflected by early topographic features on the west side of Portland, and the rock beneath some flows is tilted, indicating that at least some of the mountain building took place beforehand. The Willamette Valley was probably a lowland during the eruption of the Columbia River Basalt Group but acquired its distinctive topographic expression only 2 or 3 million years ago.

Today, the Coast Range may or may not still be rising, at a rate that is difficult to assess and so is the subject of much debate. Estimates range from less than 0.1 millimeter to about 1 millimeter per year, a range that seems inexorably slow when viewed on a human timescale. Over a period of 1 million years, however, a 1-millimeter-per-year uplift rate brings more than 0.6 mile (1 km) of uplift. However, as the mountains rise, erosion wears them down, and some researchers argue that the amount of erosion equals the amount of uplift. If so, then the range's elevation is staying approximately constant through time.

Superimposed on the long-term uplift rate are repeated cycles of geologically rapid rising and falling caused by earthquakes at the subduction zone. The North American and Juan de Fuca Plates are locked together, gradually building up their stress levels until they can suddenly slide, releasing the stress as an earthquake. During the time that the stress is building up, the overriding North American Plate compresses and bows upward, at rates that may exceed 4 millimeters per year. During earthquakes, the overriding plate releases the stress and falls back to its original elevation. Because of this cycle, vegetation becomes established along coastal estuaries as they rise slightly above sea level but then drowns when the land subsides during an earthquake. Indeed, sedimentary deposits along many of Oregon's estuaries reveal layers of peat, formed from the vegetation, alternating with mud-rich intertidal deposits. Oregon's coastline also hosts some ghost forests, stands of dead trees rooted only a few feet above sea level and drowned from subsidence during the most recent earthquake. They are above sea level now and visible because we are in the uplift portion of the cycle.

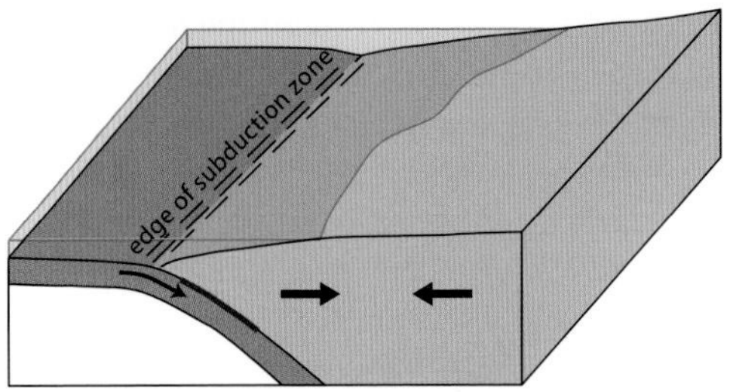

The upper part of a subduction zone is locked, causing compression of the overriding plate.

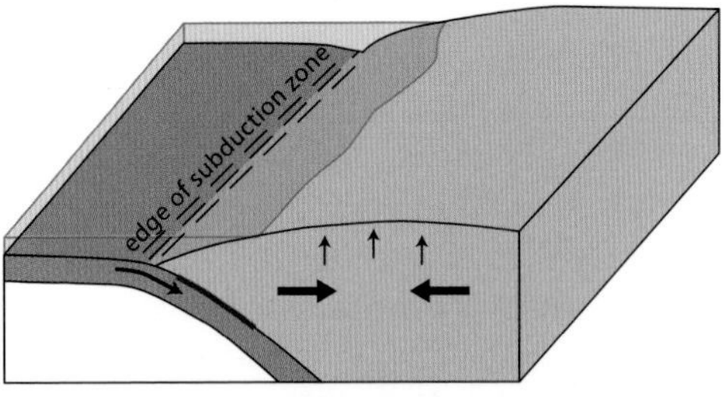

As compression increases, the plate bulges upward and the sea retreats.

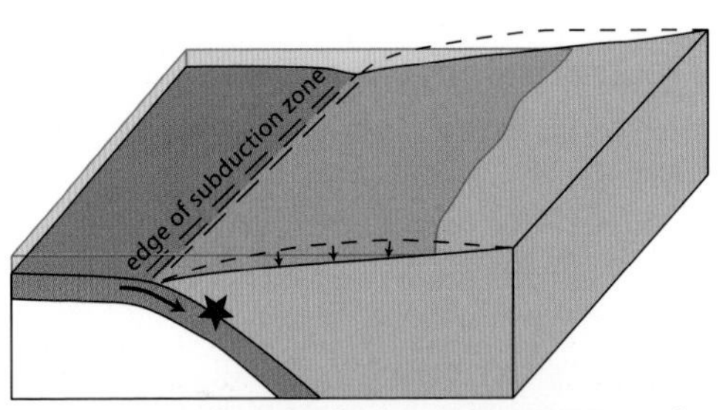

During an earthquake, built-up compression relaxes, the bulge goes back down, and the sea advances.

Because the subduction zone is locked, compression builds up in the Coast Range, which causes it to rise during an interseismic period. As the mountains rise, the sea recedes. When compressional stress becomes too great, the plates slip, releasing seismic energy as a large earthquake and allowing the Coast Range to relax back to its original position.

These ghost forests, along with sedimentary deposits from earthquake-generated tsunamis, give us a great deal of information about the potential and frequency of subduction zone earthquakes. Tsunami deposits exist up and down the coast from northern California to Washington, and by carefully applying carbon-14 dating methods to pieces of wood preserved in the deposits, researchers can tell when they formed. The ages of the deposits are consistent from one place to the next, telling us that large portions of the subduction zone periodically slip at the same time. Occasionally, the entire zone slips, releasing a correspondingly large amount of seismic energy. Indeed, most researchers argue that Oregon's subduction zone earthquakes are huge, on the order of magnitude 9 when the entire zone fails. We also know that they tend to occur once every 300 to 500 years because the ages of the sedimentary deposits are about 300 to 500 years apart. In general, northern California and southern Oregon have

Ghost forest of Sitka spruce exposed in the tidal zone of Sunset Bay near Coos Bay. These trees grew some 1,200 years ago in a forest that was just above high tide during an interseismic period. After an earthquake, the land subsided below sea level and the trees drowned. They have since been exposed and submerged over several earthquake cycles.

endured more earthquake events than the northern half of the zone. It seems the northern half of the zone experiences only the largest earthquakes every 500 years or so, while the southern half also experiences smaller ones.

While driving along the Oregon coast, you'll see tsunami warning signs along the highway wherever it dips to a low elevation. Tsunamis, or seismic sea waves, form from the same slip events that cause earthquakes. When part of the seafloor shifts upward, it displaces a great volume of water, which then rolls outward as a low-amplitude but long-wavelength wave. When it approaches a coastline and the seafloor becomes shallower, the wave increases to a height that may greatly exceed high tide, inundating low-lying areas and estuaries near the shoreline. Tsunamis devastated Crescent City, California, after the 1964 Alaska earthquake and caused more than one million dollars of damage to Brookings, Oregon, during the 2011 Tohoku earthquake in Japan. Tsunamis leave evidence in the geologic record as well. Many of the peat and intertidal deposits in the low-lying areas contain intervening thin beds of sand, deposited during tsunamis.

In the mid-1990s, Brian Atwater of the US Geological Survey and his colleagues determined the timing of the most recent subduction zone earthquake to an unusual precision: the night of January 26, 1700. This earthquake left drowned forests and tsunami deposits in its wake up and down the entire Pacific Northwest coastline, so it must have involved slip along the entire zone. From a combination of radiocarbon dating and tree-ring analysis, researchers were able to pinpoint the timing of the earthquake to within only a few years. Further study discovered records of a tsunami that hit the coast of Japan on January 27, 1700. By calculating the length of time it would take a tsunami to cross the Pacific Ocean, the researchers were able to backtrack to the actual time of the earthquake.

The Oregon Coast

The Oregon Coast stretches some 300 miles (480 km) from the California border, just south of Brookings, to Astoria, at the mouth of the Columbia River. It offers miles and miles of relatively unoccupied beaches, countless rocky headlands, a relentless surf, and a fascinating blend of ancient and modern geologic processes. Moreover, the coast is easily accessible from US 101, which closely follows the continent's edge in Oregon. Dozens of state parks allow additional access, as well as camping and picnicking.

Headlands and Sea Stacks

Some of the most dramatic elements of Oregon's coastline are its rocky headlands, sea stacks, and pocket beaches, all of which form because of erosion. Erosion takes place just about everywhere and through a variety of means. It occurs right at the shoreface, where waves break against beaches and rocks; during high tides and storms, when waves reach the hillsides and cliffs behind the beaches; and on the forested slopes away from the coast, where water-saturated material breaks loose and slides downward.

Because the headlands jut outward from the coastline and receive the most punishing wave energy, they must consist of more durable, erosion-resistant rock than areas without headlands. On the northern coast, nearly all of the headlands consist of basalt of either Siletzia, various Eocene or younger basalt flows and shallow intrusions, or the Columbia River Basalt Group. Along the southern coast, where the rocks belong to accreted terranes, the bedrock is highly variable, but the rule still holds true: where a more resistant rock type lies next to a less resistant one along a coastline, the more resistant rock will form the headland while the less resistant one will recede and form a beach.

The central part of the Oregon coast hosts only a few headlands, most notably around Cape Arago and Cape Blanco. Most of the central coast consists of long stretches of beach with an extensive coastal dune field behind it, the longest coastal dune field in North America. The central coast is developed on sandstone of the Eocene Tyee Formation, a fairly homogeneous and erodable rock. Cape Arago and Cape Blanco are unusual because, as described in the section on marine terraces, they are actively rising relative to adjacent parts of the coast and form headlands because of this uplift.

Headlands erode in irregular ways because crashing waves exploit zones of weakness—fractures, faults, or weak sedimentary laters—in the otherwise strong rock. This process, called differential erosion, causes headlands to retreat

Yaquina Head, north of Newport, is formed from Columbia River Basalt Group, a resistant rock.

A pocket beach, framed by basaltic headlands, north of Newport.

behind a wreckage of not-yet-eroded rocks, which rise above the surf as a series of sea stacks. In some places, sea arches link sea stacks to each other or to the coast, but they too eventually fail. Some sea stacks lie up to 1 mile (1.6 km) offshore. Perhaps more than anything else along the coast, these offshore sea stacks demonstrate geologic change, because just like the sea stacks close to shore, they used to be part of the coastline.

Sea caves and blowholes are two other features that form by differential erosion of headlands. Caves form through considerable erosional widening of faults or fractures while the overlying rock somehow remains intact, possibly because the fault or fracture surface is inclined. Blowholes form where the incoming surf funnels into a narrow channel eroded into the rock, encounters an abrupt dead end, and then shoots upward, resembling a geyser.

Marine Platforms and Terraces

Marine platforms are flat exposures of bedrock within the intertidal zone and often exist near headlands. They are the end product of erosion, in which wave action has reduced all the bedrock to a common level. And although they are much less dramatic than the headlands, they provide important information regarding the uplift history of the coast—as well as some of the best places for exploring tide pools!

Marine platforms are valuable for understanding relative sea level changes because their flat surfaces form at sea level. If they become uplifted, they become marine terraces, typically with a thin layer of marine sand deposited

Marine platform in the foreground, with Cape Perpetua in the background. Cape Perpetua exposes several lava flows of the approximately 30-million-year-old Yachats Basalt.

Northward view to the Whisky Run terrace from Shore Acres State Park. The terrace is the flat surface on which the forest grows, directly on top of tilted rocks of the Coaledo Formation. Note the crossbedding in the foreground.

over the bedrock. By measuring their present elevation, we can determine how much uplift has occurred since the flat surface eroded at sea level. In addition, we know that the terraces get older the higher they are in elevation; any departure from this sequence would cause earlier-formed terraces to be flooded and destroyed. In some places along the coast, such as just north of Brookings, near Cape Blanco, or in the area around Coos Bay, researchers have documented between five and seven terrace levels, reflecting different stages of uplift over the last several hundred thousand years. Where researchers are able to determine an age for a given terrace, they can infer a rate of uplift. In general, these rates range from 4 to 12 inches (10–30 cm) every 1,000 years, except near active faults, where the uplift rates reach about 3 feet (0.9 m) per 1,000 years.

Not only do the terraces reveal uplift along the coast, but several of them slope gently to the north or south, showing some degree of recent folding. Cape Arago is one such example. There, the lowest, youngest terrace reaches a high point at an elevation of about 80 feet (24 m), but gradually slopes down to sea level at the mouth of Coos Bay.

Landslides

The dramatic sculpting of rock by waves notwithstanding, landslides probably accomplish more erosion than anything along the coast because they can move a huge volume of material all at once. The conditions along the coast are

Landslide near Cape Meares, west of Tillamook. The landslide likely occurred in the last twenty to thirty years, because it still contains a host of broken trees and its toe has not eroded back from the surf zone.

perfect for landsliding: steep slopes and plenty of water. Moreover, some localities consist of sedimentary rock with bedding that slopes outward, away from the land, which encourages the overlying rock to slide. Many coastal landslides occur after storm surges undercut the slopes behind the beaches, but some landslides are triggered by human activity, such as road building, that locally oversteepens slopes. These roads are not limited to paved roads; logging roads wind through nearly all the mountains along the coast, including throughout the Coast Range and Klamaths—and those countless stands of clear-cut forests are especially prone to landsliding.

Beach Sand

Sand consists of any mineral or rock fragment between 0.06 and 2 millimeters in diameter. Initially a product of weathering and erosion, a great deal of sand comes to rest, at least temporarily, on Oregon's beaches and dunes. Finer, silt-sized particles as well as coarser pebbles, cobbles, and even boulders are also deposited in these places, but most of the material by far is sand. Oregon's beaches come in all shapes and sizes, from the one between Coos Bay and Florence that stretches continuously for some 50 miles (80 km), to sand spits at river mouths, to pocket beaches isolated between headlands.

Oregon beach sand comes from four sources: the Columbia River to north, the Klamath Mountains to the south, the Coast Range in the middle, and modern erosion of hills and cliffs behind the beaches. Detailed studies by Komar (1998) and Clemens and Komar (1988) show that even in central Oregon, there is a large component of Klamath-derived sand. The sand contains minerals such as glaucophane, pink garnet, and olivine, which, for the region, are unique to the Klamaths. However, today's headlands block lateral movement of sand, so this transport must have taken place some 20,000 years ago or earlier, during the last ice age of the Pleistocene Epoch when sea levels were much lower and the headlands as we know them now did not exist.

Sandy Beach and headlands at Carl. G. Washburne Memorial State Park, just north of Florence. View toward the south.

What did exist during the last ice age was a broad coastal plain, some 20 to 30 miles (32–48 km) wide, that extended to the edge of today's continental shelf. The Columbia, Siuslaw, Umpqua, Rogue, and Klamath Rivers transitioned from high-gradient mountain rivers to low-relief meandering ones as they carried sediment across the plain and dumped it at the water's edge. The sand migrated mostly northward along the coast, mixing Klamath-derived grains with those from the Coast Range. Northern Oregon beaches have a much smaller component of Klamath sand and a large component of material from the Columbia River.

Melting of glaciers at the close of the Pleistocene Epoch caused sea level to rise, inundate the coastal plain, and push the sand and the coastline eastward. Some 300 feet (90 m) of sea level rise occurred since the end of the last ice age, about 10,000 years ago. As the rising sea reached exposures of basalt

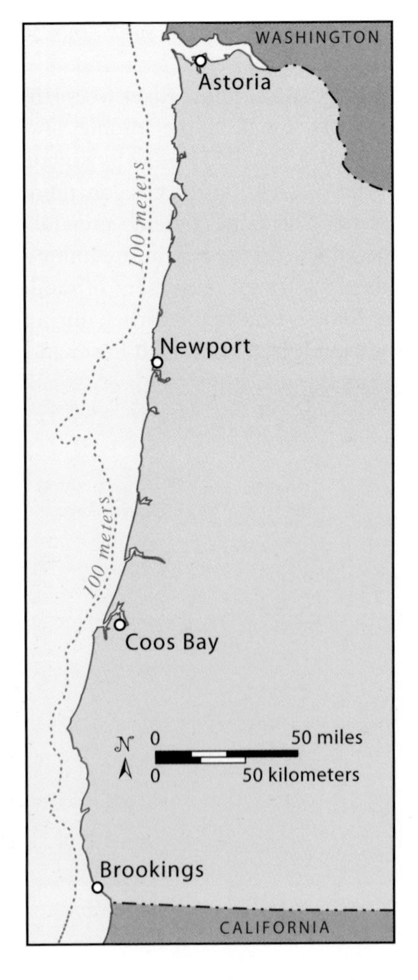

The approximate location of the Pleistocene shoreline is the 100 meter (330 feet) depth contour (blue) off the Oregon coast. The area between that shoreline and today's shoreline was a relatively flat coastal plain.
–From Walker and MacLeod, 1991

and other more resistant rock, they became headlands, effectively trapping the sand from northward transport. The rising waters also flooded low-lying river valleys in the Coast Range. Today, the lower stretches of these rivers are estuaries, with tidal influences that reach as far as 15 miles (24 km) upstream from their mouths.

Spits and Salt Marshes

Although today's headlands prevent sand from traveling a great distance parallel to the shoreline, sand still moves within the confines of its own system. Summer waves approach from the northwest and drive sand southward along the beach, while winter waves approach from the southwest and drive sand northward. The overall effect is a relatively balanced system of sand migration, with a slight northward-directed tendency because of especially strong winter storms during El Niño years. A result of this balanced longshore drift is the accumulation of numerous sand spits at river mouths along the Oregon Coast that point either northward or southward. The reason why a given spit points the way it does seems to be a combination of factors, such as the angle of the shoreline relative to incoming waves or the type of bedrock on either side of the river.

Aerial view southward of spits, headlands, bays, and lagoons at the entrance to Tillamook Bay. Cape Meares forms the prominent headland in the background. Cape Lookout forms the prominent headland in the far background.

Many of the low relief parts of the Oregon coast contain salt marshes, especially near some of the larger bays along the northern coast. For the most part, these areas are the slightly elevated parts of the Pleistocene coastal plain that were not inundated during sea level rise. They are at low enough elevations, and close enough to the shoreline, that the groundwater, which feeds the marshes, is brackish.

Coastal Sand Dunes

Oregon's central coast hosts North America's most extensive tract of coastal sand dunes. They extend continuously for some 50 miles (80 km) from Coos Bay to just north of Florence and lie directly behind the beach, from which they receive the windblown, fine-grained fraction of the sand. Other extensive fields lie between Bandon and Cape Blanco and on the Clatsop Plains, from Seaside up to the mouth of the Columbia River. In 1934, the Soil Conservation Service introduced European beach grass to help stabilize the Clatsop dunes. The grass spread quickly and now covers most of the frontal dunes along the entire Oregon coastline. The grass catches windblown sand, which causes these frontal dunes to grow in height—which in turn decreases the sand supply to the more inland dunes. Many areas on the inland side of the frontal dunes have deflated considerably, some all the way down to the groundwater table, and many have become colonized by other invasive species.

The Willamette Valley

The Willamette Valley is a part of the Coast Range that has subsided to relatively low elevations—approximately sea level in Portland and only 500 feet (150 m) in Eugene. In plate tectonic terms, the Willamette Valley is a fore-arc basin, because it lies between the subduction zone to the west and the Cascade magmatic arc to the east. It consists of three sections, divided by the Portland Hills in the north and the Salem Hills in the south. These hilly areas consist mostly of marine sandstone of the Coast Range capped by basalt of the Columbia River Basalt Group. Other spurs of bedrock, formed by volcanic or intrusive rock of the Western Cascades, protrude into the valley from the east. The southern part of the valley contains as much as 700 feet (210 m) of river-deposited sediment that overlies volcanic rock. This sediment dates back no more than 2 to 3 million years, suggesting the present valley is no older than that.

Some younger deposits include gravel and sand carried by the Missoula Floods between 18,000 and 15,000 years ago. Glacial Lake Missoula formed behind a large ice dam and filled the valleys of western Montana. Floods occurred repeatedly as the dam broke and reformed. They scoured the Channeled Scablands of Washington, rushed down the Columbia River Gorge, and inundated the Willamette Valley.

Deposits of the flood can be traced from south of Harrisburg all the way to Portland. In many places, these deposits contain boulders, some of which weigh several tons. These glacial erratics apparently rode the floodwaters encased in slabs of ice. Perhaps the most famous of these boulders is the Willamette meteorite, discovered near West Linn and now on display at the American Museum of Natural History. A copy is on display at the Museum of Natural and Cultural History in Eugene. This rock from space, made of iron and nickel, is the largest known meteorite in North America, at about 15.5 tons (14,000 kg) and more than 6 feet (1.8 m) long. It likely landed somewhere in Montana or southern Canada before being carried into Oregon by the floods. Another well-known Willamette erratic is a metamorphic rock preserved at Erratic Rock State Natural Site, near McMinnville.

Today, the Willamette River rises in the Cascade Range and flows northward the length of the Willamette Valley, picking up numerous tributaries along the way from the Cascades and the eastern side of the Coast Range. Through time, the Willamette River has swept across its floodplain, leaving former meandering channels in its wake. Just northwest of downtown Portland, the Willamette River joins the Columbia.

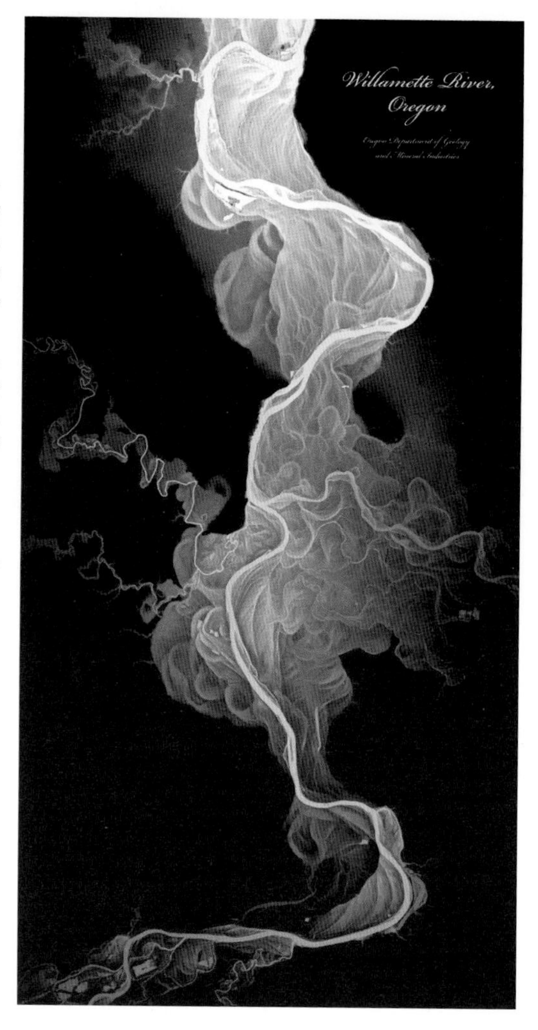

Aerial image of the Willamette River using a digital elevation model based on LIDAR, a remote sensing technique that can show subtle changes in elevation. The total elevation range shown here is only 50 feet (15 m); lower elevations are portrayed in lighter shades of blue, while higher elevations are shown in darker shades. The image shows the numerous former meander channels of the Willamette and its tributaries between Albany, near the bottom, northward to Monmouth and Independence. Note the two tributaries that join the river near the center. The Santiam, which drains part of the Cascades, joins from the east, and the Luckiamute River, which drains part of the Coast Range, joins from the west. –Image by Daniel Coe, Oregon Department of Geology and Mineral Industries

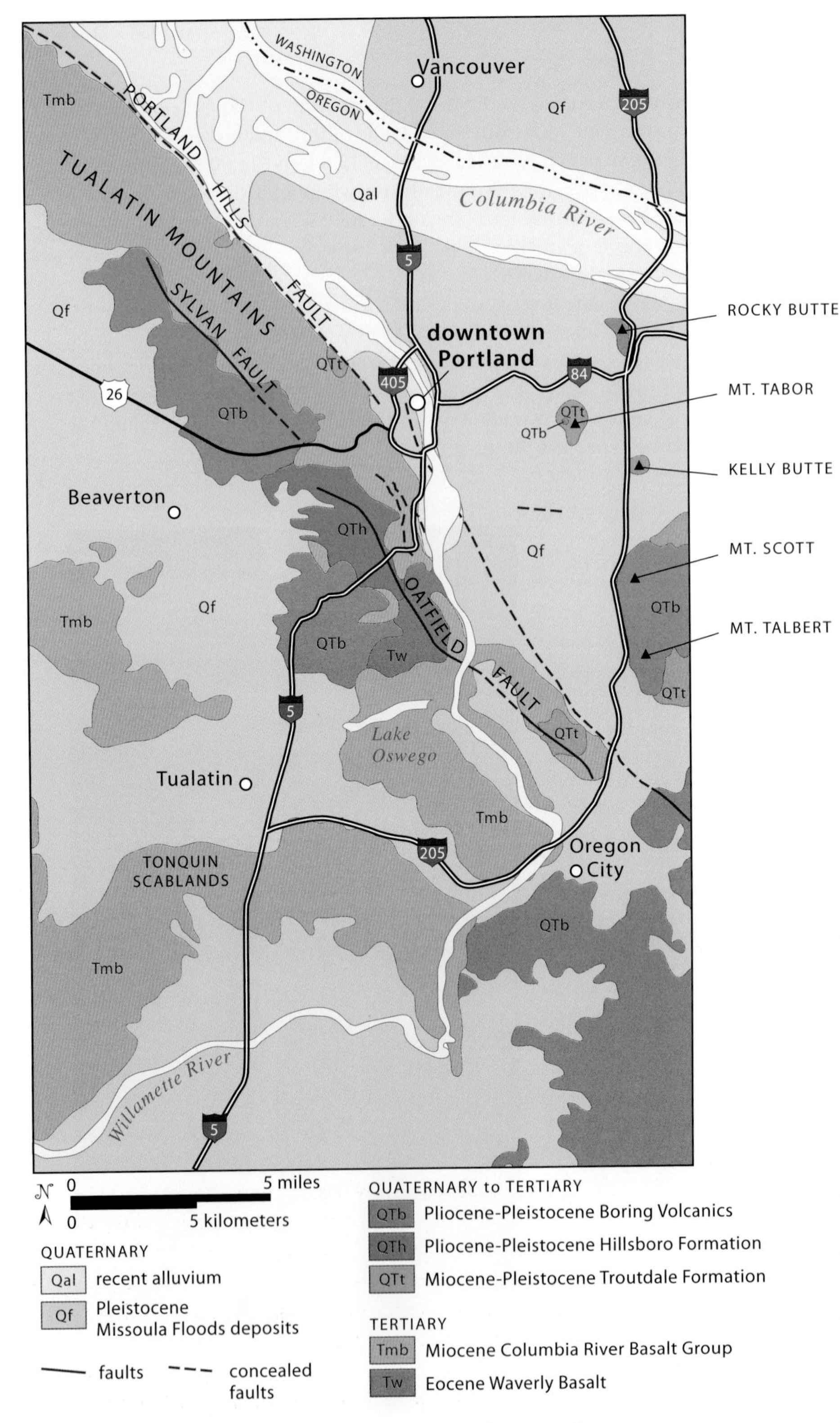

Geologic map of the Portland metropolitan area.

GUIDES TO THE COAST RANGE

PORTLAND

It can be easy to overlook Portland's geology as you concentrate on navigating the busy highways. Portland's bedrock is mostly hidden by urbanization and thick vegetation. Still, Oregon's largest city is a gem geologically as well as culturally. It offers an amazing geologic story in a relatively small area, one that reflects Oregon's plate tectonic setting and pulls together some of Oregon's most important geologic events over the past 20 million years.

Portland lies along the western edge of the Portland Basin, a low-lying area that has subsided and accumulated sedimentary and volcanic rocks for the past 20 to 15 million years. Most of these deposits are related in one way or another to the Columbia River, which has fed sediment or channeled lava flows to the basin since its inception. The city includes much of the Portland Hills, also called the Tualatin Mountains, an anticlinal ridge on the west that exposes much of the area's older rock and separates the Portland and Tualatin Basins, both part of a much larger fore-arc basin. Stretching north-south between the Cascade magmatic arc and the Cascadia subduction zone, the fore-arc basin includes the Willamette Valley to the south and the Puget Lowland in Washington.

The Portland area consists mostly of Miocene and younger formations, most notably the Columbia River Basalt Group, the Troutdale Formation, the Boring Volcanics, and deposits of the Missoula Floods. Some less extensive units include the Waverly Basalt, Hillsboro Formation, and Portland Hills Silt. In general, gravels and sands of the Missoula Floods blanket the low-lying areas of the entire region, whereas older rock units can be found in the upland areas that were not inundated by the floods.

The Portland Hills Silt accumulated in the Portland Hills as a windblown deposit called loess during the ice ages of the Pleistocene Epoch. The loess, which ranges in thickness from 1 to 100 feet (0.3–30 m), is relatively strong when dry but becomes weak when saturated with water. During the storms of February 1996, some 250 landslides occurred in the steep areas of the Portland Hills. LIDAR, a relatively new imaging technology that strips away vegetation, allows geologists to recognize many of the landslides and areas that are likely to cause future problems.

The most prominent outcrops in Portland are basalt, but the basalt comes in three varieties and ages. The Waverly Basalt is the oldest rock unit in Portland, with an age of about 43 million years, and closely resembles parts of the Tillamook Volcanics of the Coast Range. It is exposed in the area of Waverly Heights south of downtown but likely underlies much of the Portland Basin at depth. Flows of the Columbia River Basalt Group are exposed primarily in the Portland Hills and include the 16- to 15-million-year-old Grande Ronde Basalt, as well as the 15- to 14-million-year-old Wanapum Basalt. The Boring Volcanics, erupted between about 2.6 million and 50,000 years ago, show up as prominent hills throughout the region.

The Troutdale Formation, deposited between about 15 and 2 million years ago, consists of conglomerate and sandstone deposited by the Columbia River. Most of the cobbles in the conglomerate are basalt, derived from exposures of the Columbia River Basalt Group elsewhere in Oregon, but many consist of metamorphic and granitic rocks that originated even farther upriver in eastern Washington or Idaho. The best accessible exposures of the Troutdale Formation are on the sides of Mt. Tabor and Kelly Butte, on the east side of Portland, as well as along US 30 in the Columbia River Gorge. The Hillsboro Formation was deposited on the western side of the Portland Hills at about the same time as the Troutdale Formation, but it consists of material derived from the Coast Range.

The Boring Volcanics, named for exposures near the town of Boring on the east side of Portland, consist mostly of basaltic flows and cinder cones. They make up numerous buttes in the Portland area, including Rocky Butte, near the junction of I-84 and I-205, Mt. Tabor, Mt. Scott, and the hills of southwest

Cobble conglomerate of the Troutdale Formation exposed along the edge of the Willamette River beneath the Steel Bridge in downtown Portland. One of the towers of the Oregon Convention Center is visible in the background.

Aerial view of Portland and the Willamette River. The view is toward the northwest along the Portland Hills Fault. The Portland Hills, on the left side (west), end abruptly at the fault.

Portland. The Boring Volcanics seem to be related to the Cascades but are enigmatic, because they erupted in the fore-arc basin instead of along the volcanic arc.

The rocks in the Portland area are cut by a complex array of faults, many of which are potentially active, and several of which have caused historic earthquakes. Several earthquakes of magnitudes 5 and greater have recently shaken the area, including a magnitude 5 earthquake in Portland in 1962 and the 1993 Scotts Mills earthquake (magnitude 5.7) to the southeast. Most of Portland's faults, however, lie hidden beneath either urbanization or vegetation.

Portland's most dramatic fault is unquestionably the Portland Hills Fault, which shows evidence of dominantly right-lateral movement with a less significant reverse component, although some researchers suggest the minor component might be normal movement. It runs through downtown Portland and is expressed topographically where the low-elevation Portland Basin meets the Portland Hills. It continues this expression for some 30 miles (48 km) toward the northwest but loses it to the southeast where it crosses the Willamette River and extends out in the basin. A trench exposure of the fault to the southeast, however, shows that it has folded deposits of the Missoula Floods, indicating it was active in the last 15,000 years. Other potentially active faults in the Portland Hills are the Sylvan and Oatfield Faults.

The Portland Hills define a large anticline, and the Portland Basin defines a large syncline. These folds date back to before the arrival of the first flood basalts, some 16 million years ago. Studies of the Columbia River Basalt Group in Portland show that the earliest Grande Ronde flows were restricted to a relatively narrow zone northeast of the anticline, probably because the nascent fold formed an obstacle to the flow. Later flows, however, appear to have overtopped the anticline, because they can be found on its west side as well as within the Tualatin Basin. Continued growth of the Portland Hills through folding and faulting eventually led to the complete separation of the Tualatin and Portland Basins. The Portland Hills also became a long-standing barrier to the Columbia River, forcing it to turn northward for some 40 miles (64 km) before continuing its westward journey to the sea.

The faulted basalt and young volcanics were all in place by the time the Missoula Floods came charging down the Columbia, leaving behind a thick deposit of silt, sand, and gravel over all the low-lying areas around Portland and Tualatin. The floods also deposited the Portland delta, where the floodwaters emerged from the narrow Columbia River Gorge, slowed, and dropped much of their coarse sediment load. Numerous gravel bars form ridges, especially just down-current of the various buttes that rise above the delta and which obstructed the flow. The floodwaters also scoured lakes and eroded scablands in the underlying bedrock. Lake Oswego, for example, fills an abandoned channel of the Tualatin River, which was scoured down to bedrock by the floods. Just west of I-5 south of Tualatin, the Tonquin Scablands attest to some of the erosion in the Tualatin Basin.

INTERSTATE 5
Portland—Eugene
108 miles (174 km)

Nothing reinforces the concept of a fore-arc basin like driving the length of one, and this stretch of busy interstate through the Willamette Valley gives that opportunity. A fore-arc basin is the linear valley between a subduction zone and a volcanic arc. In Oregon, the fore-arc basin consists of several individual basins separated by ridges of resistant rock. From north to south, I-5 traverses first the Portland Basin, then part of the Tualatin Basin, and then finally the length of the Willamette Valley. Good outcrops are rare, but occasional roadcuts, quarries, and landforms illustrate the geologic history of this important part of the state.

Crossing the Columbia River into Oregon, I-5 passes over the flat surface of the Portland Basin, which here consists of Columbia River floodplain sediments

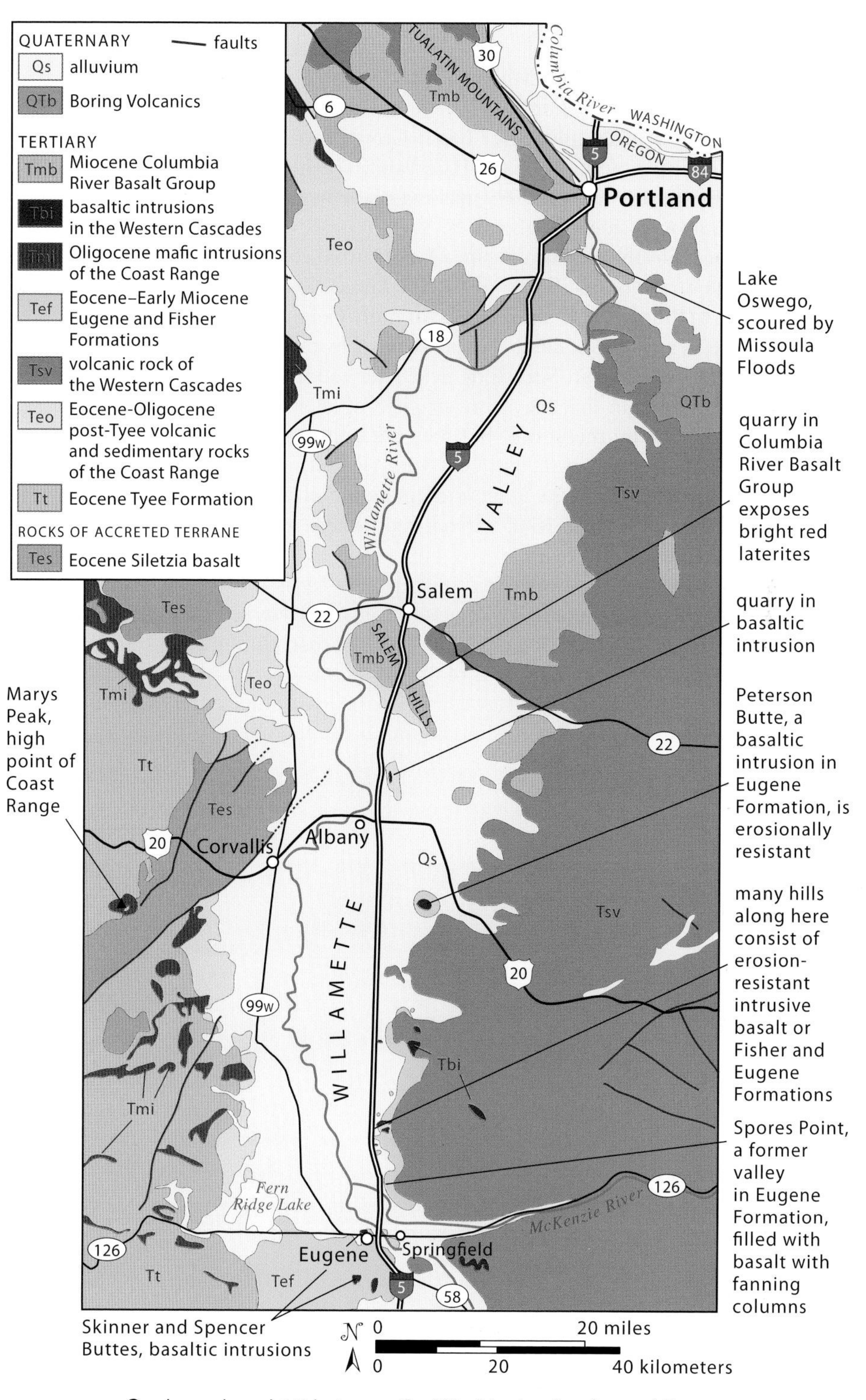

Geology along I-15 between the Washington border and Eugene.

mantled by deposits of the Missoula Floods. The Portland Hills, also called the Tualatin Mountains, rise steeply behind Portland's downtown area. They consist mostly of Columbia River Basalt Group, broadly folded into an anticline, and uplifted by the Portland Hills Fault. The highway climbs out of the Portland Basin near milepost 299, passing roadcuts of Columbia River Basalt Group on the west side of the road.

I-5 drops into the Tualatin Basin, a highly urbanized area. Low hills on the north edge of the basin consist of basaltic lavas of the Boring volcanic field, while hills to the west and south consist of more Columbia River Basalt Group. Lake Oswego, at exit 290, fills a channel scoured by the Missoula Floods, which also eroded the Tonquin Scablands near Sherwood. The Tonquin Scablands are true scablands in that they were stripped of all soil and eroded into deep winding canyons by the floods, but they are best viewed in aerial images or Google Earth because of subsequent forest growth and urbanization. The highway leaves the Tualatin Basin and enters the northern Willamette Valley a few miles south of the I-205 interchange, where it crosses a low ridge of Columbia River Basalt Group.

North of Salem, the northern Willamette Valley is mostly flat farmland, drained by the north-flowing Willamette River, as well as several other north-flowing tributaries. South of Salem, however, the road climbs the Salem Hills, another large ridge formed of flows of the Columbia River Basalt Group. Occasional exposures of deep red soil are laterites, formed from such deep weathering of the underlying basalt that most nonaluminum components of the rock were

The red laterite overlying the quarried basalt developed on the surface of the Grande Ronde Basalt of the Columbia River Basalt Group during the middle part of Miocene time. These laterites cover much of the Salem Hills and are also found in Portland and as far north as Vancouver, Washington.

leached away. In some cases, laterites form highly enriched aluminum ores called bauxite. This laterite, which formed on top this lava flow throughout much of northern Oregon, qualifies as bauxite and was commercially mined during the 1930s and 1940s. The deep weathering that creates laterites only occurs under tropical conditions, indicating a hot, wet climate prevailed here during the middle part of Miocene time. A basalt quarry on the east side of the highway about halfway between mileposts 247 and 246, exposes this laterite on top of the basalt. Southbound travelers gain a glimpse of it through the quarry gates; the view for northbound travelers is largely blocked by a wall lining the road. It is much more satisfying and safer to view this quarry from the frontage road. After crushing, the basalt from the quarry is used for a variety of purposes, including road construction and landscaping.

The laterites of the Salem Hills differ from most Jory soils, the better known and more widespread state soil of Oregon. Laterites are weathered much more deeply, so they are far more depleted in elements other than iron and aluminum than most Jory soils. In Oregon, both laterites and Jory soils form on flows of the Columbia River Basalt Group.

South of the Santiam River, which drains the Western Cascades and empties into the Willamette River, the Willamette Valley marks a clear separation between the Oregon Coast Range to the west and the Western Cascades to the east. Geologically, these ranges are very different. The Coast Range consists of Early Eocene basalt of the accreted Siletz terrane below a cover of mostly Middle to Late Eocene marine sedimentary rock with some volcanic and intrusive rock. The Western Cascades are younger and formed as a volcanic arc, the precursor to the modern High Cascades. They are mostly volcanic, with numerous intrusive bodies and relatively minor amounts of sedimentary rock.

Along the eastern edge of the Willamette Valley, older sedimentary rocks crop out between about Albany and Creswell, south of Eugene. These rock units consist of the Spencer and overlying Eugene Formations, which were deposited in an ocean between about 50 and 35 million years ago, and the Fisher Formation, which was deposited on land during the same time. An ancient coastline probably oscillated between Albany and Cottage Grove, with the Fisher Formation deposited on land to the south of the shoreline. Above these sedimentary rocks lie a series of lava flows called the Little Butte Volcanics, which were erupted between about 35 and 25 milli on years ago, during early phases of the Western Cascades volcanic arc. Because the Little Butte Volcanics tend to be fairly resistant to erosion, they cap many of the hills along the eastern edge of the Willamette Valley, especially along the highway south of Brownsville.

Just south of the Santiam River, I-5 crosses a smaller ridge. Another quarry, between mileposts 239 and 238, shows sandstone resting on top of intrusive basalt that exhibits a crude columnar jointing pattern suggestive of intrusion at a shallow level. Part of a series of mafic intrusive rocks that are resistant to erosion, the basalt forms hills along the east side of the Willamette Valley from here all the way to Creswell, some 15 miles (24 km) south of Eugene. These hills include Peterson Butte, Lone Pine Butte, and Diamond Hill. The known ages of about 31 million years and chemical compositions of these intrusions resemble those of the Little Butte Volcanics, suggesting they are related.

Between about Albany and Coburg, the highway affords a good view of the Coast Range to the west, and Marys Peak, its highest point at 4,098 feet (1,249 m). The slopes of Marys Peak consist primarily of Eocene seafloor basalt of the accreted Siletz terrane overlain by Eocene marine sandstone. Its summit is made of Oligocene-age gabbro that intruded these older rocks as a sill. These gabbro sills intrude much of the Coast Range but are older and unrelated to those on the east side of the Willamette Valley.

Immediately north of where the highway crosses the McKenzie River, between mileposts 198 and 197 at a place called Spores Point, a roadcut exists on the east side of the road. There, a basalt flow of the Little Butte Volcanics occupies the edge of an ancient valley cut into gently folded sandstone and shale of the Eugene Formation. Moreover, the basalt flows exhibits a radial pattern of columns, likely a product of irregular cooling along the valley edge.

Basalt of Little Butte Volcanics (right and above) and sandstone and shale of the Eugene Formation (left) at Spores Point. Notice how the lava flow fills the edge of an ancient valley cut into the older Eugene Formation, and how on the left (north) side of the roadcut, the basalt overlies the gently folded Eugene Formation. The basalt exhibits radial cooling fractures, possibly because of thermal irregularities caused by cooling against the valley wall.

Skinner and Spencer Buttes

Skinner and Spencer Buttes, isolated hills just north and south of Eugene, respectively, are mafic intrusions, related to the 33- to 30-million-year-old Little Butte Volcanics. They intrude marine sandstone of the Eugene Formation. Skinner Butte offers a wonderful exposure of its basaltic bedrock, broken into columns as it cooled. If you look closely at the rock, you will see that it consists of interlocking crystals that, while small, are easily visible with a hand lens, and that it completely lacks air bubbles. Both these features suggest the rock cooled beneath Earth's surface rather than as a lava flow, which would have smaller crystals and probably contain air bubbles. To get to Skinner Butte from I-5, follow I-105 westward from exit 194 for 3.4 miles (5.5 km) to its termination at Seventh Avenue. Turn east (left) on Seventh Avenue for 0.2 miles (320 m) to Lincoln Street. Turn north (left) on Lincoln and follow it for 0.4 miles (640 m) to a small parking lot at the base of the columns. From there, a road leads 0.5 mile (800 m) to the top of the butte with views of the surrounding area, including Spencer Butte, directly to the south.

Known as the Columns, the popular rock climbing cliff at Skinner Butte Park features vertical columns that formed by the cooling of intrusive basaltic magma.

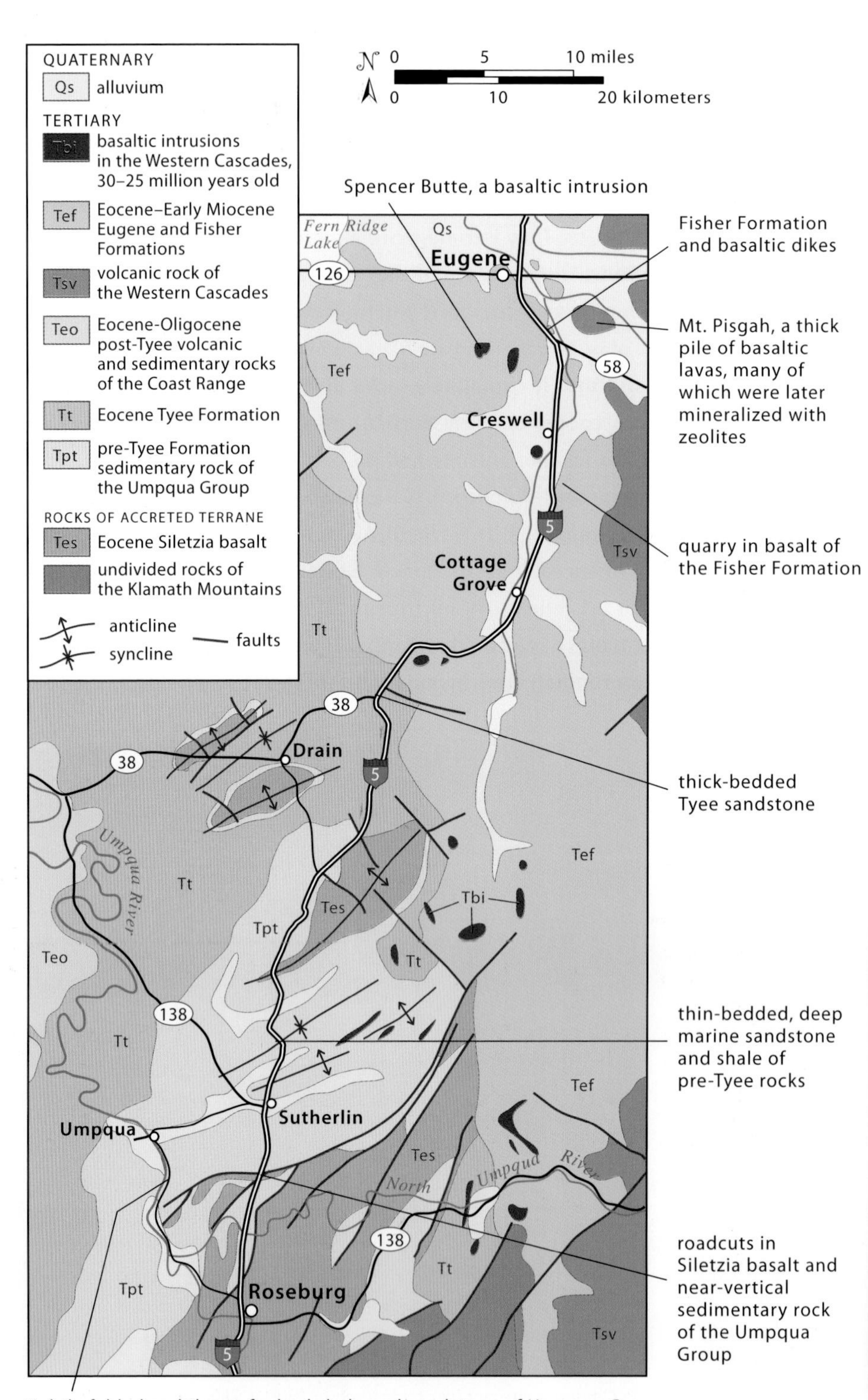

Geology along I-5 between Eugene and Roseburg.

INTERSTATE 5
Eugene—Roseburg
70 miles (113 km)

On the south side of Eugene, I-5 climbs through hills with sparse bedrock exposures. The hills consist of Eocene and Oligocene sedimentary and volcanic rocks that were intruded at shallow levels by basaltic magma in some places. The resulting rock, although technically intrusive, is fine-grained and frequently called basalt. The sedimentary rock belongs to the Eugene and Fisher Formations, deposited at roughly the same time but in marine and continental environments, respectively. An east-west shoreline oscillated back and forth between the vicinities of Albany, some 45 miles (72 km) north of Eugene, and Cottage Grove, 23 miles (37 km) south of Eugene. A prominent cliff of the intrusive basalt exhibits beautiful columnar jointing on OR 126 (Franklin Blvd.) just before the south-directed highway entrance. From the highway, this cliff is visible just past the south edge of the Whilamut Passage Bridge. On either side of milepost 191 the Springfield Quarry, which mines basaltic lava flows, is visible to the east. Mt. Pisgah, an elongate ridge immediately south of the quarry, consists of more basalt flows, with a probable age of about 30 million years.

The roadcut at milepost 188 offers an opportunity to inspect the Fisher Formation. Here, floodplain and river deposits form prominent thick sandstone and conglomerate beds that overlie more thinly bedded lake deposits. Some of these lake deposits contain fossil leaves; others display beautiful examples of graded beds, which are single layers in which coarse sediment at the bottom grades into finer material at the top. A conspicuous white ash layer lies near the base, and several basaltic dikes cut steeply through the exposure. A less obvious ash layer near the top of the roadcut yielded an age of 30.6 million years. To reach this roadcut, take exit 188 northbound and park on the wide shoulder at the north end of the exit ramp.

Upon rejoining the Willamette River floodplain, I-5 maintains a low gradient all the way to Cottage Grove. The Fisher Formation, with numerous basaltic lava flows, forms the surrounding hills. Spencer Butte, a basaltic intrusion, is clearly visible to the west between mileposts 188 and 182. The prominent hill west of the highway just south of Creswell is another basaltic intrusion, and a large quarry exposes basalt flows within the Fisher Formation on the east.

South of Cottage Grove, the valley narrows and rock exposures become more frequent. Some outstanding roadcuts at exit 162, the interchange with OR 99 and OR 38 toward Drain and Reedsport, expose thick-bedded sandstone of the Eocene Tyee Formation. For the next 6 miles (10 km) or so, scattered exposures of the Tyee show up on both sides of the road. Just south of exit 154, the road crosses a fault and onto older rock. Sedimentary rock of the Eocene Umpqua Group lies on the west side of I-5 but is poorly exposed, and basalt of Siletzia makes up the ridge on the east. This ridge, which follows the highway for the next 5 miles (8 km) or so, is the erosional remains of a large anticline, cored by the basalt. The interstate crosses through the core and a low, double roadcut of basalt and basaltic sandstone of Siletzia between mileposts 145 and 144.

Roadcut of Fisher Formation at milepost 188 along the northbound exit ramp. Graded lake deposits are the darker beds near the bottom. The dark, nearly vertical features are basaltic dikes, and the prominent white layer is an undated ash. The dated ash, which is 30.6 million years old, lies near the very top of the roadcut.

Turbidite deposits of the Umpqua Group near milepost 141. The more resistant brown layers are sandstone; the darker, more easily eroded layers are shale.

Where well exposed, this part of the Umpqua Group is much finer-grained and more thinly bedded than most of the younger Tyee Formation. Some of the roadcuts consist of uniformly thin-bedded shale, and some contain turbidites, alternating thin beds of sandstone and shale deposited from deep-sea currents of turbid, sediment-charged water. In further contrast to the Tyee, which was deposited after the final stages of the accretion of the Siletzia terrane, these rocks were deposited during the final stages, and so are more faulted and folded than the Tyee. Perhaps the best exposures of the turbidites along the interstate are on the west side of the road just south of milepost 141, about 0.25 mile (400 m) north of the turnoff to Oakland. A detour from Sutherlin to Roseburg takes one to a truly spectacular exposure of folded and thrust-faulted sandstone and shale of the Umpqua Group.

Detour to Folded Umpqua Group

This detour through beautiful countryside is approximately 10 miles (16 km) longer than driving directly to Roseburg on the interstate, but the roads are easy, and halfway there, the roadside geologist can pull off and inspect some amazing rocks. From the Sutherlin exit (136), take OR 138 about 0.25 mile (400 m) west, then bear left on the road to Umpqua and follow that road westward about

Geology students inspect folded turbidites of the Umpqua Group along Garden Valley Road near Umpqua.

6 miles (10 km). At Garden Valley Road, go left and follow the east bank of the Umpqua River southward. After 2.75 miles (4.4 km), a long roadcut on the east side exposes folded rocks of interbedded shale and sandstone turbidites of the Umpqua Group. In many places, the rocks contain graded beds. Toward the south end of the roadcut, there is a big pull-out on the right, and just beyond that, two prominent southward-directed thrust faults. Looking across the Umpqua River to the west, you can see a bird's-eye view of the folds where they outcrop over a small, flat area.

Near milepost 132, I-5 crosses a thrust fault that places basaltic rock of Siletzia over the sedimentary Umpqua Group. The approximate location of the fault runs along the base of the more resistant ridge that was uplifted along the fault and crosses the highway at a high angle. Immediately south, roadcuts on both sides of the road contain some faulted basalt but mostly overlying sandstone and shale that is in a near-vertical orientation. From here to Roseburg, roadcuts expose either more of the Siletzia basalt, or remnants of the Umpqua Group deposited on top of the basalt. Roadcuts at exit 121 within Roseburg display spectacular examples of the pillow basalt.

US 20
Corvallis—Newport
53 miles (85 km)

Driving US 20 to the coast gives a sense of the anticlinal nature of the Coast Range: good exposures of east-dipping rocks lie near the center of the range, and predominantly west-dipping rocks lie to the west. Between Corvallis and Philomath, US 20 leaves behind the Willamette Valley and passes over the buried trace of the Corvallis Fault, which likely originated as a thrust fault during the later part of Eocene time, bringing the Siletzia basalt eastward over the Tyee Formation at the eastern edge of the nascent Coast Range. The topography it created in turn provided source material for some of the sedimentary rocks deposited during and afterward in Late Eocene time. Since it originally formed, however, the fault likely had a complicated history, including normal and strike-slip fault movement, some of which may have continued into Quaternary time. Just to the north of Corvallis, the fault marks the topographic boundary between the Willamette Valley and the Coast Range.

Four miles (6.4 km) west of Philomath, and only 0.25 mile (400 m) east of the junction with OR 223, a low-lying cut on the south side of US 20 exposes basalt of the Siletz terrane. A closer look, afforded by the small pull-out, reveals several distinct pillow shapes, created when the basalt erupted underwater.

Near milepost 27, gently east-dipping beds of the Tyee Formation mark the first in a series of large roadcuts for the next 10 miles (16 km) to the west.

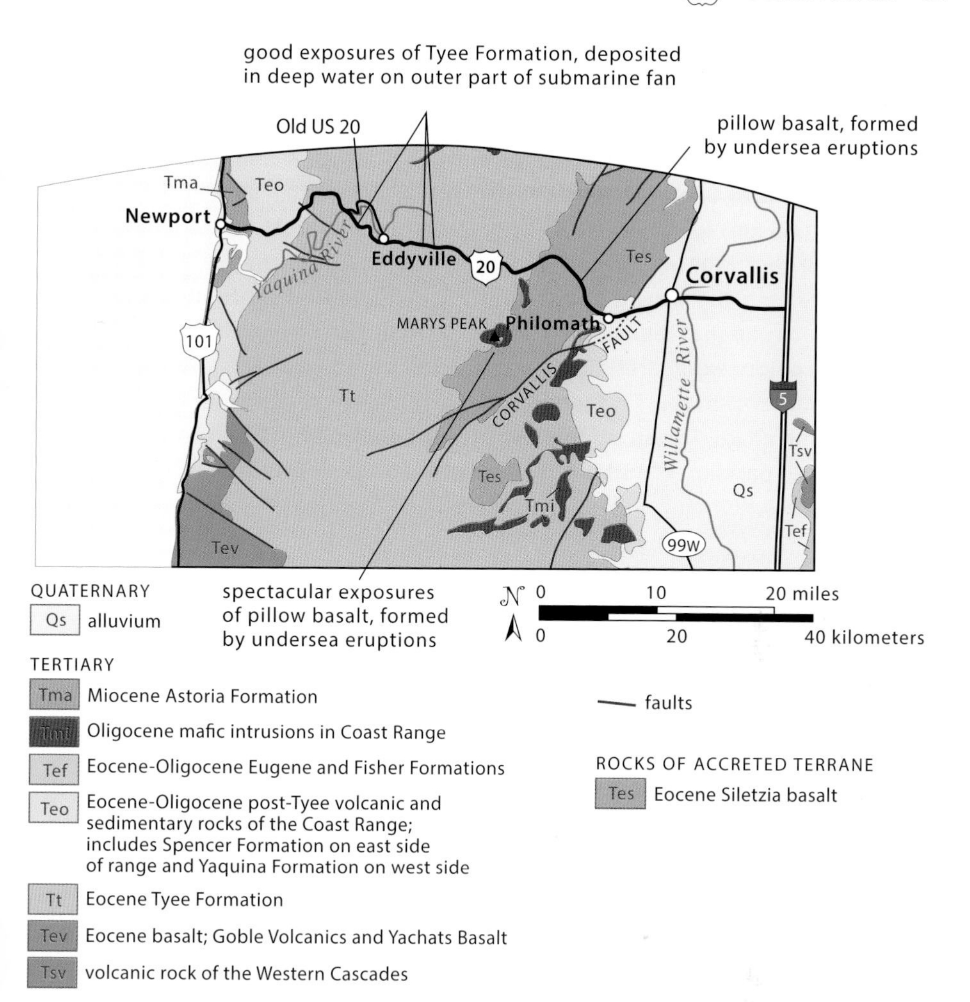

Geology along US 20 between Corvallis and Newport.

These exposures contain abundant shale as well as sandstone. They are also more thinly bedded than the coarser, more thickly bedded Tyee sandstones to the south. In many places the beds here display turbidites, indicating deposition in deep water. This interpretation fits nicely with the concept of the Tyee as being a submarine fan complex that deepened northward from a delta at the edge of the continent. These exposures also demonstrate folding of the rock, as the Tyee bedding dips to the west in the vicinity of Eddyville, near milepost 23.

Some 0.75 mile west of milepost 25, a turn-off to Eddyville crosses the Yaquina River and follows its tight meanders to rejoin the highway some 10 miles downstream. The new main highway, mostly completed in 2016, bypasses those meanders to offer a shorter, more direct, and much safer route. However, the highway project lasted seven years longer than expected and suffered cost overruns in excess of $200 million because the route passes through extremely

Marys Peak

A side trip to Marys Peak, some 10 miles (16 km) to the southwest offers great views, a chance to inspect one of the Coast Range's mafic sills, and quarry exposures of Siletzia pillow basalt. Take OR 34 south from Philomath for 9 miles (14.5 km) to the Marys Peak road and turn right. The road climbs almost 3,000 feet (900 m) in just under 10 miles (16 km) to the summit parking lot. Along the way, a parking lot 3.7 miles (6 km) up from OR 34 marks a gated road that leads a short distance to an abandoned quarry with outstanding pillow exposures. The roadcuts just past this parking lot offer some good pillows as well. Farther up, the road passes exposures of Eocene sandstone and shale. To cap off the visit, a short walk along the gravel road from the upper parking lot follows the mafic sill to the peak's summit and a 360-degree view.

View from the top of Marys Peak, looking southwestward, with an exposure of a mafic sill along the left side.

Pillow basalt of Siletzia, exposed in cross section in the quarry wall, accessed from the parking lot 3.7 miles (6 km) up Marys Peak Road.

West-dipping, interbedded deep water sandstone (thicker, light gray layers) and shale (dark eroding layers) of the Tyee Formation. A fault zone cuts through the rock near the center of the photo.

rugged terrain that includes nearly a dozen old landslides. These slides began to move again during construction, forcing multiple delays and eventually the demolition of some partially completed bridges.

The new route passes by some spectacular roadcuts along the north side of the road. One of these cuts, less than a quarter mile west of the turn-off for the Chitwood Covered Bridge, beautifully displays a fault zone. You can't match specific beds across the fault, implying either dip slip motion on the fault greater than the scale of the exposure or a significant strike-slip component to its motion.

Just west of Pioneer Summit, US 20 crosses a hidden fault onto unexposed Yamhill and younger rocks. Newport, described in more detail in the US 101 guide, occupies an uplifted marine terrace.

US 26
PORTLAND—SEASIDE
79 miles (127 km)

Called Oregon's Sunset Highway, US 26 provides a fast but comparatively long route to the coast through deep forestland as it diagonally crosses the northern Coast Range. Similar to other routes across the range, it displays the Coast Range's anticlinal nature, crossing progressively older rocks as it approaches the core and then progressively younger ones out toward the coast.

Turning west onto US 26 from I-405 in Portland, the road crosses the Portland Hills Fault and almost immediately passes into the Vista Ridge Tunnel. From there, it climbs steeply through the Portland Hills, a northwest-trending anticline made mostly of Columbia River Basalt Group lava flows and overlying Boring Volcanics. Some poor roadcuts of the Columbia River Basalt Group line the highway just west of the tunnel, but these are inaccessible given

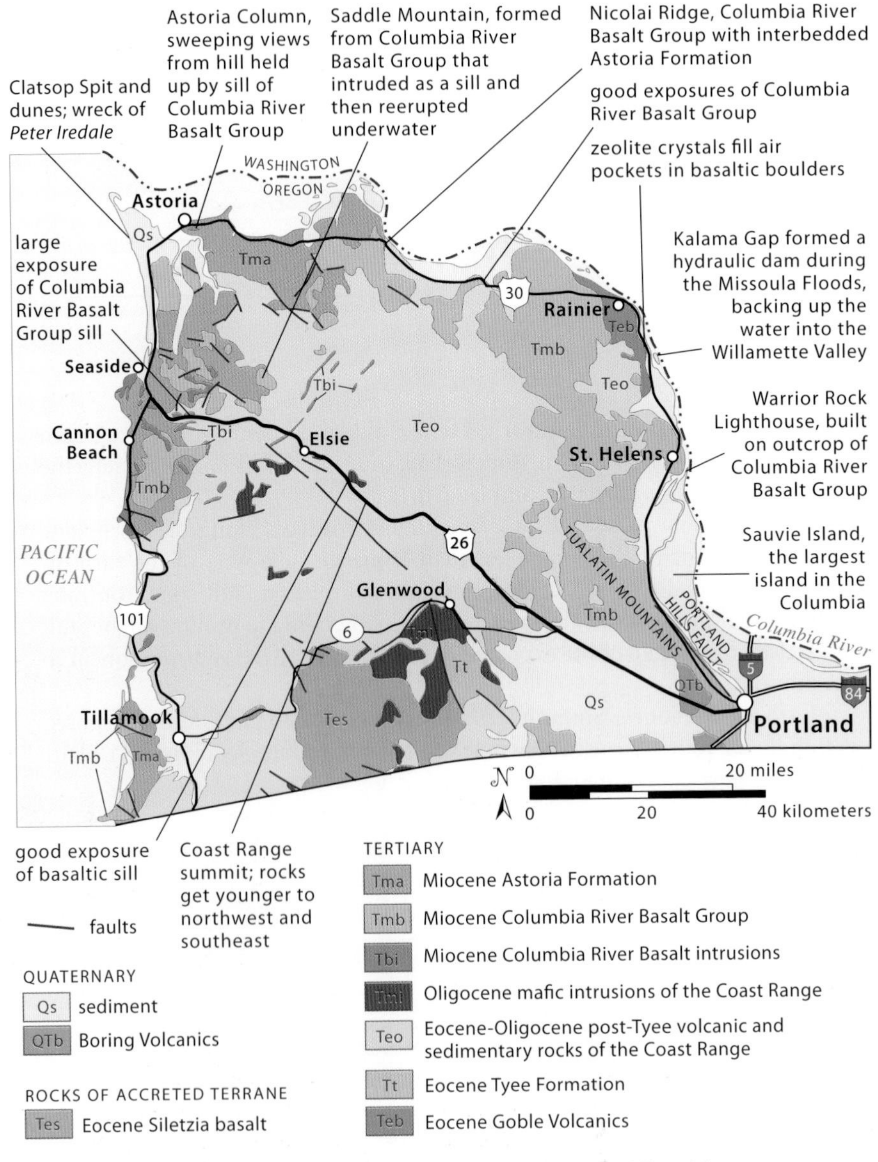

Geology along US 26 between Portland and Seaside.

the intensity of the traffic. A few miles farther west, overgrown roadcuts on the Boring Volcanics lie on the north side of the highway just east of its intersection with Oregon 217. For 16 miles (26 km) between the outskirts of Portland and the turnoff for OR 6, US 26 travels in a straight line across alluvium deposited on the Columbia River Basalt Group. Good views northward show the crest of the Portland Hills anticline.

About 2 miles (3.2 km) west of the turnoff for OR 6, the road travels between steep forested hills without any rock exposure along the road. From exposures off the road, however, geologists have determined that these hills consist of Columbia River Basalt Group. From here to about the Coast Range summit, 24 miles (39 km) away, the rocks mostly dip eastward and, for the most part, become progressively older as you head west. The Columbia River Basalt Group is Miocene age; sandstone of the marine Scappoose Formation exposed between mileposts 49 and 50 is Late Oligocene; deltaic mudstones of the Pittsburgh Bluff Formation are Middle Oligocene; tuff-rich mudstones of the Keasey Formation are Late Eocene to Early Oligocene; and marine sandstones of the Cowlitz Formation and the Tillamook Volcanics near the Coast Range summit are Middle to Late Eocene.

Some of the better exposures of these rocks exist near mileposts 45 and 41, just east of the Dennis L. Edwards Tunnel, known as the Sunset Tunnel prior to 2002. At milepost 45, you can see good outcrops of east-dipping Pittsburgh Bluff Formation. Some of the beds contain fossil clams and carbonized wood, as well as tiny shards of volcanic glass. At milepost 41, you can see the upper part of the Keasey Formation, which contains numerous lime-rich concretions. These concretions make the rock fairly resistant to erosion so that it stands out as a ridge. Between mileposts 33 and 32, a good exposure of Tillamook Volcanics lies on the south side of the road.

-ial view of the Portland Hills anticline, looking eastward toward downtown Portland. The Port- -d Hills Fault runs along the north side of the anticline, just south of the Willamette River. US leads northwestward from the downtown area across the anticline and into the Tualatin Basin.

Saddle Mountain State Natural Area

Near milepost 10, the Saddle Mountain Road leads 7 miles (11 km) north to a parking area and trailhead for Saddle Mountain State Natural Area. At an elevation of 3,267 feet (996 m), Saddle Mountain marks the highest point in the Coast Range northwest of Portland. The rock that makes Saddle Mountain consists of breccias and highly fractured pillow basalt of the Columbia River Basalt Group, a common rock in Oregon but one that took an incredible path to get this far west. The Columbia River Basalt Group flows originated in northeastern Oregon and flowed to the Pacific shoreline, where they poured down submarine canyons and managed to intrude into the existing sedimentary deposits as gigantic sills. Saddle Mountain is one place where the flows entered the submarine canyons. Humbug Mountain, just north of US 26, consists of similar rock, as do Sugarloaf Mountain and Onion Peak a few miles to the south. Basaltic dikes that cut through these flows can be seen from the Saddle Mountain parking lot as the narrow, vertical, gray lines on the south face of the mountain. You can gain a much better view of the mountain by walking 0.25 mile (400 m) up the main trail and then taking the short spur trail to the Humbug Mountain viewpoint.

Saddle Mountain, as seen from Humbug Mountain viewpoint. Three basaltic dikes appear as vertical lines, cutting through the basalt near each of the summits and in the saddle between.

Detailed view of a fractured pillow (the large feature draped by the hand lens) and surrounding black and brown hyaloclastite, an accumulation of fragmented volcanic glass formed as the lava erupted underwater.

Slightly more than 2 miles (3.2 km) west of the Coast Range summit, the road passes a cliff made of a sill of Oligocene basaltic rock, which displays an interesting fracture pattern, probably as a result of cooling underground within the surrounding Cowlitz Formation. Being much less resistant to erosion, the Cowlitz Formation is entirely obscured by vegetation. Some good exposures of its shale and sandstone do show up along the road a couple of miles farther west, between mileposts 24 and 23. In contrast to the rocks on the east side of the summit, these ones dip gently westward, consistent with the anticlinal nature of the Coast Range.

Near the community of Elsie, the road passes back into the younger Keasey Formation, although here it lacks the limy concretions, so it is poorly exposed. Still younger sedimentary rocks are exposed farther west, but along the road they are completely covered by forest.

Rock exposures along US 26 between Saddle Mountain Road and the intersection with US 101 are almost all sills of the intrusive Columbia River Basalt Group. Some of these exhibit a crude columnar jointing, caused during cooling. An especially large exposure exists on the north side of the highway between mileposts 7 and 6, behind a maintenance station of the Oregon Department of Transportation.

US 30
Portland—Astoria
90 miles (145 km)

See map on page 68

US 30 hugs the Columbia River along most of its length between Portland and Astoria, following the same route taken to the Pacific Ocean by Lewis and Clark's Corps of Discovery in late 1805, the path of the Missoula Floods 18,000 to 15,000 years ago, and the approximate path of the Columbia River Basalt Group between about 15 and 11 million years ago. From its confluence with the Willamette River on the west side of Portland, the Columbia River drops only 10 feet (3 m) in elevation over this route. By the end of the last ice age, though, the Columbia had carved a canyon more than 300 feet (90 m) deep at its present mouth at Astoria and extended some 40 miles (64 km) westward to the coastline at the edge of the continental shelf. As the ice melted and sea level rose, the coastline moved east to Astoria, and the river began depositing huge volumes of sediment in its channel, which is now marked by numerous mid-channel islands.

Between Portland and Burlington, US 30 follows the trace of the Portland Hills Fault, which uplifted the Portland Hills to the southwest relative to the Columbia River valley. Roadcuts of uplifted Columbia River Basalt Group lie immediately southwest of the fault. To the northeast, the combined floodplain of the Willamette and Columbia Rivers occupies the low ground. The rivers join at Kelly Point Park, about the same latitude of Burlington, but some 5 miles (8 km) across the floodplain to the east. Sauvie Island, surrounded by channels of the Columbia, occupies much of the floodplain immediately downstream from the confluence. With an area greater than 30 square miles (78 km^2), Sauvie Island ranks as the largest island in the Columbia. Its size appears to be dictated by a combination of its location and the likely presence of bedrock at shallow depth. Its location near the confluence of two major rivers ensures a high sediment load, and the shallow bedrock likely decreases flow velocities, resulting in deposition. The Columbia River Basalt Group crops out beneath the Warrior Rock Lighthouse at the northern tip of the island.

Just north of Burlington, US 30 turns due north, away from the trace of the Portland Hills Fault, and follows the edge of the river's Multnomah Channel to Scappoose and then St. Helens. In many places, the abrupt edge of the adjacent hills makes it appear that the fault continues parallel to the highway, but this edge is likely a product of erosion from the Missoula Floods between about 18,000 and 15,000 years ago. Numerous roadcuts of mostly Columbia River Basalt Group line the west side of the highway until just south of Scappoose, where the floodplain widens. The basalt reappears just north of St. Helens.

Between milepost 37 and just west of the Lewis and Clark Bridge at Rainier, roadcuts expose basalt flows of Late Eocene age, substantially older than the Columbia River Basalt Group. This basalt is called the Goble Volcanics. At Jaquish Road, 0.5 mile (0.8 km) south of Goble, a seemingly random pile of basaltic boulders, probably left by road construction crews, are rich in white

Radial fibers of thomsonite, a zeolite mineral, in basalt from near Goble. Photo is about 1.5 inches (4 cm) across. –Rock sample from the collection of Tyler Gass

and gray zeolite crystals precipitated in air pockets by fluids circulating through the rock long after it erupted.

Goble sits on the Columbia River at Kalama Gap, the narrowest part of the river's course between Portland and the ocean. During the Missoula Floods, this constriction created a hydraulic dam, causing the floodwaters to back up and flood the entire Willamette Valley all the way to Eugene. This idea seems inconceivable, but consider that some of the floods exceeded 500 cubic miles (2,100 km^3) in volume. If only a third of the water temporarily ponded behind the gap, it would fill the Willamette Valley to depths greater than 300 feet (90 m).

Just west of the Lewis and Clark Bridge, US 30 ascends steeply through the Eocene basalt and up into the Columbia River Basalt Group, which caps the hill. Between here and Clatskanie, the road passes over poorly exposed Columbia River Basalt Group. At Clatskanie, the highway rejoins the floodplain, marked by a marshland with a dramatic array of meandering channels that empty into the river. The steep hills west of Clatskanie offer numerous exposures of Columbia River Basalt Group; a particularly good one exists by milepost 63. Farther west, between mileposts 67 and 68, small slides in the steep hillside reveal bedrock of Oligocene or Miocene sandstone.

Just west of Westport, US 30 climbs steeply up Nicolai Ridge, a prominent cliff of Columbia River Basalt Group left standing after the Missoula Floods. Roadcuts expose several flows of basalt and, toward the top, interbedded sandstone that is considered to be part of the marine Astoria Formation. This interbedding shows that deposition of the sandstone and eruption of the basalt was at least partly concurrent. The Bradley State Scenic Viewpoint sits at the

View northwestward over Astoria and the Columbia River from the Astoria Column on top Coxcomb Hill. The Astoria-Megler Bridge, in the middle ground, is 14 miles (23 km) from the Columbia's mouth.

top of the cliff, on more Columbia River Basalt Group, and affords a nice view toward the east. The long roadcut just west of the viewpoint consists of younger Miocene or Pliocene sandstone that was deposited on land.

Astoria occupies a peninsula of Astoria Formation that separates the Columbia River from the Youngs River, another estuary that merges with the Columbia from the south. Very little bedrock crops out, however, because it is mostly covered by buildings and roads. The Astoria Column, on top of Coxcomb Hill at an elevation of more than 500 feet (150 m), gives a sweeping panorama of the region. Coxcomb Hill is held up by a resistant sill of Columbia River Basalt Group.

US 101
Astoria—Lincoln City
107 miles (172 km)

Oregon's northern coast hosts an especially varied landscape, from long sandy beaches to imposing headlands and sea cliffs to wide bays and estuaries. The most prominent headlands are held up by basaltic flows and intrusions, whereas less resistant sandstone, mudstone, and shale tend to form the low-lying areas.

The northern stretch of US 101 highlights intrusive features, including basalt dikes and sills intruding the Astoria Formation and older rocks. Most

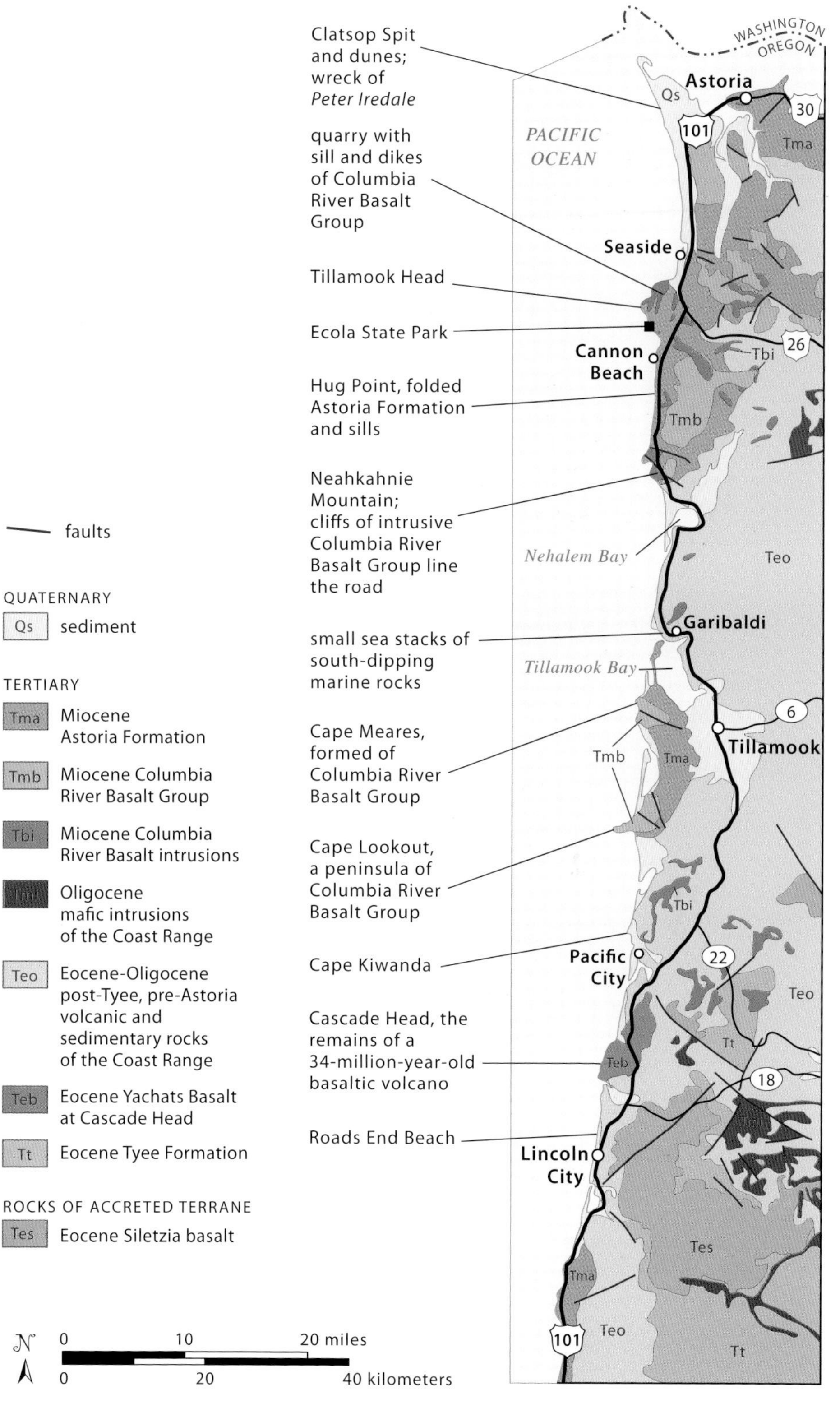

Geology along US 101 between Astoria and Lincoln City.

early geologists in Oregon recognized these relations and naturally inferred that the basalt flows originated near the coast because most intrusive magmas come directly from magma chambers deep below the surface. As a result, they argued the basalt must be unrelated to the Columbia River Basalt Group flows farther to the east. However, both sets of basalt are of the same age and have identical chemistry, and mapping has shown them to be connected via the Columbia River corridor. It is now generally accepted that many of the basalts along the coastline, including those that form dikes and sills in older rocks on the coast, are in fact flows of the Columbia River Basalt Group that made it all the way to the Pacific Ocean.

After reaching the Miocene coastline, the lavas formed massive deltas of pillow basalts, which continued to grow through addition of lava from lava tubes along their bases, as well as from flows on their tops. As the deltas grew, they increased pressure on the lava flowing out of their bases and caused it to intrude at sloping angles downward along faults, fractures, and bedding surfaces. In some places, the intruded rock exhibits highly irregular folding, suggesting it was still fairly soft when the lava intruded.

Some of the intrusions grew large enough to form their own small magma chambers, some of which crop out today as megasills, greater than 1,000 feet (300 m) in thickness. The large sills fed the upward injection of smaller dikes and sills, some of which even erupted as small basaltic volcanoes! Many of the dikes and sills consist of mixed sedimentary and fragmental basaltic material and are called peperites, which form when intruding magma encounters wet sediment and explodes.

After crossing the Youngs River, an estuary that drains much of the northern Coast Range, US 101 encounters the Clatsop coastal plain, which extends nearly 20 miles (32 km) from the mouth of the Columbia southward to Seaside. The plain hosts another of Oregon's extensive dune fields, with a several-mile-long sand spit at its northern tip and multiple beach ridges and intervening swales, many of which are filled by shallow, elongate lakes. The area was the site of numerous shipwrecks from the mid-1800s to the early 1900s as ships ran aground attempting to maneuver into the Columbia River. You can see remnants of one such shipwreck, the *Peter Iredale*, which came aground in 1906, in the beach sand on the spit at Fort Stevens State Park.

The sand contains abundant dark heavy minerals, particularly magnetite and ilmenite, which appear to be originally derived from a basaltic source somewhere up the Columbia River. Since construction of the numerous dams on the Columbia River, however, little additional sand is reaching to the coast.

Tillamook Head juts into the ocean immediately south of Seaside. It consists of several thick sills of Columbia River Basalt Group intruding the Astoria Formation. Just south of the bridge across Necanicum River by milepost 24, a quarry exposes one of these sills, with dikes extending into deep marine mudstone of the overlying Astoria Formation. This part of the Astoria is typically a dark gray color, but much of it was bleached to light gray or even white from the heat of the intrusion.

Ecola State Park

Ecola State Park offers the easiest and certainly most instructive access to Tillamook Head's geology. Where bedrock consists of the Columbia River Basalt Group, it is relatively stable and forms prominent features, such as headlands and sea stacks, but where it consists of interbedded sandstone and shale of the Astoria Formation, it tends to erode easily and even form landslides. The parking lot for Ecola Point rests on top of a recently active landslide, and a prominent scar just uphill marks the head scarp of its most recent sliding event in 1961. From Ecola Point, you can see other, smaller slides, some of which go straight down to the beach. To the north, Indian Beach lies between two headlands of intrusive basalt.

The cliff at the north end of Crescent Beach, below and just south of Ecola Point, contains exposures of tightly folded sandstone and shale of the Astoria Formation. This folding appears to have been caused by the intrusion of basalt when the rocks were still soft sedimentary deposits. In some places where the basalt comes into contact with the sediment, the basalt is brecciated and locally contains fragments of the folded rock. The Astoria Formation consists of turbidites, deposited from turbid flows of mixed sediment. In marine environments, these flows occur where the ocean floor slopes steeply basinward, past the continental shelf. A close inspection of these rocks shows that they contain some single beds whose grain size decreases gradually from coarse at the bottom to fine at the top, suggesting the sediment was deposited as it settled from suspension. At the south end of the beach, the rocks consist of basalt.

Folded Astoria Formation and intrusive basalt at the north end of Crescent Beach at Ecola State Park. It is likely that the folding resulted from intrusion of the basalt, which is part of a large sill of Columbia River Basalt Group.

Wreck of the Peter Iredale *on the spit at Fort Stevens State Park.*

Offshore from Cannon Beach is Haystack Rock, one of Oregon's iconic landmarks. This sea stack originated as an undersea volcano, where the magma of an invasive sill of Columbia River Basalt Group had erupted on the seafloor. Smaller sea stacks just to the south, called the Needles, are part of the same complex. Close inspection of Haystack Rock shows that it contains mostly basaltic breccia and pillows overlying mudstone and sandstone of the Astoria Formation, cut through by basaltic sills and dikes. The base of Haystack Rock is accessible at low tide, but scrambling or climbing on it is prohibited to protect the seabird habitat.

Silver Point, immediately north of milepost 32, offers a chance to inspect mudstone of the Astoria Formation at its type locality. The mudstone is appropriately named the Silver Point Member of the Astoria Formation because it was first described in formal detail in the sea cliffs below. The thinly bedded and soft mudstones erode easily, as demonstrated by the amount of landsliding apparent in the slopes above the point. Notice the large number of small head scarps. Sea stacks to the west consist of the resistant remnants of a basaltic sill.

Just 0.5 mile (0.8 km) to the south, Astoria sandstone is exposed in roadcuts on both sides of the road, with dramatically different orientations from one side to the other. It dips steeply to the northwest on the west side of the road and gently east on the east side. Moreover, a dike cuts the sandstone on the west side of the road. Like the folding at Ecola Point, the folding here may reflect the basaltic intrusions.

Hug Point State Recreation Site, between mileposts 33 and 34, shows tilted sandstone of the Astoria Formation intruded by a basaltic sill and overlain by sands of an uplifted marine terrace. Numerous alcoves and sea caves have been carved into the rock by the surf. Apparently, its name derives from people having to hug the point to get around it at high tide. Short walks north and south along the beach allow inspection of more sea cliffs and sea caves, with more basaltic and peperite dikes, as well as highly deformed Astoria sandstone. Alan Niem provided an outstanding guide to the area, published in 1975.

Just south of Arch Cape, US 101 enters Oswald West State Park, named for Oregon's fourteenth governor. The road passes through a tunnel cut through a large basaltic intrusion. Both Cape Falcon and Neahkahnie Mountain, immediately south of Arch Cape, also consist of Columbia River Basalt Group intrusions into the Astoria Formation. South-dipping Astoria Formation holds up the sea cliffs behind Short Sand Beach, easily accessed via a short, beautiful walk through old-growth forest between mileposts 39 and 40. One mile (1.6 km) to the south, cliffs of Neahkahnie Mountain line the east side of the road with numerous pull-outs and breathtaking views on the other side.

Nehalem Bay marks the mouth of the Nehalem River, the first major estuary south of the Columbia River. It is extremely shallow, with depths at mean low tide of only 1 to 2 feet (0.3–0.6 m); narrow channels with depths greater than 10 feet (3 m) traverse most of the shallows. The river flows southwestward toward the ocean along the contact between largely Miocene-age marine

Astoria Formation at Hug Point State Recreation Site illustrates complex relations with intrusions of the Columbia River Basalt Group. A basaltic sill (dark line) in this photo intrudes the rock and thickens considerably toward the top of the outcrop. Below, the sandstone is faulted against the basalt.

sandstone, principally the Astoria Formation, to the north and mostly Eocene marine sandstone to the south. Some of these older sandstones crop out near milepost 45. About 4 miles (6.4 km) farther south, the community of Rockaway Beach occupies a 7-mile (11 km) stretch of sand, easily accessed from side roads within the town. From the beach, you can see Twin Rocks, two large sea stacks, one of which has eroded into an arch.

At Barview, US 101 curves eastward, following the north shore of Tillamook Bay. Just north of Garibaldi, small sea stacks are Oligocene-age marine sedimentary rocks, with bedding dipping southward beneath the younger Astoria Formation cropping out on the other side of the bay. The tall smokestack in Garibaldi marks the site of the town's first lumber mill, built in the 1920s.

Like the other bays along the northern coast, Tillamook Bay is shallow, with an average depth of only about 6 feet (1.8 m) during high tides. At low tides, about half the bay emerges as tidal flats. The bay receives water and sediment from five different rivers and receives a great deal of sediment from the long spit and beach that separate it from the ocean. Studies of the bay by Paul Komar and others in 2004 found that the sediment accumulates at rates of between 8 and 16 inches (20–40 cm) per century. Numerous exposures of basalt of the Oligocene-age Tillamook Volcanics frame the eastern edge of the bay.

From Garibaldi to Lincoln City, US 101 mostly follows a low-relief inland route, continuing east of Tillamook Bay, up the Tillamook River, over a divide, and then onto the floodplain of the Nestucca River, which empties into Nestucca Bay. The highway passes by numerous lakes, low-lying swampy areas, and small dune fields, bordered in places by steep landslide-prone hills of Oligocene-age rock. This stretch of road illustrates the north coast's coastal plain, the remnant of the much larger plain that, during the low sea levels of Pleistocene time, extended westward several tens of miles to the edge of the continental shelf. The Three Capes Road offers a more dramatic, albeit slower, alternative route for the section of US 101 between Tillamook and Pacific City. It accesses the coastline near Cape Meares and Cape Lookout, both promontories of Columbia River Basalt Group flows, and Cape Kiwanda, which consists of Astoria Formation.

Neskowin and Cascade Head to the south occupy a 34-million-year-old basaltic volcano. Many of its older flows are interbedded with marine sandstone and siltstone and show evidence for erupting underwater, but the bulk of the flows were erupted on land. Some of the youngest erupted flows lie along the east edge of Nestucca Bay, where they are highly brecciated and even contain fragments of the sandstone. These rocks erupted at the same time as the Yachats Basalt, which forms much of the coastline farther south near Yachats.

Lincoln City is mostly built on an uplifted marine terrace developed on Eocene marine sandstone somewhat younger than the Tyee Formation. The rock is soft and prone to erosion, so does not crop out along US 101. It does show up at the north end of Roads End Beach, 1 mile (1.6 km) down Logan Road, which intersects US 101 near milepost 113 in Lincoln City.

Just below the parking lot at Roads End Beach, sand caps a beautifully exposed dark gray paleosol, an ancient soil formed from a former spruce forest. The paleosol developed on top of older marine sands and slopes upward from beach level to just below the parking lot elevation, reflecting the topography at the time. Close inspection of the marine sands, which are orange from oxidation, reveals some highly folded bedding, a product of slumping or sliding, possibly triggered by an earthquake.

Sloping paleosol (dark) beneath recently deposited beach sand and overlying oxidized Eocene sandstone, originally deposited in shallow ocean water. The exposure is directly below the parking lot at Roads End Beach.

US 101
Lincoln City—Bandon
145 miles (233 km)

US 101 between Lincoln City and Bandon passes from headlands of resistant basalt, with sea cliffs and pocket beaches, south into a low-relief area that boasts one of the longest continuous beaches of the west coast as well as the largest set of coastal sand dunes in the United States. The dunes end at Coos Bay, where US 101 moves inland to avoid Cape Arago, another rugged upland.

Between Lincoln City and Gleneden Beach, the road follows Siletz Bay at the mouth of the Siletz River. Salishan Spit extends several miles north from Gleneden Beach, protecting the bay entrance. Like the other bays to the north and south, Siletz Bay is gradually filling with sediment, and at low tide, you can see extensive tidal flats. Between mileposts 118 and 119, three small sea stacks form a line near the shoreline, marking the remains of a basaltic dike related to the Columbia River Basalt Group. The road travels over Quaternary terrace deposits almost as far as Depoe Bay.

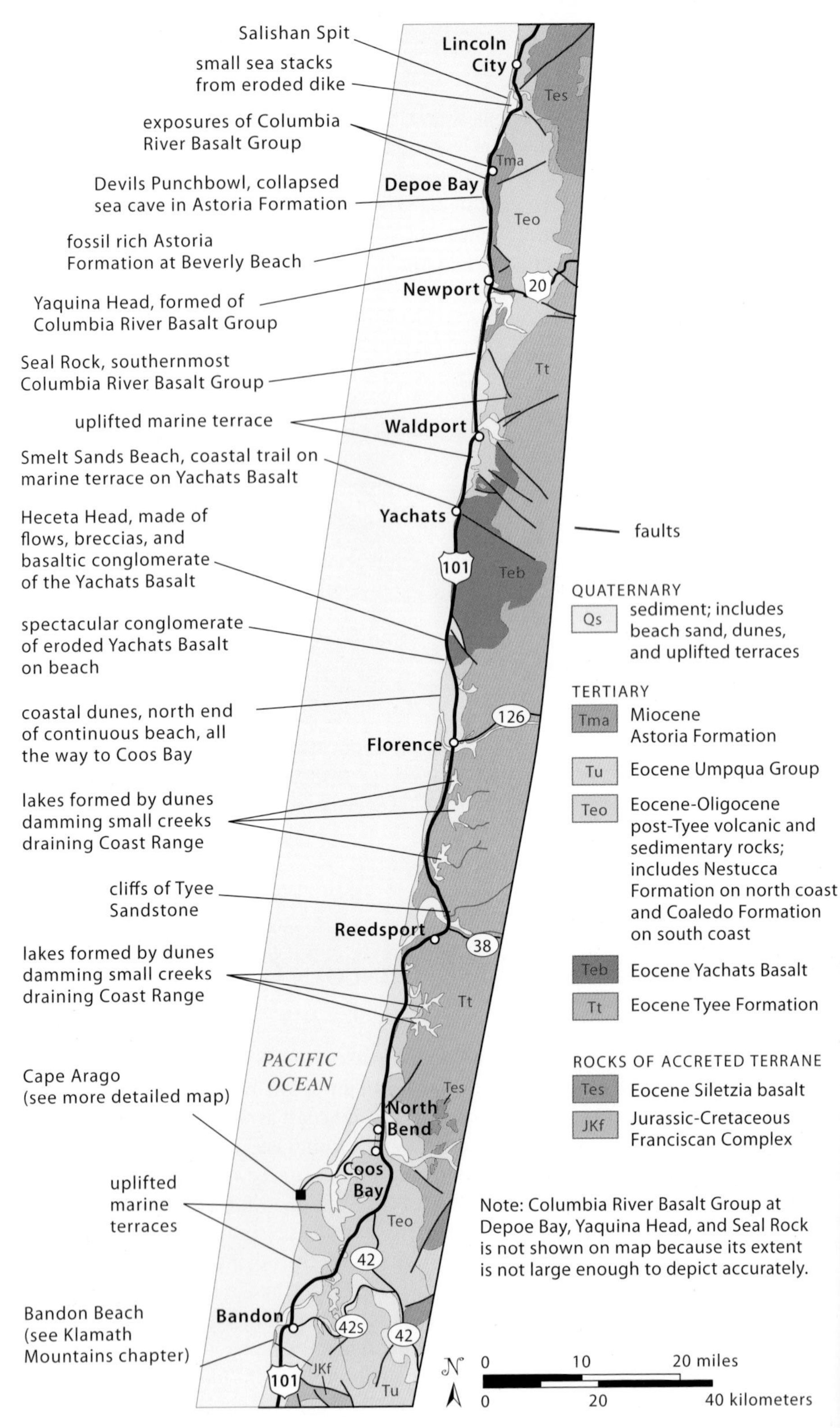

Geology along US 101 between Lincoln City and Bandon.

Immediately north and south of Depoe Bay, basalt flows of the Columbia River Basalt Group form the inner harbor and coastline: the older Grande Ronde Basalt forms the inner harbor and the younger Wanapum Basalt forms the coastline. Astoria Formation sandstone, also called the Sandstone of Whale Cove, crops out in the harbor between the basalts and in the hills to the east. Prior to the recognition of their origin as part of the Columbia River Basalt Group, the flows along the coast were called the Cape Foulweather Basalts, while those in the bay were called the Depoe Bay Basalt. The coastal Wanapum flows are mostly fragmental, probably from flowing into the ocean and breaking up in the ensuing explosions. Some 4 miles (6.4 km) to the south, more intact flows of the Wanapum Basalt make up Cape Foulweather. Within the inner harbor, the Grande Ronde Basalt shows off some beautiful pillow lavas. To the north, at Boiler Bay, these flows intrude the Astoria Formation and form peperite dikes.

It's easy to forget that these flows erupted some 15 million years ago in eastern Oregon and are far more extensive than these narrow coastal exposures. Offshore drilling projects have provided information on their actual extent. Ten miles (16 km) offshore, a drilling rig encountered a 50-foot-thick (15 m) flow of the Grande Ronde Basalt at a depth of more than 3,500 feet (1,070) and a basaltic sill at a depth of 8,250 feet (2,500 m).

Just south of milepost 134, Beverly Beach fronts exposures of the Astoria Formation, with clam-rich fossil beds in an outcrop just south of the beach access. Just beyond that, another outcrop yields fossils of the tusklike mollusc dentalium. Between mileposts 135 and 136, the road passes an older portion of the highway, broken up by a landslide.

Between mileposts 137 and 138, a well-marked road accesses Yaquina Head, a promontory of the Wanapum Basalt of the Columbia River Basalt Group. There, visitors can see a variety of flows and flow features, including columnar joints and breccias, and also dikes, small normal faults, wind-deposited sand, and a dramatic marine terrace. Many of the features are beautifully exposed in either of two former quarries, where the basalt was mined for road construction. The area is now protected as the Yaquina Head Outstanding Natural Area, and the park charges a nominal fee for parking. The Yaquina Head Lighthouse sits at the end of the peninsula.

South of Yaquina Head, Agate Beach offers one of many excellent opportunities to hunt for agates, which are brightly colored pebbles and cobbles of extremely fine-grained silica. Most of the agates of the Oregon coast formed by precipitation of silica within cavities in basalt, which then separated from the original rock through weathering and erosion. Many of the agates now reside in terrace gravels that erode during storms, providing an ongoing supply to the beaches, both to the north and south of Newport.

Much of Newport rests on an uplifted marine terrace, developed on marine sand deposited over the Astoria Formation. The terrace erodes through frequent landslides along its edge. The older section of town occupies the area in and around Yaquina Bay, which supports a commercial fishing industry. Frequent dredging keeps the estuary navigable as far as Toledo, about 10 miles (16 km)

Devils Punchbowl

Devils Punchbowl, 1 mile (1.6 km) off US 101, offers spectacular views of crashing waves at high tide. The surf explodes through a small sea arch, spilling over scoured-out Astoria sandstone. The arch appears to be the remnants of a collapsed sea cave. A well-developed soil profile is beautifully exposed at the top of the sandstone immediately below the overlook parking lot. At the beach at Devils Punchbowl State Natural Area, accessed by a side road about 0.25 mile (400 m) east of the overlook, you can see sandstone of the Astoria Formation up close, inspect several small normal faults that offset its bedding, and, at low tide, walk through a second arch into the punchbowl (use caution!). The beach trail traverses a small landslide, developed in the seaward-inclined Astoria Formation. The slide is best seen from the beach. Also from the beach, the view northward shows three levels of marine terraces, one at beach level and two older ones that have been uplifted.

View at low tide into the Devils Punchbowl through an arch made of Astoria Formation.

Seal Rock, an intrusion of the Columbia River Basalt Group, overlies and intrudes the Astoria Formation, which is visible underneath the overhang at the bottom of the cliff.

upriver. In 2011, the National Oceanographic and Atmospheric Administration relocated its home port to Newport.

South of Newport, US 101 travels over a marine terrace for much of the way to Yachats. Numerous cuts expose the marine sands deposited over the bedrock. South Beach State Park, just south of Newport, contains exposures of Astoria Formation, while Seal Rock State Recreation Site presents a dramatic exposure of Columbia River Basalt Group, intruded as part dike and part sill into the Astoria Formation. Remnants of the intrusion form a linear array of sea stacks that extend for nearly 0.5 mile (0.8 km) toward the north-northwest. The southernmost sea stack forms a high ridge displaying spectacular columnar jointing. It sits above well-bedded Astoria Formation and shows some small-scale intrusions into the Astoria near the north edge of the beach.

At Yachats, US 101 first encounters exposures of Yachats Basalt, a Late Eocene series of underwater and on-land lava flows. It dominates the coastline from here to a few miles south of Heceta Head. It also formed at about the same time and is compositionally similar to the basalt of Cascade Head. Smelt Sands State Recreation Site provides easy access to the coast on the north side of town. A trail follows the contact between marine sands of the terrace and the underlying bedrock, which consists of different parts of the Yachats Basalt, including flows, breccias, and basaltic conglomerate, formed by the erosion and

Exposure of Yachats Basalt and overlying basaltic conglomerate at Smelt Sands State Recreation Site. The man's hand is touching the base of the conglomerate.

redeposition of some of the basalt. A beautiful soil profile is exposed in places at the top of the terrace.

South of Yachats, US 101 winds in and out of headlands and small bays, over a bedrock of Yachats Basalt. Cape Perpetua exposes a series of lava flows, best viewed from below at Devils Churn, a narrow channel that funnels and concentrates the wave energy. Devils Churn has eroded along a small fault zone that trends perpendicular to the coastline. The fault offsets several small basaltic dikes that cut through the basaltic bedrock, likely feeder dikes in the volcano. Strawberry Hill, just south of milepost 169, also exhibits basaltic dikes, and one particularly obvious one forms much of the footpath down to the coast.

Heceta Head Lighthouse State Scenic Viewpoint, formerly Devils Elbow State Park, is just south of milepost 178 and provides instructive exposures of the Yachats Basalt, best reached during low to medium tides. The two prominent sea stacks exhibit bedding that dips about 30 degrees toward the southwest. The rocks consist of basaltic lava flows, flow breccias, and sedimentary deposits made of eroded basalt, which in some places is so closely related to the lava that it's difficult to tell them apart. Many of the beach cobbles here contain large crystals of the mineral plagioclase. The Heceta Head Lighthouse sits on the marine terrace above the beach. Constructed in 1894, it is more than 50 feet (15 m) high and boasts Oregon's strongest beam.

Just north of milepost 179, US 101 passes through the Cape Cove Landslide, which broke loose during a particularly heavy rainy season in winter 2000 and closed the road for several months. Several stabilization efforts are visible here, including shotcrete, bolts, netting, and drainage pipes. Just south of milepost 179 lies the entrance to Sea Lion Caves, eroded by the ocean surf along a fault zone. One half mile (0.8 km) south of the caves, small pull-outs afford outstanding views of Oregon's coastal dunes to the south. A trail from one of the pull-outs, some 300 feet (90 m) south of milepost 181, leads directly to the beach and outstanding exposures of basaltic conglomerates, shed from volcanoes of the Yachats Basalt.

View southward along coastal dunes from a headland of Yachats Basalt just north of Florence. The beach and dunes continue nearly uninterrupted all the way to Coos Bay, about 50 miles (80 km) to the south.

Aerial view of the Cape Cove Landslide taken in September 2000. The landslide occurred the previous winter.

By milepost 182, US 101 is south of the headlands and onto the coastal plain, which from here south to Coos Bay is mostly occupied by dunes, lakes, and tidally influenced rivers and stream channels. In Florence, the dunes come right up to some of the buildings. The Siuslaw River turns abruptly north at Florence, following behind the dunes until it empties into the ocean. South of Florence, occasional roadcuts expose natural cross sections of the sand deposits. The numerous lakes on the east side of the road formed because the dunes blocked small rivers draining the Coast Range. The bedrock, which consists of sandstone of the Tyee Formation, is not exposed along most of US 101, but you can see several large, south-dipping exposures near Gardiner just north of the Smith River, which joins the Umpqua River at Reedsport.

Between Reedsport and North Bend, US 101 mostly follows sand dunes that form a prominent ridge along the western edge of the road in some places. At about milepost 230, the highway reaches the northernmost arm of Coos Bay, which gradually increases in size southward. Coos Bay, the largest of Oregon's estuaries, covers more than 12,000 acres (50 km^2) at high tide, about 1.5 times the size of Tillamook Bay, Oregon's second-largest estuary. At milepost 237, between North Bend and the town of Coos Bay, north-dipping, thin-bedded sandstone of the upper part of the Coaledo Formation crops out in the cliff face on the west. These rocks are slightly younger than the Tyee Formation. Downtown Coos Bay occupies a former salt marsh, a reclaimed part of the bay.

From Coos Bay to Bandon, the road follows Isthmus and then Davis Slough of the Coos Bay estuary for nearly 10 miles (16 km) before it reaches a low divide and then drops onto the approximately 80,000-year-old Whisky Run terrace north of Bandon. Old Town Bandon occupies low ground along the

Coquille River, but most of the newer development sits on top of the Whisky Run terrace, about 100 feet (30 m) up on the south side. You can also reach Bandon via the Seven Devils Road, which leads south from Charleston along the west edge of the South Slough of Coos Bay and rises over the uplifted terraces before rejoining US 101 just north of Bandon. Bandon Beach offers a host of amazing sea stacks and cliff exposures of the Jurassic-age Otter Point Formation, part of the accreted basement of the Klamath region. They are discussed in the Klamath Mountains road guide that describes US 101 between Bandon and California.

Cape Arago

At the mouth of Coos Bay, the coastline abruptly changes from the sandy beaches of the Oregon coastal dune sheet in the north to the steep and rocky headlands of Cape Arago. Home to three state parks, this area offers up sea cliffs, sea stacks, pocket beaches, a wave-cut platform, uplifted marine terraces, and drowned trees—not to mention hiking trails, accessible tide pools, wonderful views, and a botanical garden. Perhaps more than any other coastal area in Oregon, this one demonstrates the Coast Range's interplay of active uplift and erosion.

Unlike the predominantly basaltic headlands of the northern Coast Range, the headlands of Cape Arago consist of tilted, layered sedimentary rock, giving a striped appearance to the headlands and sea stacks. In sedimentary rocks, erosion can exploit bedding surfaces rather than just the highly variable fractures and faults available in basalt. Because the bedding of the headlands tilts consistently eastward, the sea stacks tend to consist of ridges of resistant rock sloping eastward to the mainland and facing the ocean with a vertical cliff.

This stretch of the coast also beautifully displays a series of uplifted marine terraces, which originally formed as flat surfaces in the tidal zone, such as the wave-cut bench at Sunset Bay. With uplift, they now appear as low-relief sandy areas perched on top the bedrock at different elevations. Researchers have documented five different terrace levels in this area, the oldest residing at the highest elevations, having been uplifted the most. Using the estimated age of a given terrace, they can estimate the rate of uplift of parts of the coast, which farther south near Cape Blanco reaches nearly 3 feet (0.9 m) per 1,000 years. The lowest terrace, named the Whisky Run terrace, formed about 80,000 years ago. It forms the flat tops of many of the sea stacks, as well as the prominent bench just above many of the coastal cliffs.

Outstanding exposures of the tilted sedimentary rock exist at Shore Acres State Park and Cape Arago State Park, as well as the wave-cut bench

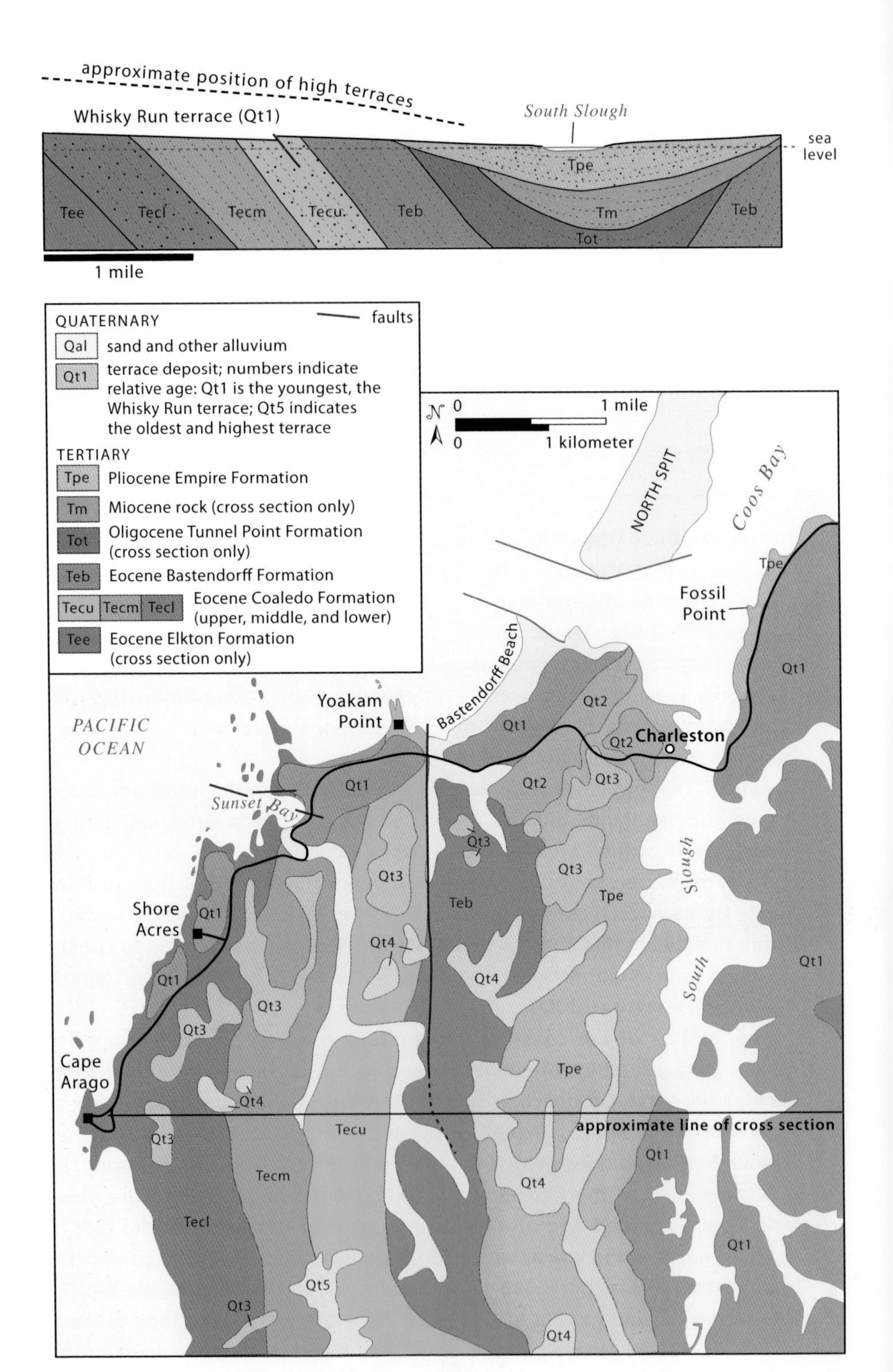

Geologic map of the Cape Arago Peninsula. Geologic cross section from the Cape Arago area (left) across South Slough. Note how folding is tightest in the older rocks and becomes progressively less tight in the younger rocks and features, including the Whisky Run terrace. This variation indicates the folding has been going on since before the Miocene rock was deposited. —*Map modified from Baldwin and Beaulieu, 1973; cross section modified from Armentrout, 1980*

at Sunset Bay. These rocks mostly belong to the lower part of the Coaledo Formation, which consists mostly of sandstone and lesser amounts of shale. The Coaledo was deposited during Middle Eocene time, about 52 to 44 million years ago, after accretion of Siletzia and deposition of the Tyee Formation. Its name derives from the presence of coal seams, found in other parts of the formation. The Coaledo hosts innumerable nearly spherical concretions, many of which are the size of bowling balls. Concretions form because groundwater may precipitate extra amounts of cement around irregularities in the rock, such as stray shells or pebbles. The additional cement then makes that part of the rock especially resistant to erosion. Some of the best examples of these concretions are at Shore Acres State Park.

The most accessible low-tide exposures, on the bench at Sunset Bay, display a variety of sedimentary features that indicate shallow marine environments of deposition, probably near the front of a delta. These features include a variety of crossbedding, ripples, fragments of fossil wood, and some marine fossils. They also include sequences of strata 100 to 200 feet thick (30–60 m) that show a progression from mostly offshore shale near the base to mostly nearshore and river-mouth sandstone at the top. These sequences suggest gradually increasing water energies and sediment supplies, followed by sudden drop-offs in both, which might be expected

This wave-cut bench at Sunset Bay features outstanding exposures of steeply dipping Coaledo Formation and wonderful tide pools at low tides. Note how the prominent sandstone rib at lower right is offset along two right-lateral faults, running from the lower left corner to the upper right.

as distributary channels move back and forth over a delta. The coal seams reflect swampy conditions, also typical of many deltas.

Younger parts of the Coaledo Formation lie east of Sunset Bay. The middle member, being finer-grained, erodes more easily than the rest of the formation and so occupies the low-relief area on the east side of Sunset Bay to just west of Yoakam Point. It likely reflects deeper water conditions. The upper member largely resembles the lower member in its details and even displays a coal seam at Yoakam Point. It probably reflects a return to shallow marine and deltaic conditions. The Elkton Formation, which lies below the Coaledo Formation, forms beautiful cliffs immediately north and south of Cape Arago. These exposures consist largely of turbidites, deposited in deeper water.

Rocks younger than the Coaledo Formation lie just to the east and are poorly exposed. The Bastendorff Formation, deposited soon after the Coaledo, consists mostly of shale; the Empire Formation consists almost entirely of sandstone, deposited during the Pliocene Epoch, perhaps 4 million years ago. It is exposed in some places, most notably along South Slough and at Fossil Point.

South Slough, which separates the Cape Arago peninsula from the rest of the mainland, fills the trough of a syncline; bedding in the rock units on

East-dipping beds of the Coaledo Formation at Sunset Bay State Park. View toward the south.

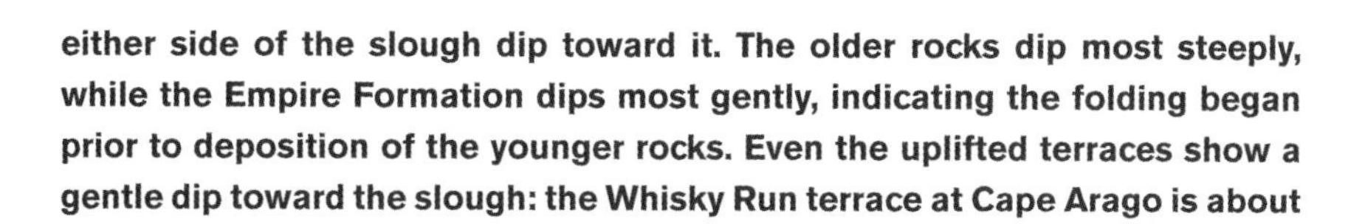

either side of the slough dip toward it. The older rocks dip most steeply, while the Empire Formation dips most gently, indicating the folding began prior to deposition of the younger rocks. Even the uplifted terraces show a gentle dip toward the slough: the Whisky Run terrace at Cape Arago is about 20 feet (6 m) higher than it is at Sunset Bay.

Several faults cut the peninsula, the most visible of which offset the beds exposed at low tide at Sunset Bay. These faults show an apparent strike-slip motion and have even caused some folding of the rock next to the faults. They also cut the Whisky Run terrace, indicating they have moved since the terrace formed 80,000 years ago. Other faults that cut the terrace include reverse faults that might be related to the folding.

While these faults and folds are clear expressions of the area's proximity to the Cascadia subduction zone, the stumps of drowned trees on the south side of Sunset Bay speak volumes about the region's earthquakes. Look for these stumps during low tides, though they may be hidden by sand. They drowned during land subsidence that accompanied a great earthquake some 1,200 years ago. Since then, several cycles of land emergence during interseismic periods have alternated with subsidence during earthquakes. The most recent earthquake occurred in 1700, just over 300 years ago.

OR 6
US 26—Tillamook
52 miles (84 km)

OR 6 gives a wonderful taste of the northern Oregon Coast Range, with its steep forested valleys carved by meandering rivers. Beginning at an elevation of 200 feet (60 m) on the outskirts of Portland, OR 6 rises steadily to an elevation of about 1,600 feet (490 m), at which point it drops precipitously to the Wilson River, which it follows the rest of the way to Tillamook.

The first rock exposures west of US 26 are partly overgrown roadcuts of 30-million-year-old basaltic intrusive rock near milepost 37, a few miles west of Glenwood. Although mostly obscured along this route, these basaltic rocks are important because they intrude many parts of the Coast Range and, being resistant to erosion, form many of the higher peaks.

On the steeper, west side of the divide, a variety of good exposures reveal the Tillamook Volcanics, Yamhill Formation, and Siletzia basalt, each formed during a progressively older time in the Eocene Epoch. These rocks can be difficult to distinguish in outcrop, let alone through the car window. The Tillamook Volcanics are mostly basalt, some of which erupted underwater. The Yamhill Formation consists mostly of fine-grained marine sediments, although it includes some basaltic sandstones and flows. The accreted Siletzia basalt, which consists of oceanic basalt, forms the basement of the Coast Range and, as such,

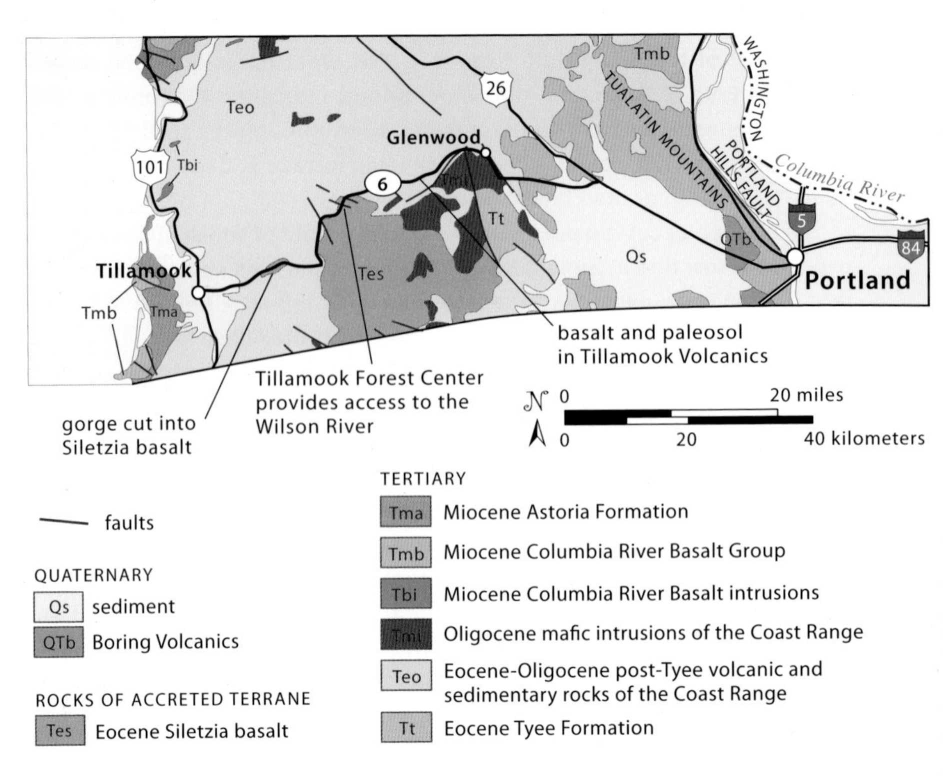

Geology along OR 6 between US 26 west of Portland and Tillamook.

is the core of the Coast Range anticline. Good roadcuts of Tillamook Volcanics lie between mileposts 32 and 30, one of which, near milepost 31 exhibits a thin but distinctly red paleosol between flows.

Beginning near milepost 24, the Wilson River cuts into the Siletzia basalt; 2 miles (3.2 km) farther west, the Tillamook Forest Center offers easy access to bedrock exposures of Siletzia basalt in the river channel, as well as extensive exhibits on the natural history of Tillamook State Forest. From here to about milepost 13, the road skirts the western edge of the Siletzia basalt but only crosses over the rock for the next 1 mile (1.6 km). For the most part, the road passes sedimentary and volcanic rock of the Yamhill Formation. The best exposures of these rocks are in the river channel, parts of which can be seen from the many pull-outs along the road.

One last exposure of Siletzia basalt exists between mileposts 10 and 9 on the north side of the highway. The river cuts a stunning narrow gorge through these rocks just east of milepost 10. These exposures contain pillows, the rounded blobs that form when lava erupts underwater. The pillows are one of the hallmarks of the Siletzia basalt. For the next several miles past here, the road passes over Tillamook Volcanics, although they crop out only in the river channel.

Just east of milepost 4, OR 6 leaves the canyon and emerges onto Tillamook Bay's coastal plain.

View looking up the Wilson River over an outcrop of Siletzia basalt from the bridge at Tillamook Forest Center. The gravel bar in middle ground includes cobbles of Tillamook Volcanics and some Yamhill Formation, younger rock units that crop out in the channel farther upstream.

Aerial view of Tillamook and its coastal plain, bordered by the Coast Range. The Wilson River canyon empties onto the coastal plain in the back center of the photo.

OR 18

Tualatin—US 101

72 miles (116 km)

OR 18 provides the most direct route from the southwestern side of Portland to Oregon's central coast. To reach OR 18, take OR 99W southwest through Tigard. The flat, highly urban landscape typifies much of the Tualatin Basin, separated from the larger Portland Basin by the Portland Hills. The basin is covered mostly by alluvial sediments that overlie lava flows of the Columbia River Basalt Group. Farther southeast, the road passes through some low hills made of Columbia River Basalt Group a few miles north of Newberg, but very little of these rocks actually show up. The Newberg Hills divide the Tualatin Basin from the northern

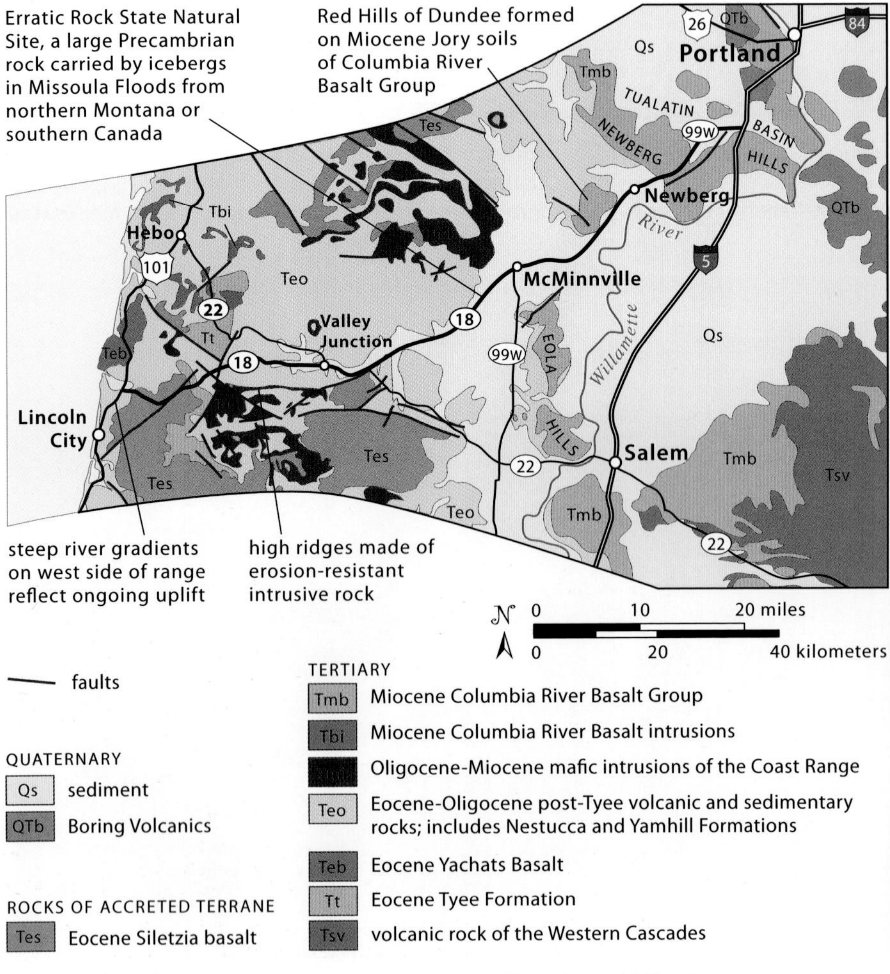

Geology along OR 18 between Tualatin and US 101.

end of the Willamette Valley. This area is one of Oregon's prime grape-growing regions, with dozens of small vineyards located within easy reach of the highway.

The junction of OR 99 and OR 18 is just south of vineyard-covered hills called the Red Hills of Dundee. The red soils derive from the Jory soils that formed on the Columbia River Basalt Group in a near-tropical climate about 15 million years ago in Miocene time. Now, the poor soils provide ideal conditions for wine grapes. One mile (1.6 km) southwest of the junction, OR 18 crosses the Yamhill River, which it follows deep into the Coast Range. The Yamhill empties into the Willamette River about 4 miles (6.4 km) downstream.

Erratic Rock State Natural Site

About 6 miles (10 km) southwest of McMinnville, the highway passes Erratic Rock State Natural Site, which highlights a large, ice-rafted glacial erratic left from the Missoula Floods. The erratic rests about 150 feet (45 m) off the valley floor, so the water must have been at least that deep when the ice-encased erratic came to rest here. What's especially striking about this erratic is its large size and rock type. It weighs some 40 tons (36,000 kg) and consists of a slightly metamorphosed shale called argillite from a Precambrian rock unit found only in northern Montana and southern Alberta. To get to the erratic, take Southwest Sauter Road west to Southwest Oldsville Road and take a right for 0.25 mile (400 m). A paved walkway climbs alongside a vineyard about 0.2 mile (320 m) to the erratic.

This glacial erratic is set among vineyards a couple hundred feet off the valley floor at Erratic Rock State Natural Site, southwest of McMinnville. It rafted here within an iceberg carried by the Missoula Floods about 15,000 years ago.

Between Dayton and Valley Junction, OR 18 is on the floodplain of the South Yamhill River.

West of Valley Junction, OR 18 follows the South Yamhill floodplain for a few more miles, framed between hills made of poorly exposed marine sandstone of Eocene age. To the south, high ridges made of Oligocene-age basaltic sills dominate the landscape. Near milepost 17, the highway leaves the floodplain and rises over more poorly exposed sandstone to a low divide that drops into the watershed of the Nestucca River. About 1 mile (1.6 km) beyond that, the road crosses another divide into that of the Salmon River. In general, the rivers that drain the west side of the Coast Range show steep gradients and expose bedrock in their channels, characteristics of active uplift. By contrast, rivers on the east side have low gradients developed on channels of deep accumulations of alluvium. This asymmetry suggests that uplift in the western part of the Coast Range is occurring faster than in the eastern part.

The only real rock exposures along this stretch of road show up west of milepost 7. Just east of Rose Lodge, a dark-colored outcrop of sandy conglomerate lies on the south side, a sedimentary part of the otherwise basaltic Siletzia terrane. Just east of the town of Otis, some of the Eocene-age marine sandstones of the Yamhill Formation are well exposed.

OR 22
Salem—US 101
55 miles (89 km)

See map on page 96

OR 22 allows easy access to the coast from the Willamette Valley, but offers clear exposures of bedrock only over its westernmost 10 miles (16 km). Heading west from Salem, the road follows the north bank of the Willamette where the river cuts through the southern Eola Hills, which consist of Eocene-age marine sedimentary rock capped by flows of the Grande Ronde Basalt of the Columbia River Basalt Group. None of these rocks are exposed near the highway. Baskett Slough, a national wildlife refuge near milepost 14, preserves some riparian features of the western Willamette Valley, including seasonal wetlands, forest, and some open water. It provides winter habitat for a variety of waterfowl, especially geese, swans, and ducks. The rolling farmland for the next 15 miles (24 km) or so is developed mostly on Eocene marine sandstone of the Yamhill Formation. The high hills to the south and west consist mostly of basalt of the accreted Siletz terrane.

West of Valley Junction, OR 22 rises along the Yamhill River and its tributaries to an almost imperceptible divide, where it passes into the Nestucca watershed, which empties into the Pacific Ocean at Nestucca Bay. If you look southward from the Yamhill Valley, you'll see the high ridge of Saddleback Mountain, held up by a basaltic intrusion at an elevation above 3,000 feet (900 m). West of the divide, OR 22 winds down a canyon cut into Eocene sandstone but overgrown by lovely dense forest. Roads descending westward from divides in the Coast Range are steeper than those descending to the east. This asymmetry most likely results from increased rates of uplift to the west than to the east.

At the intersection with OR 130 (the Little Nestucca Highway), OR 22 begins to exhibit roadcuts, as well as mileposts, which mark the distance remaining to US 101. Nearly all of these roadcuts consist of resistant Miocene basalt that intruded older, and much less resistant, marine sandstone. Age dating and chemical analysis confirm that these intrusions originated as Columbia River Basalt Group lava flows in eastern Oregon and, after flowing across the state, managed to intrude into these older rocks along the Oregon Coast!

OR 38
I-5—Reedsport
56 miles (90 km)

The western two-thirds of OR 38 follows the Umpqua River through the Coast Range. The Umpqua, which begins in the Cascades northwest of Crater Lake, is one of only three rivers that cut completely through the Coast Range in Oregon, the other two being the Rogue and Columbia Rivers. In terms of discharge, it's Oregon's fifth-largest river, behind only the Columbia, Snake, Willamette, and Santiam.

As you travel between I-5 and Drain, you'll see good exposures of flat-lying Tyee sandstone and shale, especially at the interstate exit, near milepost 56, and between mileposts 54 and 53 on OR 99. The Drain anticline, made mostly of Siletzia basalt, forms a high ridge just south of Drain. OR 99 passes through the anticline on its way south to Yoncalla. West of Drain, OR 38 passes interbedded sandstone and shale of the Tyee Formation. About 6 miles (10 km) west of town, the road passes through another anticline, also cored by Siletzia basalt. Scattered, weathered exposures of these rocks exist along the road between mileposts 44 and 43. A quarry in the rock can be seen just north of the road 0.25 mile (400 m) west of milepost 44.

One of the most interesting rock exposures along this route is a double roadcut between mileposts 38 and 37, just over 2 miles (3.2 km) west of a highway tunnel drilled through thick sandstone beds of the Tyee Formation. There, fine sandstone and shale of the Tyee are cut by a variety of small thrust and normal faults.

At Elkton, OR 38 joins the Umpqua River. Just west of Elkton, near milepost 31, the road passes exposures of Yamhill Formation, a marine deposit just slightly younger than the Tyee Formation. It has thinner-bedded, darker sandstones than the Tyee and contains some interbedded shale. By milepost 27, however, the road is back within thick-bedded sandstone of the Tyee and stays within the Tyee until Reedsport.

When the Umpqua River is not flowing too high, you can see exposures of Tyee Formation in the river channel; good examples occur near milepost 18. The presence of bedrock in river channels suggests that a river is cutting

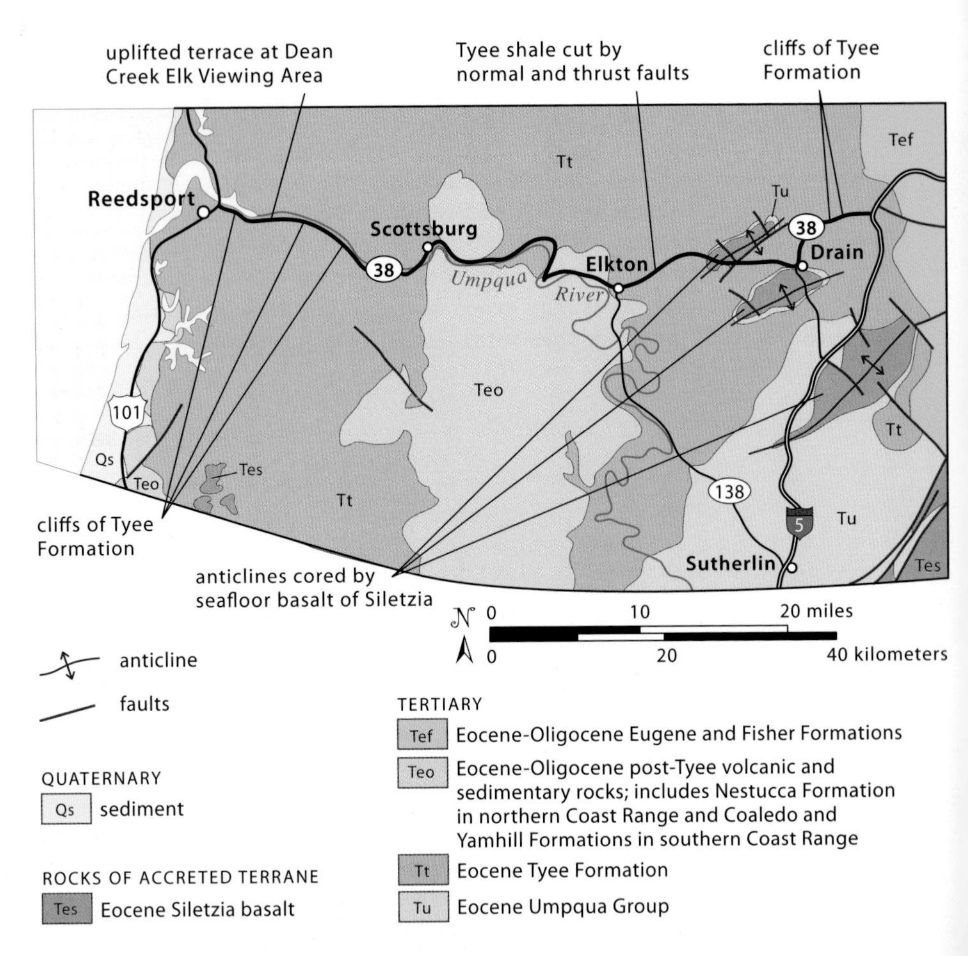

Geology along OR 38 between Interstate 5 and Reedsport.

downward, a result of uplift of the land. The Umpqua River, however, meanders lazily through the Coast Range. It's cutting downward at a pace that keeps up with the uplift. At Scottsburg, between mileposts 16 and 17, the river reaches its tidal head—the point farthest upstream at which it is influenced by tides. From there to the coast, the Umpqua is an estuary, flowing through a deep valley created when sea level was some 300 feet (90 m) lower during the Pleistocene Epoch. Cliffs of Tyee sandstone follow the road most of the way to Reedsport.

At the Dean Creek Elk Viewing Area, between mileposts 4 and 3, look for flat areas along the ridgetops north and the south of the road. The flat surface is a marine terrace, formed at sea level and since uplifted. Rising sea levels, brought on by glacial melting at the end of Pleistocene time, drowned the Umpqua River valley, turning it into an estuary. The highway through this stretch is built on a dike that separates the active river channel from the floodplain. Uplift and

Sandstone of the Tyee Formation near milepost 56, likely deposited on the middle to upper reaches of a submarine fan that extended northward into deep ocean water.

Bedrock exposures of Tyee Formation in the Umpqua River channel, a few miles upstream from Elkton on OR 138.

Elk viewing area at Dean Creek on the floodplain of the Umpqua River. The long, somewhat flat ridge in the background is the eroded edge of an uplifted marine terrace.

submergence may seem contradictory until you consider their different timescales and the timing of their occurrence. Over the long term, the Coast Range rises at about 0.1 millimeter per year, and so marine terraces take tens of thousands of years to reach their present elevations. Sea level rise after Pleistocene time occurred independent of this uplift, was comparatively rapid, and so was superimposed over the longer-term uplift process.

OR 42

Roseburg—Coos Bay

82 miles (132 km)

OR 42, a beautiful winding drive, primarily follows the Middle Fork of the Coquille River, after cresting the Coast Range divide southwest of Roseburg. Just 2 miles (3.2 km) west of I-5, OR 42 crosses the South Umpqua River, which drains parts of the Western Cascades as well as the accreted terranes near Canyonville. Immediately east of the bridge, deformed Franciscan Complex rocks make up the hill to the north, and you can see additional bedrock exposures in the river channel. One mile (1.6 km) south of Winston on OR 99, you can see the Dillard Landslide, which cut loose in February 2004, at least in part because of oversteepening of the slope brought on by basalt quarrying near the base.

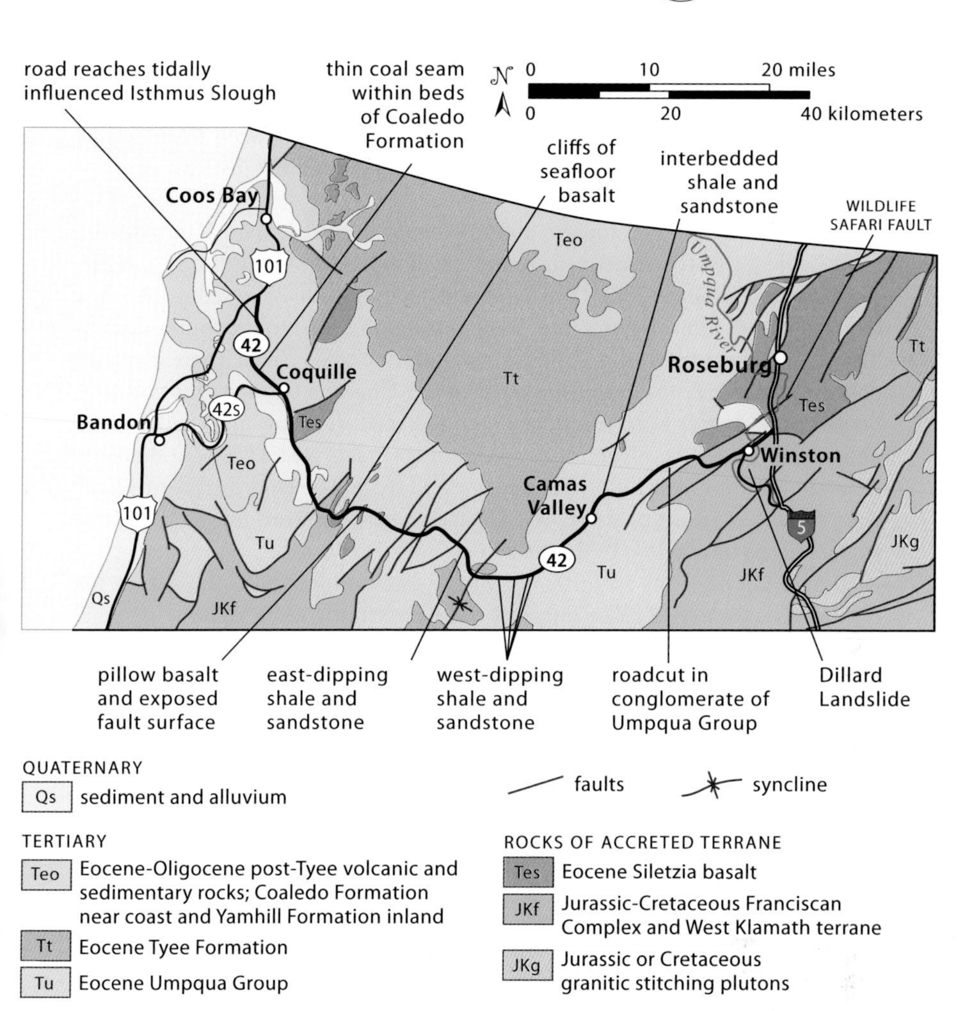

Geology along OR 42 between Winston and Coos Bay.

For several miles west of Winston, OR 42 parallels the Wildlife Safari Fault, which places the Franciscan Complex northward over the Siletzia basalt and Umpqua Group. The fault lies near the base of the ridge just to the north. About 5.5 miles (8.9 km) west of Winston, at milepost 68, a long roadcut exposes conglomerate of the Umpqua Group. These rocks are part of the Bushnell Rock Formation, which lies near the base of the Umpqua and was probably deposited along a northeast-trending shoreline just north of an early Klamath highland. The Umpqua Group, which consists mostly of Early Eocene, deep-marine mudstone, shale, sandstone, and minor conglomerate, is the oldest sedimentary rock of the Coast Range and was deposited in part during the accretion of Siletzia.

For the next 12 miles (19 km) west of milepost 68, the highway passes over marine sandstone and shale of the Umpqua Group, which is well exposed in a roadcut less than 0.25 mile (400 m) up Benedict Road between mileposts 64 and 63 and in a large roadcut on both sides of the highway just east of milepost 57.

Camas Valley, named for the abundance of camas plants that provided a food staple for Native Americans in the area, is eroded into sandstone and siltstone of the Umpqua Group, visible in scattered roadcuts. The sediments were deposited in shallow ocean water. About 3 miles (4.8 km) to the southwest, OR 42 enters the canyon of the Middle Fork of the Coquille River. There, the road passes abundant exposures of somewhat older Umpqua Group rocks that were deposited in deeper water. Particularly good exposures show up at milepost 50 and 49 and at several localities between mileposts 45 and 43, where there are imposing cliffs of west-dipping sandstone and shale. The sandstone forms thick beds of 10 feet (3 m) or more that appear to grade upward into thin-bedded shale, only to be overlain by another thick bed of sandstone.

Where OR 42 turns abruptly northward between mileposts 39 and 38, the road crosses the axis of a syncline. Just north and south of the highway, younger rocks of the Tyee Formation occupy higher elevations. At milepost 37, just west of the covered bridge at Sandy Creek, sandstone and shale beds of the Umpqua Group dip eastward.

Sandstone and shale of the Umpqua Group on OR 42 just east of milepost 57.

Less than 0.5 mile (0.8 km) west of milepost 29, on the north side of the road directly across from McMullen Creek Road, a spectacular exposure of Siletzia basalt rises more than 100 feet (30 m) above the highway. You can find good examples of pillows if you take the time to park and poke around. A small inactive quarry cuts into the exposure on its east side. Behind it, a large fault surface presents subhorizontal striations, indicating an episode of oblique-slip faulting. More exposures of the Siletzia basalt exist farther down the canyon, most notably between mileposts 26 and 25 and between mileposts 25 and 24, but these cliffs contain little in the way of obvious pillows.

Close-up view of pillow basalt.

Exposure of Siletzia pillow basalt across from the intersection with McMullen Creek Road, with near-horizontal striations on a fault surface near the upper center of the cliff exposure.

The small town of Myrtle Point lies on a wide floodplain 2 miles (3.2 km) to the north of where the Middle Fork of the Coquille River exits its canyon and joins the South Fork. Several large roadcuts and quarries of Siletzia basalt exist between mileposts 17 and 15 just a few miles north of Myrtle Point. On close inspection, these rocks show some poorly defined pillow shapes, indicative of their submarine origin. The quarries mined the rock for road construction because of its hardness.

Near Coquille, the road crosses a fault and into the Late Eocene Coaledo Formation, which makes up the headlands of much of the Cape Arago peninsula near Coos Bay. For those going to Bandon, the alternate road, OR 42S, branches off at Coquille and follows the floodplain of the Coquille River past scattered exposures of the Coaledo Formation.

Coquille lies on deltaic deposits of the Coaledo Formation, which is comparatively soft and outcrops poorly compared to the other rocks along this route. A roadcut of Coaledo Formation with a coal seam lies on the east side of the road between mileposts 9 and 8. At about milepost 3, the road encounters the upper reaches of Isthmus Slough, which flows northward into Coos Bay. Even this far from the coast the creek provides a great example of tidal effects because at low tides its water level is noticeably low, with a lot of exposed mud.

OR 126
Eugene—Florence
61 miles (98 km)

OR 126, one of the more direct routes to the coast, passes mostly through the Tyee Formation of Eocene age. The Late Eocene Eugene Formation and Oligocene-age intrusive bodies also form important components of this part of Oregon but do not appear along the route in outcrop.

In Eugene, OR 126 passes north of hills made of Eugene Formation and through parts of the West Eugene Wetlands. Most of the wetlands lie just north of OR 126 and offer trails and viewing areas. You can reach the largest area, called Meadowlark Prairie, by driving 1 mile (1.6 km) north on Greenhill Road, which intersects OR 126 west of the heavily urbanized area. An impermeable layer of clay prevents the water from easily draining from the wetlands. According to mineralogical and chemistry studies by Eugene-area geologist Michael James and geomorphologist Karin Baitis, the clay is altered volcanic ash erupted from Mt. Mazama. The ash occurs in many places in Oregon, but the impermeability here was likely caused by extensive alteration in this lowland area. The West Eugene Wetlands must have formed relatively recently, after the Mazama eruption 7,700 years ago. Three miles (4.8 km) west, the road reaches more wetlands along the south edge of Fern Ridge Lake. The Army Corps of Engineers built Fern Ridge Dam in 1942, impounding the reservoir.

At Veneta, OR 126 crosses into the Tyee Formation, although significant exposures are absent until midway between mileposts 26 and 25. There, a roadcut exposes thick-bedded Tyee sandstone along the south side of the road

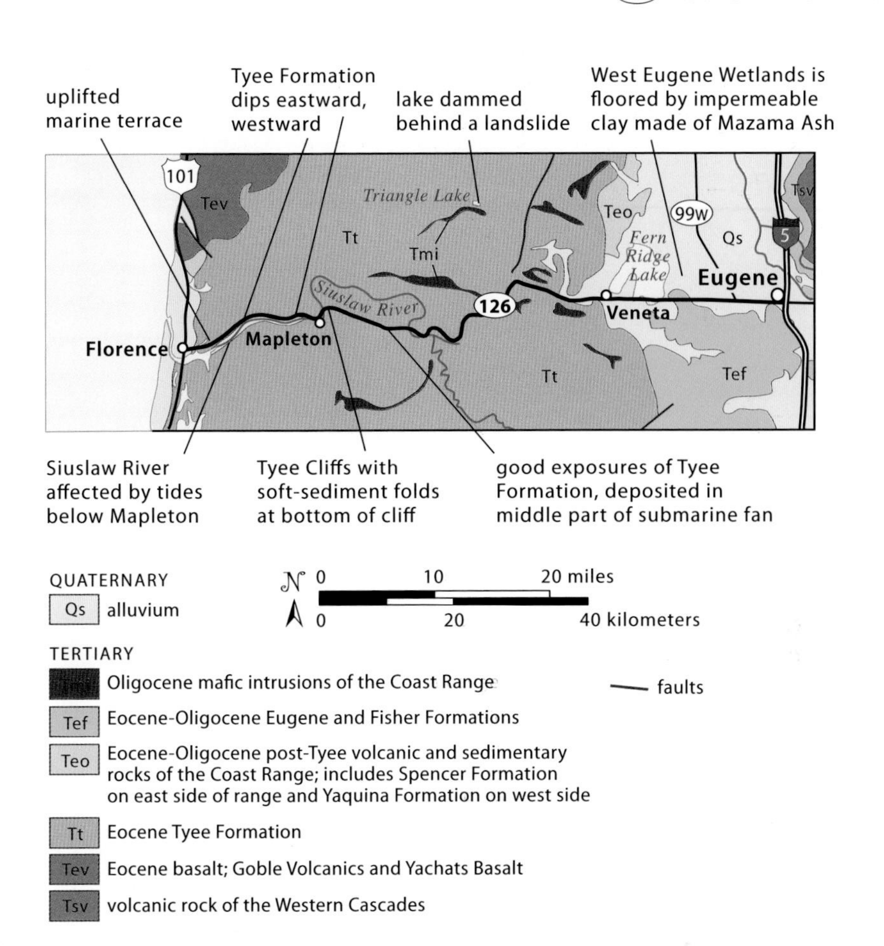

Geology along OR 126 between Eugene and Florence.

for more than 0.5 mile (0.8 km). This stretch of road lacks pull-outs, however, so even the most desperate roadside geologist should pass it by and just enjoy the view from the window. Other good exposures of the Tyee exist farther to the west.

A small, curved normal fault in Tyee sandstone occurs across from a pull-out 0.25 mile (400 m) past milepost 23. Good exposures of Tyee Formation exist on both sides of the tunnel between mileposts 20 and 19, and there is an imposing roadcut on the south side of the road just west of milepost 15, on the east side of the Siuslaw River at Mapleton. This latter exposure displays some unusual examples of folding near the base of the roadcut. As the folds don't seem to affect any other part of the rock, they are likely a product of slumping or sliding while the sediment was still soft, soon after it was deposited.

At Mapleton the highway crosses the Siuslaw River, which even here is an estuary, affected by ocean tides more than 15 miles (24 km) from its mouth.

West Eugene Wetlands stay wet because an impermeable layer of clay, probably derived from Mazama Ash, prevents the water from draining easily.

Typical exposure of Tyee Formation along OR 126. Note the small curved fault near the center of the photo.

This part of the Siuslaw formed when rising sea levels flooded the river valley. Its width and flatness reflect the huge amounts of sediment deposited after sea level rise, framed in by the old valley walls. OR 36, which heads north and east from Mapleton toward Junction City, passes through the town of Triangle Lake, named for one of the few large natural lakes in the Coast Range. The lake formed during Pleistocene time when a landslide blocked a river drainage.

West of Mapleton, the Tyee forms cliffs on the north side of the highway. Similar to sandstones along the rest of this route, these are thick bedded. The Tyee Formation is more thin bedded and more variable along the highways to the north and south. Between mileposts 13 and 12, the Tyee sedimentary layers dip gently eastward, whereas they dip westward near milepost 3. Directly across from milepost 11, an especially good exposure contains some interesting minor faults.

Between mileposts 5 and 4, the road follows a dike, built to control flooding along the river, and passes uplifted marine terraces that cap the hills on either side of the river. Recent deposits of dune sand line the north side of the road west of the North Fork of the Siuslaw River at about milepost 1.

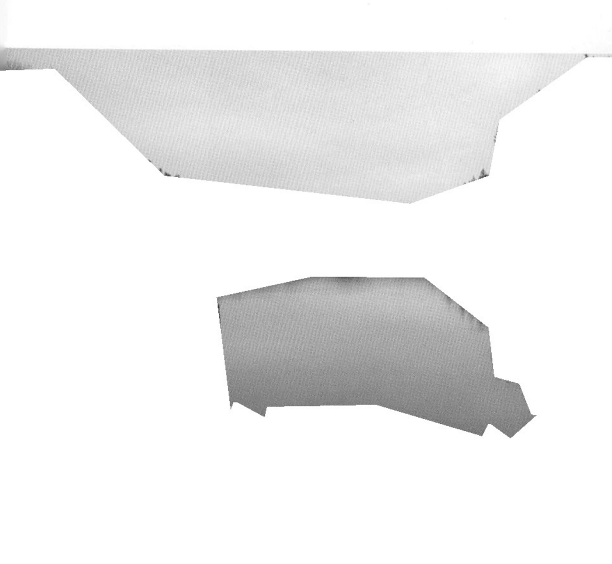

The Siuslaw River is affected by ocean tides for 15 miles (24 km) upstream from its mouth to the town of Mapleton. This photo looks upstream from a point along the highway between Mapleton and US 101.

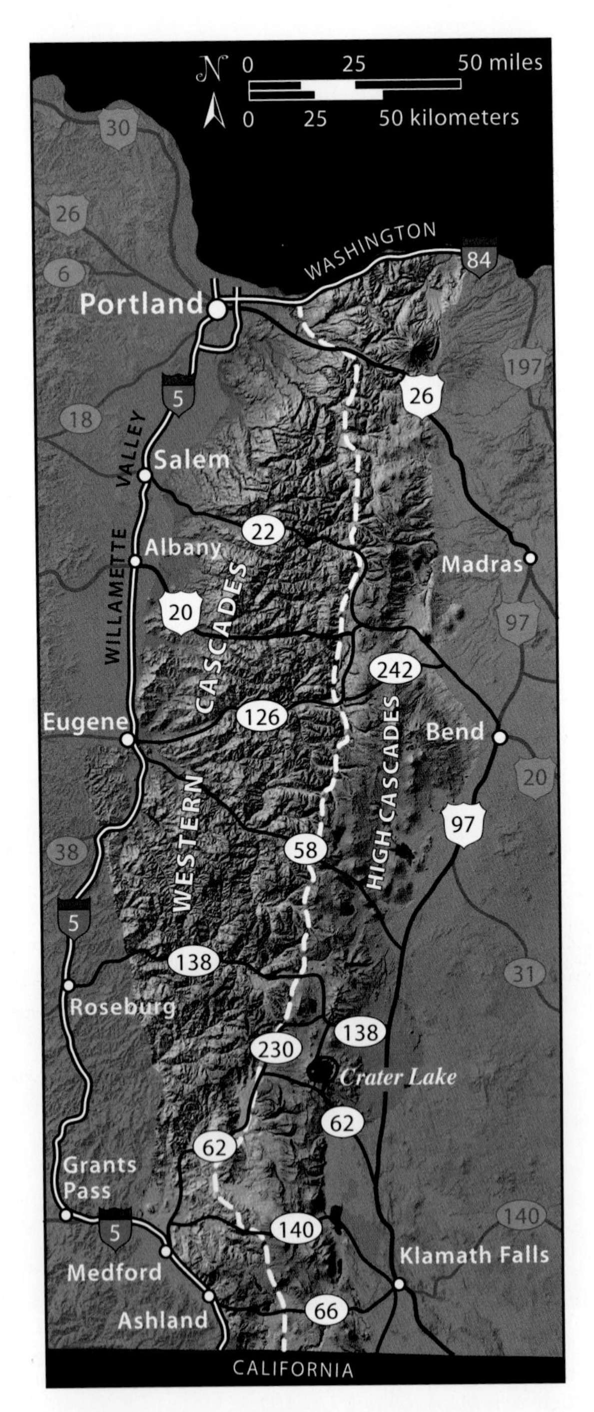

Physiographic map of the Cascade Range. The dashed line marks the boundary between the High Cascades and Western Cascades. Note the relative smoothness of the topography of the two areas: the inactive Western Cascades is deeply eroded, while recent eruptions smoothed over the irregularities in the High Cascades. –Base image from US Geological Survey, National Elevation Data Set Shaded Relief of Oregon

Cascade Range

Oregon's Cascade Range offers some of the most spectacular scenery of the Lower Forty-Eight, with magnificent waterfalls, densely forested valleys, and glaciated volcanoes, several of which exceed 10,000 feet (3,000 m) in elevation. Some nationally known landmarks include Crater Lake National Park, Mt. Hood, and Multnomah Falls. The Cascade Range, originating entirely through volcanic activity, reflects a wonderful interplay of creation and destruction. Volcanic eruptions in the High Cascades occur every 50 to 100 years or so, nearly continuously at the geologic timescale. Meanwhile, weathering and erosional processes work away at new and old material alike, sculpting the peaks and ridges and cutting the valleys and waterfalls.

The Cascade volcanoes extend as a narrow belt some 600 miles (970 km) from northern California to southern British Columbia, Canada. They define a volcanic arc, an arcuate chain of volcanoes above a subduction zone. The Andes of South America and the Aleutian Islands of Alaska are also volcanic arcs. During the last 2 million years, magma has erupted from more than two thousand sites within the Cascades, more than one thousand of which are in Oregon. These sites include small volcanic cones, fissures, and shield volcanoes, as well as the much larger stratovolcanoes, such as Mt. Hood. In Oregon, most of the lava is basalt and basaltic andesite, but in some places, it contains more silica and forms andesite, dacite, and even rhyolite. Because lavas with greater silica content tend to be less fluid, they form steeper cones. In contrast, basalt lava has the least silica and flows long distances. Some volcanoes of basaltic andesite, as well as andesite and dacite-rich volcanoes, rise as peaks above the low-lying basalt flows. The introduction to this book presents more detailed information on the chemistry and behavior of different types of lava.

The magma that fuels volcanic activity in the Cascades, as well as other volcanic arcs, originates at the subduction zone. There, the cold oceanic plate gradually heats as it sinks beneath the continent, and as it heats, it causes water-rich minerals to dehydrate. The released water rises into the overlying continental lithosphere, which is hotter than the subducting plate (remember, the subducting plate was cold to begin with), so the continental lithosphere begins to melt as it interacts with the water.

The reason for the melting is relatively simple: wet rock melts at lower temperatures than dry rock, and the continental lithosphere at these depths is hot enough for wet melting but not hot enough for dry melting. The magma, which begins with a variety of basaltic compositions, can evolve into more silicic ones as it rises, through either engulfing smaller silica-rich bodies of

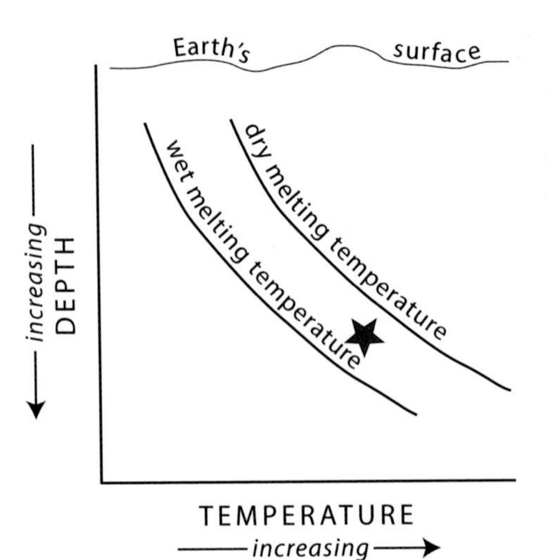

In melting curves for wet and dry rock, the temperature of melting increases with depth (and pressure), but for wet rock, the temperature of melting is less than for dry rock. A dry rock at the conditions represented by the star, therefore, would begin to melt if combined with water.

Southward view over Mt. Washington to the Three Sisters. Belknap Crater, a shield volcano, occupies the right middle ground, and you can see several recent basaltic lava flows between it and North Sister. McKenzie Pass lies in the low area on the south (far) side of Belknap Crater.

rock, mixing with other magmas, or crystal fractionation. In this latter process, crystallization of early-formed minerals, which tend to be proportionately low in silica, results in a remaining magma that is richer in silica.

In Oregon, the Cascade Range consists of two parts: the High Cascades and the Western Cascades. The High Cascades are the recently active volcanoes that define the crest of the range, whereas the much older Western Cascades lie just to the west. In some places, such as at Salt Creek Falls along OR 58, recent lava flows from the High Cascades flowed down valleys cut into older rock of the Western Cascades.

The High Cascades form an effective north-south barrier to Pacific storms as they move eastward across Oregon, resulting in a classic rain shadow to the east and an unusually high amount of precipitation to the west. The Western Cascades receive about 4 to 10 feet (1.2–3 m) of precipitation each year; the east side receives only a quarter of that.

The High Cascades

The High Cascades mostly consist of rock erupted during the latest Pliocene and Quaternary Epochs, beginning about 3 million years ago. The most visible features however, such as the volcanoes along the range crest, are younger than a half million years, and many are much younger. The oldest rock exposed in Crater Lake National Park, for example, is 420,000 years old, but the eruption that blew the mountaintop off and created Crater Lake occurred only 7,700 years ago. At Mt. Hood, most of the early volcano-building eruptions are younger than a half million years.

It is likely that parts of the High Cascades were active as far back as 8 million years ago. The Deschutes Formation, which consists of mafic and silicic lavas, ash-flow and ash-fall deposits, debris flows, and volcanic-derived sedimentary rocks, covers much of the area north of Bend and immediately east of the volcanoes. Although these rocks are, for the most part, in the Lava Plateaus region, they appear to be derived from the Cascades between about 7.5 and 4.5 million years ago. In addition, just east of Mt. Jefferson, the andesitic volcano at Castle Rocks yields an age of 8 million years; west and south of Mt. Jefferson, several basaltic volcanoes give ages between 7 and 5 million years.

There is not much evidence of the early activity in the High Cascades because the older rocks have been buried beneath younger lava flows. They are further hidden because much of the range is built upon crustal blocks down-dropped between inwardly dipping normal faults. Together, these blocks define the High Cascades graben, which probably first formed about 5 million years ago, although many of its faults have continued to slip throughout much of Quaternary time. Many of the smaller volcanic vents of the High Cascades are aligned in north-south directions, suggesting their locations were influenced by faults in the subsurface.

The graben is most distinctive in the area between Bend and Mt. Jefferson, where it likely dropped more than 2,000 feet (610 m) throughout its history. Faults help define the McKenzie River drainage near Belknap Hot Springs on the west side, while on the east side, they trend northward through Camp

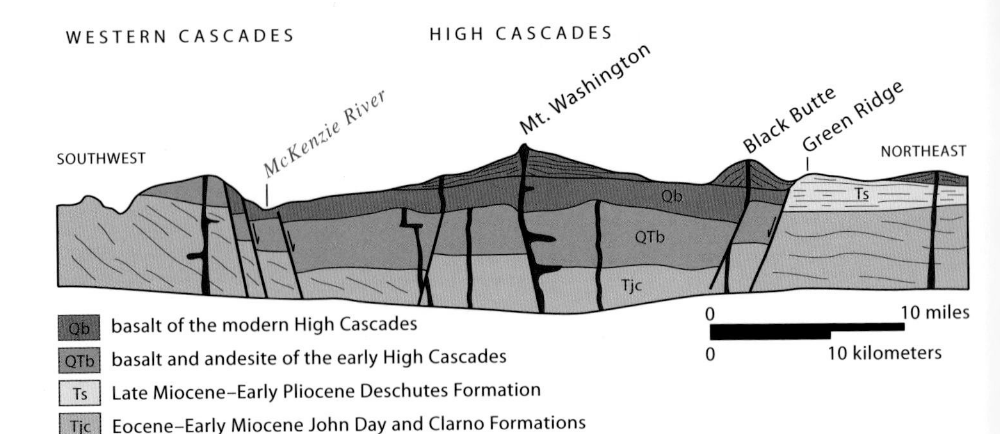

Cross-section from the southwest to the northeast across the High Cascades at Mt. Washington. Note that the High Cascades occupy a graben, and most of the early High Cascades lavas have been dropped down and buried beneath the younger volcanic material. –Modified from Taylor, 1990

Sherman and along the edge of Green Ridge. Extensional stresses form grabens, but what caused the extension here is not entirely clear. It might be linked to slight changes in plate motions, related to Basin and Range extension, or caused by the volcanic arc foundering under its own weight.

In general, the longer a volcano goes without erupting, the more dissected by erosion it becomes. Recently active volcanoes, such as South Sister and Mt. Hood, are much more conical and smooth than North Sister or Mt. Jefferson, which haven't erupted for tens of thousands of years. Catastrophic eruptions make important exceptions to this rule, of course, because they can decimate the volcanic edifice altogether. Crater Lake occupies the caldera where Mt. Mazama erupted and collapsed, and Mt. St. Helens, just across the border in Washington, consists of the emptied shell of itself after its eruption in 1980. These exceptions notwithstanding, this rule also explains important differences in the landscapes of the Western and High Cascades. The Western Cascades, which have been inactive for millions of years, show far deeper erosion than even the oldest elements of the High Cascades.

Most of the peaks of the High Cascades were eroded by glaciers during the last ice age of the Pleistocene Epoch. The upper elevations contain numerous cirques, which are bowl-shaped regions scoured by the upper reaches of a glacier. Many of the deep canyons display U-shaped profiles, characteristic of glacial valleys. These glaciers deposited much of the eroded material as moraines, many of which can be seen at the higher elevations. Today, Mt. Hood, Mt. Jefferson, and each of the Sisters still host small, remnant glaciers, but the rest of the ice succumbed to the warming climate at the end of the ice age 10,000 years ago.

The Western Cascades

The Western Cascades formed from about 40 to 5 million years ago. They no longer host any active volcanoes, so their landscape is deeply eroded. Because they lie on the wet side of the crest, dense forest covers the bedrock. Exposed rocks, mostly basaltic to andesitic lava flows, tuffs, and lahars, resemble the younger ones of the High Cascades. Erosion has exposed some shallow intrusions in the range, some of which likely fed volcanic activity at the surface. The Western Cascades also contain some sedimentary rocks derived from the erosion of their volcanoes.

The Western Cascades demonstrate that the volcanic arc we see today extended back about 40 million years into the past. A slight flattening of the angle of subduction likely caused of the eastward shift of eruptions to the modern location. The descending plate needed to be farther east under the continent to reach the depth and temperature conditions necessary to cause the overlying lithosphere to melt.

View over the Western Cascades, looking east toward the High Cascades. Mt. Thielsen, just north of Crater Lake, forms the spike-shaped peak near the center of the horizon.

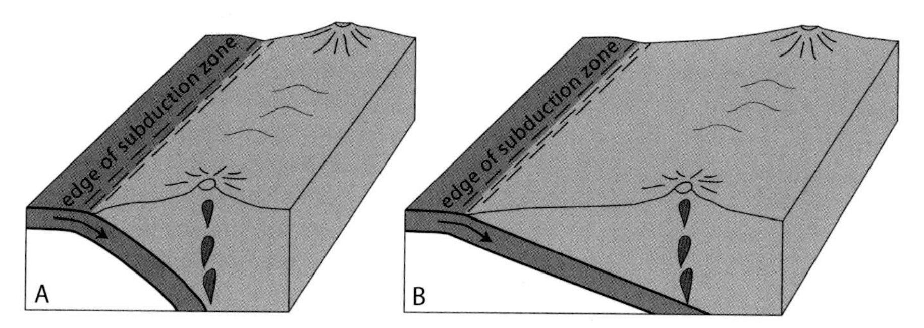

A decrease in the angle of subduction (B) shifted the volcanic activity eastward.

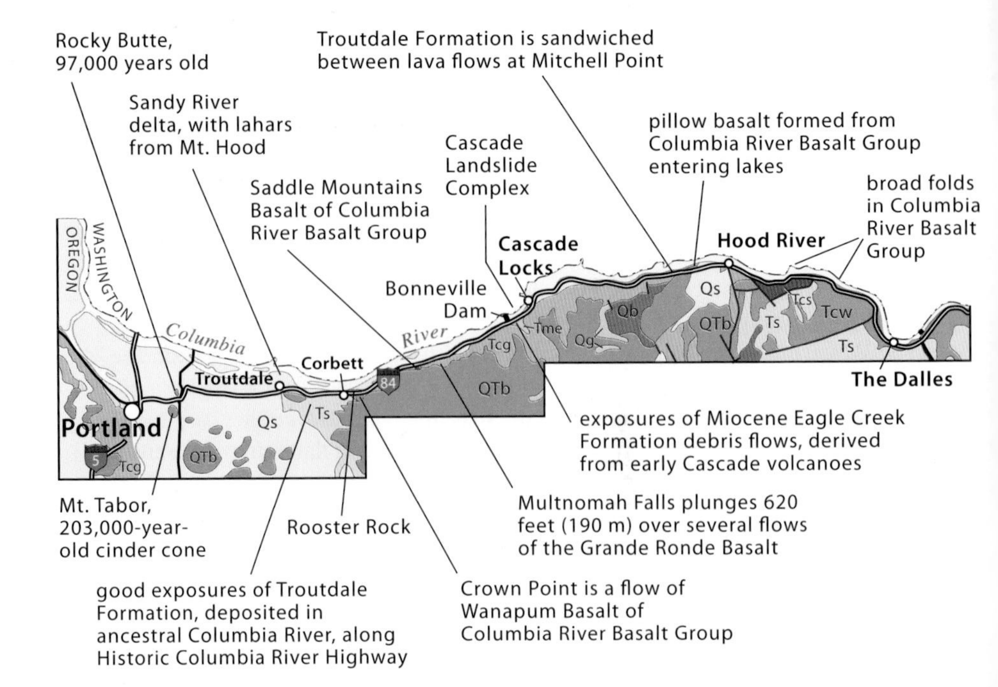

Reconstruction of former channels of the Columbia River during eruptions of Wanapum and Saddle Mountains Basalts of the Columbia River Basalt Group

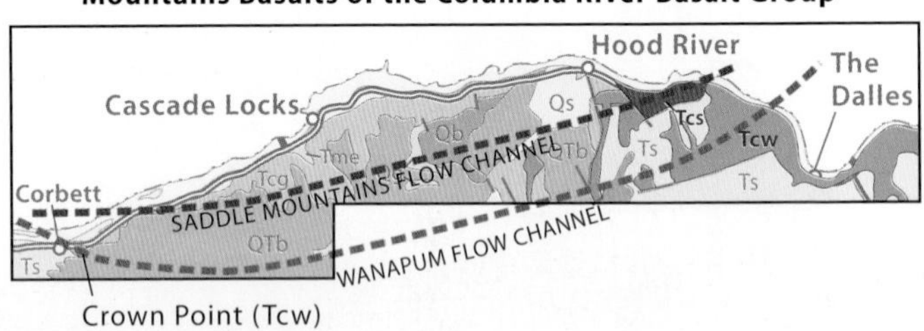

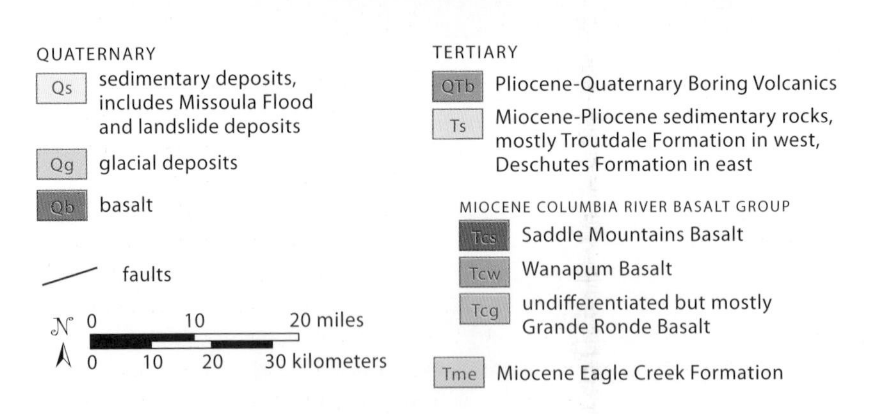

Generalized geology along I-84 between Portland and The Dalles. The lower map shows approximate locations of two former channels of the Columbia River, now occupied by Columbia River Basalt Group lava that flowed down the channels and forced the river to seek a new path.

GUIDES TO THE CASCADE RANGE

INTERSTATE 84
Portland—The Dalles
84 miles (135 km)

The stretch from Portland to The Dalles must be one of the most dramatic stretches of interstate highway anywhere. I-84 follows the Columbia River Gorge through the Cascade Range, passing by soaring cliffs, waterfalls, and busy locks on the river. The gorge is relatively young, created by downcutting of the Columbia River over the last 3 to 2 million years during uplift of the Cascade Range. The Historic Columbia River Highway, the original highway through the Columbia River Gorge, was constructed between 1913 and 1920. Remaining parts of it provide an alternate, slower route to sightseers and geologists.

Columbia River Gorge

The rocks of the Columbia River Gorge consist mostly of different flows of the Columbia River Basalt Group and underlying Eagle Creek Formation, a sequence of debris flows and river-deposited conglomerate and sandstone deposited about 20 million years ago. In some places, the formation contains fossilized tree logs. Deposits of the Miocene-Pliocene Troutdale Formation, deposited by an early Columbia River, figure prominently from Mitchell Point westward. In addition, basaltic lava flows of the Boring Volcanics exist at higher elevations on top of the Columbia River Basalt Group. Larch Mountain, a shield volcano some 5 miles (8 km) southeast of Multnomah Falls, erupted many of the Boring lava flows. Another rock unit, the volcanic Ohanapecosh Formation, lies beneath the Eagle Creek Formation. It crops out on the Oregon side of the river in only a few places but plays a huge role in the landslide story.

Although lava flows of the 15-million-year-old Grande Ronde Basalt, the most voluminous member of the Columbia River Basalt Group, make up most of the basaltic cliffs of the gorge, younger flows of the Columbia River Basalt Group form many of the cliffs just west of Multnomah Falls. The Vista House at Crown Point, for example, sits on top the 14-million-year-old Wanapum Basalt, and Bridal Veil Falls flows over the 12-million-year-old Saddle Mountains Basalt. What's more, these flows are unusually thick: the Wanapum Basalt at Crown Point measures greater than 650 feet (200 m). Detailed studies of these flows indicate that they mark former channels of the Columbia River, eroded into bedrock of the Grande Ronde Basalt. The Columbia River today cuts obliquely across those old channels, exposing them in the cliffs.

Perhaps the most immediately visible aspect of the Columbia River Gorge is that it is asymmetric: the Oregon side is extremely steep, with countless cliffs and waterfalls, whereas the Washington side is not nearly as imposing. This asymmetry results from the near-ubiquitous landslides on the north side, which carry enormous masses of land toward the river and greatly decrease the overall slope gradient on that side. Perhaps the most striking example of these landslides is the Bonneville Slide across the river from Cascade Locks. It covers some 5.5 square miles (14 km^2), pushed part of the channel of the Columbia River southeastward, and temporarily dammed the river until it failed in a large flood some years later.

The landslides tend to form on the river's north side because the underlying bedrock tilts between 2 and 10 degrees southeastward, which encourages sliding along bedding surfaces toward the river. On the south side, rocks tilt

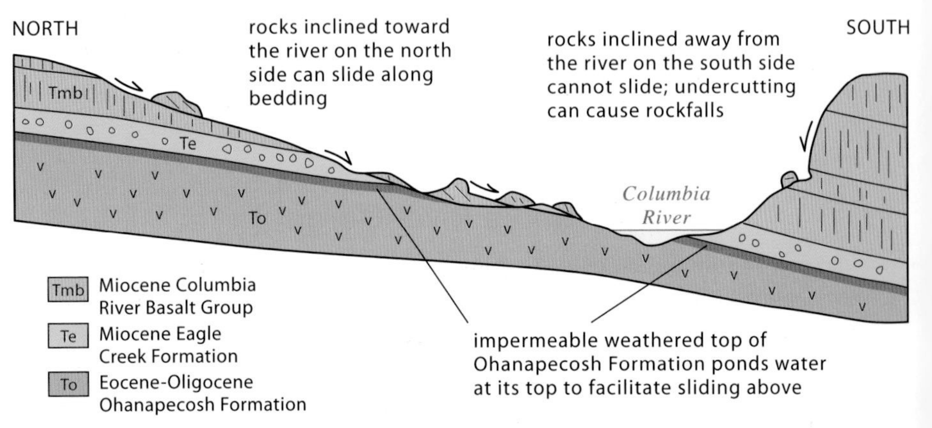

Schematic cross section from north to south across the Columbia River Gorge. On the north side of the river, rocks can slide easily toward the river because they are tilted southward; on the south side, erosion takes place more often by rockfalls, which tend to maintain the steep slopes.

away from the river, an orientation that discourages sliding. Contributing to the landslides is the underlying Ohanapecosh Formation, which is deeply weathered and contains a great deal of clay. Water passes relatively easily down through the fractured basalt of the Columbia River Basalt Group and permeable gravels of the Eagle Creek Formation but collects at the clay-rich top of the Ohanapecosh, creating a slick surface, ripe for sliding. Near Hood River, where the Columbia River Basalt Group is at river level, the landslides are far less prevalent and the gorge is more symmetrical.

The rocks tilt gently southward in the Columbia River Gorge because they are part of the Yakima Fold Belt, a series of anticlines and synclines that stretches from near Yakima, Washington, to Pendleton, Oregon. Perhaps the most obvious folds along this stretch of I-84 are the anticline between Cascade Locks and Hood River and the syncline between Hood River and The Dalles. The fold belt greatly affects the landscape, as the anticlines tend to form ridges and the synclines form valleys. Many of the anticlines are cored by thrust faults that may be potentially active. Researchers suggest that the fold belt may be a result of the ongoing clockwise rotation of Oregon into the more stationary Washington.

Water and cliffs abound on the Oregon side of the Columbia River Gorge, so it is no wonder that the area contains so many waterfalls. Just as the southeastward tilt of the rocks promotes landsliding on the north side, the tilt promotes cliff formation on the south side. With little sliding, the more resistant rocks erode piecemeal by breaking off cliff faces along steeply inclined fractures after being undercut by faster erosion of softer material below. The fracture surfaces then become the new cliff face. Slides do occur on the south side, but they tend to be relatively small when compared to the north side.

View upriver from Crown Point. Note how much steeper the slopes are on the south side of the river.

The Missoula Floods also played an important role in maintaining the cliffs of the Columbia River Gorge. Their incredible erosive power swept out the existing erosional debris on both sides of the river, exposing the bare rock below. The floods also undercut some of the material at higher elevations, which caused both rockfalls and landslides.

Portland to The Dalles

East of downtown Portland and the Willamette River, I-84 first climbs through an old flood channel scoured by the Missoula Floods about 18,000 to 15,000 years ago. Just west of the intersection with I-205, you can see Rocky Butte to the north and Mt. Tabor to the south. The steep eastern side of Rocky Butte was eroded by the Missoula Floods, which deposited the long gravel bar on the butte's downstream, western side. The bar, called Alameda Ridge, has long been quarried for gravel and sand. Both buttes consist of bedrock of volcanic rocks of the Boring volcanic field. Mt. Tabor is a cinder cone that formed 203,000 years ago on top of the Troutdale Formation. Both I-84 and I-205 pass good exposures of the Boring Volcanics on the flood-eroded east side of Rocky Butte, less than 1 mile (1.6 km) north of their intersection. Rocky Butte, at 97,000 years old, is Portland's youngest volcano.

Just east of Troutdale, near milepost 17, the interstate crosses the Sandy River and passes beneath the first cliffs of the Columbia River Gorge. These densely vegetated cliffs consist of Troutdale Formation, deposited by the Columbia River in Miocene time and overlain by Boring Volcanics. Immediately north, the low-lying area is the delta of the Sandy River. It consists of lahar sediments from the last two major eruptions of Mt. Hood, some 30 miles (48 km) to the south. These eruptions occurred about 1,500 years ago and just over 200 years ago, just before Lewis and Clark arrived. In their journal from 1805, Lewis and Clark called this river the Quicksand River because of the easily fluidized material in the recent lahar deposit.

At exit 18, take the Historic Columbia River Highway toward Oxbow Regional Park to inspect the Troutdale Formation in a safer and quieter environment than the interstate. Multiple exposures show up on the east side of the road for a distance of some 2.5 miles (4 km) and include mostly sandstone and gravel with crossbeds and channels, features typical of river-deposited sediments of the ancestral Columbia River. Ages on the Troutdale Formation range from about 15 to 2 million years ago.

About 2 miles (3.2 km) east of exit 18, I-84 encounters the Columbia River and hugs the banks all the way to The Dalles. Exit 22, for Corbett, offers a slower but scenic alternative on the Historic Columbia River Highway. Just 1 mile (1.6 km) to the east of Corbett, it passes by expansive views at the Portland Women's Forum State Scenic Viewpoint and the Vista House at Crown Point. Between the two is a landslide complex, manifest on the old roadway by numerous cracks and rough spots.

Rooster Rock, which appears from the interstate as a spire set out from the cliffs beneath Crown Point, is part of the landslide. The landslide broke

Outcrop of Troutdale Formation along the Historic Columbia River Highway.

off Crown Point during the Missoula Floods and slid toward the river. The hummocky topography of the valley immediately west of Crown Point is part of the landslide. It continues to be somewhat active, causing breakage and bending of the Historic Columbia River Highway directly behind it.

Rooster Rock and Crown Point consist of Wanapum Basalt of the Columbia River Basalt Group that flowed down and filled an ancient canyon, oriented at a high angle to the cliffs. This canyon was likely one of several early courses of the Columbia River. From Rooster Rock State Park (exit 25) you can look up to Crown Point and see a waterfall spilling gracefully over a cross section of the thick Wanapum Basalt flow. The bottom of the flow shows a distinct layering, as does the bottom of Rooster Rock, visible from the boat launch at the park. This layering is produced when basalt flows enter bodies of water and break into small glassy fragments called hyaloclastite, which is then deposited as layers. The eastern end of Rooster Rock State Park is a long, forested sand dune that runs parallel to I-84 and is actively forming today. During winter months, east winds blow sediments out of the Columbia River floodplain onto the dune in the trees. Just east of milepost 27, you can look across the river to see some cliffs of Wanapum Basalt at water level.

Latourell Falls, east of Crown Point along the Historic Columbia River Highway, flows over an example of Columbia River Basalt Group lava that flowed down an ancient canyon. There, the lava flow forms excellent columnar joints, called the colonnade, beneath a zone of closely spaced irregular columns, called the entablature.

Multnomah Falls, the highest waterfall in the Oregon, is easily the most famous waterfall in the gorge, with a combined drop of 620 feet (190 m) for its upper and lower falls. Even with the crowds, this waterfall is well worth the visit. Exit 31 to the parking lot provides access to both sides of the interstate, so travelers can turn around here as well. After parking, visitors walk through a tunnel beneath the railroad tracks and across the Historic Columbia River Highway to a visitor center, lodge, hiking trails, and the waterfall. A foot trail leads to a bridge that crosses in front of the plunge pool of the upper falls, and then continues to the top of the falls and beyond. From the bridge, you can view several flows of the Grande Ronde Basalt, including one that consists mostly of pillow basalt that formed when the lava flowed into a lake. You can also see some of the enormous boulders that have toppled from the cliff as the plunge pool cut away their support. Several of the boulders came from one larger block, about 40 feet by 20 feet by 6 feet (12 x 6 x 1.8 m) that broke free from the upper falls in September 1995 and shattered upon impact. Fortunately, nobody was badly hurt, although some twenty people who were standing on the bridge at the time received minor injuries. Most recently, a falling boulder damaged the bridge and caused its temporary closure in January 2014.

Just west of its intersection with the Historic Columbia River Highway at milepost 37, I-84 passes Beacon Rock on the Washington side of the river. Beacon Rock is a basaltic plug that was eroded by the Missoula Floods. It erupted 60,000 to 50,000 years ago, which makes it the youngest of the Boring volcanoes. A short trail switchbacks to its summit.

This stretch of road affords numerous good views of several large landslides on the Washington side. They appear as sunken areas of irregular topography, in many cases framed in by cliffs. One good example is across from milepost 34. Probably the largest and best known, however, is the Cascade Landslide Complex, across from milepost 39. This series of landslides covers 10 square miles (26 km^2) and has slipped repeatedly over the past 1,000 years. The most recent landslide, called the Bonneville Slide, occupies much of the center of the complex. It likely slipped between about AD 1425 and AD 1450. The event dammed the river, which gave rise to the Native American Bridge of the Gods legend and pushed the river's channel more than 1 mile (1.6 km) southward. The dam formed a lake that probably extended more than 50 miles (80 km) upstream. Eventually, the lake crested the dam, and in a manner reminiscent of the much larger Missoula Floods, broke through, unleashing a major flood downstream that was more than 50 feet (15 m) deep when it hit the Portland area.

Conglomerate and minor sandstone of the Miocene Eagle Creek Formation shows up in roadcuts on the south side of the highway beginning at about milepost 37; an especially prominent exposure of it exists near milepost 40. These

Multnomah Falls has a combined drop of 620 feet (190 m) from the top of its upper to the base of its lower falls. Note the pillow lavas near the top of the upper falls.

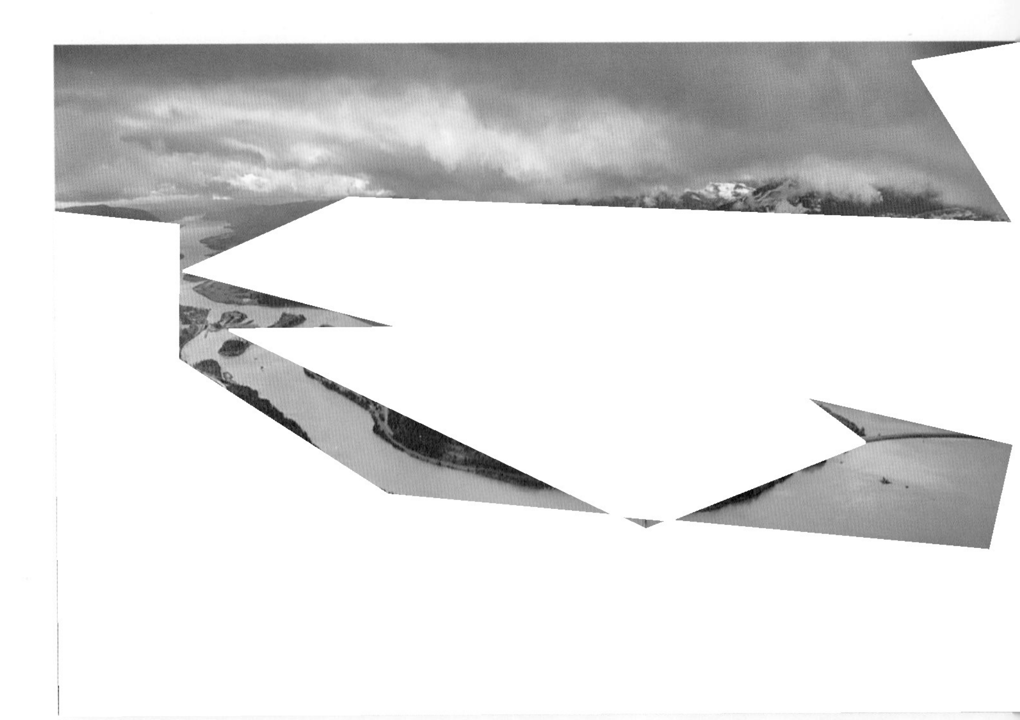

Aerial view of the Bonneville Slide, view downriver toward Cascade Locks.

rocks originated mostly as debris flows off the early Cascade volcanoes, and in many places contain petrified logs. Good off-highway exposures of this interesting formation exist at exit 41 for the Eagle Creek Trailhead (eastbound only).

Below Mitchell Point on a forested slope (exit 58), the Troutdale Formation is sandwiched between underlying Grande Ronde Basalt and overlying Wanapum Basalt, all dipping moderately to the southeast. This relationship demonstrates that the Columbia River, which deposited the Troutdale Formation, was here during the time of the Columbia River Basalt Group, and that overlying lavas likely flowed down its old channel. Geologists familiar with Columbia River Gorge geology regard this outcrop as an exposure of the southern edge of the Columbia River channel, some 15 million years old. From the overlook at Mitchell Point, you can get a close look of the Grande Ronde Basalt, including an exposed transition from the colonnade to the entablature part of the flow. Loose rocks from the cliffs above consist of solid dense varieties from flow interiors, as well as ones full of air bubbles from flow tops. A large landslide appears just downriver on the Washington side.

Just east of milepost 60, brown-colored roadcuts of pillow basalt exist on both sides of the highway. The pillows formed when the basalt flow poured into a body of water, possibly the early Columbia River. Accompanying explosions shattered much of the basalt, producing hyaloclastite, which then altered to the brownish yellow material palagonite. A much safer and equally outstanding exposure of basalt pillows and palagonite exists just east of The Dalles and off the interstate. Refer to The Dalles to US 97 road guide in the Lava Plateaus chapter for directions.

Besides the many exposures of columnar jointing and entablature, the stretch between Hood River and The Dalles offers views of the geology on the Washington side of the Columbia. Inclined basalt flows mark limbs of gentle folds of the Yakima Fold Belt, which forms a zone of deformation that extends from about Hood River to Richland, Washington. Near milepost 66, you can see flows dipping upriver, then near milepost 68 a syncline that includes the Oregon side, and near milepost 79, another anticline. Directly across the river from milepost 78, you can see some basalt flows in a vertical orientation. These flows mark the western limb of an anticline that moved westward along a thrust fault over the adjacent basalt.

The Dalles occupies a bend of the Columbia, framed by bluffs of Columbia River Basalt Group and overlying sedimentary rock of the Dalles Formation. The sedimentary rock was derived from erosion of the Cascade Range to the west about 9 to 7 million years ago and deposited in rivers. Good exposures of this formation can be found about 2 miles (3.2 km) to the south on US 197.

Mitchell Point. The rock in the foreground is Grande Ronde Basalt. Above it, the steep slopes and cliffs consist of Troutdale Formation, which was deposited in an old channel of the Columbia River. Cliffs of Mitchell Point at the top of the photo, above the Troutdale, consist of Wanapum Basalt.

US 26
Portland—Madras
112 miles (180 km)

Passing diagonally through the Western and High Cascades, US 26 provides access to Mt. Hood and also crosses several canyons cut in basalt. Heading eastward from its intersection with I-205, US 26 crosses gravel deposits from the Missoula Floods. Don't expect to see any exposures of the flood deposits though, because the area is heavily developed. You will see basaltic lava flows, small shield volcanoes, and cinder cones of the Boring volcanic field, named for the nearby town of Boring. These features erupted from some eighty small volcanoes between about 2.6 million and 50,000 years ago in and around the Portland area. Between SE 136th Avenue and SE 162nd Avenue, US 26 passes north of Powell Butte, which bears one of the younger lavas on its northwest corner. For the next several miles through Gresham, the road passes around the northern and then eastern side of the Boring Hills, which consist of more Boring Volcanics erupted between 1 million and 500,000 years ago. The town of Boring has a sister city in the United Kingdom: Dull, Scotland.

East of Boring, US 26 crosses farmland developed on ancient lahar deposits. The Jonsrud Viewpoint, only 1 mile (1.6 km) north of Sandy on Bluff Road, offers a wonderful view of a deeply incised meander bend of the Sandy River, more than 400 feet (120 m) below, as well as a view eastward to Mt. Hood. Notice the abrupt change of slope near treeline on Mt. Hood. The high part of the mountain is glacially eroded and steep, whereas the lower smoother part is made of accumulations of lahars and pyroclastic flows.

At Alder Creek, the valley becomes constricted, framed in by forested hills. From here to just a few miles west of Government Camp at the base of Mt. Hood, there are virtually no rock exposures along the highway. Geologic maps of the area show that, until a few miles east of Zigzag, the hills are made of andesitic breccias, flows, and tuffs of the Miocene Rhododendron Formation. At higher elevations, such as at Zigzag Mountain, 10- to 9-million-year-old andesite flows overlie the Rhododendron Formation. These rocks form the upper part of the series of rocks that make up the Western Cascades. East of Brightwood, the highway travels over forest-covered pyroclastic and lahar deposits, erupted from Mt. Hood in the last 1,500 years.

Approaching Government Camp from the west, US 26 passes by cliffs of 9- to 8-million-year-old diorite, beginning at about milepost 49. These cliffs continue intermittently for a good 2 miles (3.2 km), providing a dramatic example of the roots of some of the Western Cascades volcanoes. The diorite cooled from magma below the volcanoes. Unlike the Cascades farther south, where active volcanoes reside several miles to the east of the boundary between the High Cascades and the Western Cascades, the boundary here is somewhat diffuse and approximately coincident with Mt. Hood. Just east of the diorite, the road passes an exposure of a 175,000-year-old basaltic andesite that erupted from south of the highway. Near Mt. Hood Skibowl, reddish-gray, 1,500-year-old lahars and pyroclastic flow deposits form scattered roadcuts.

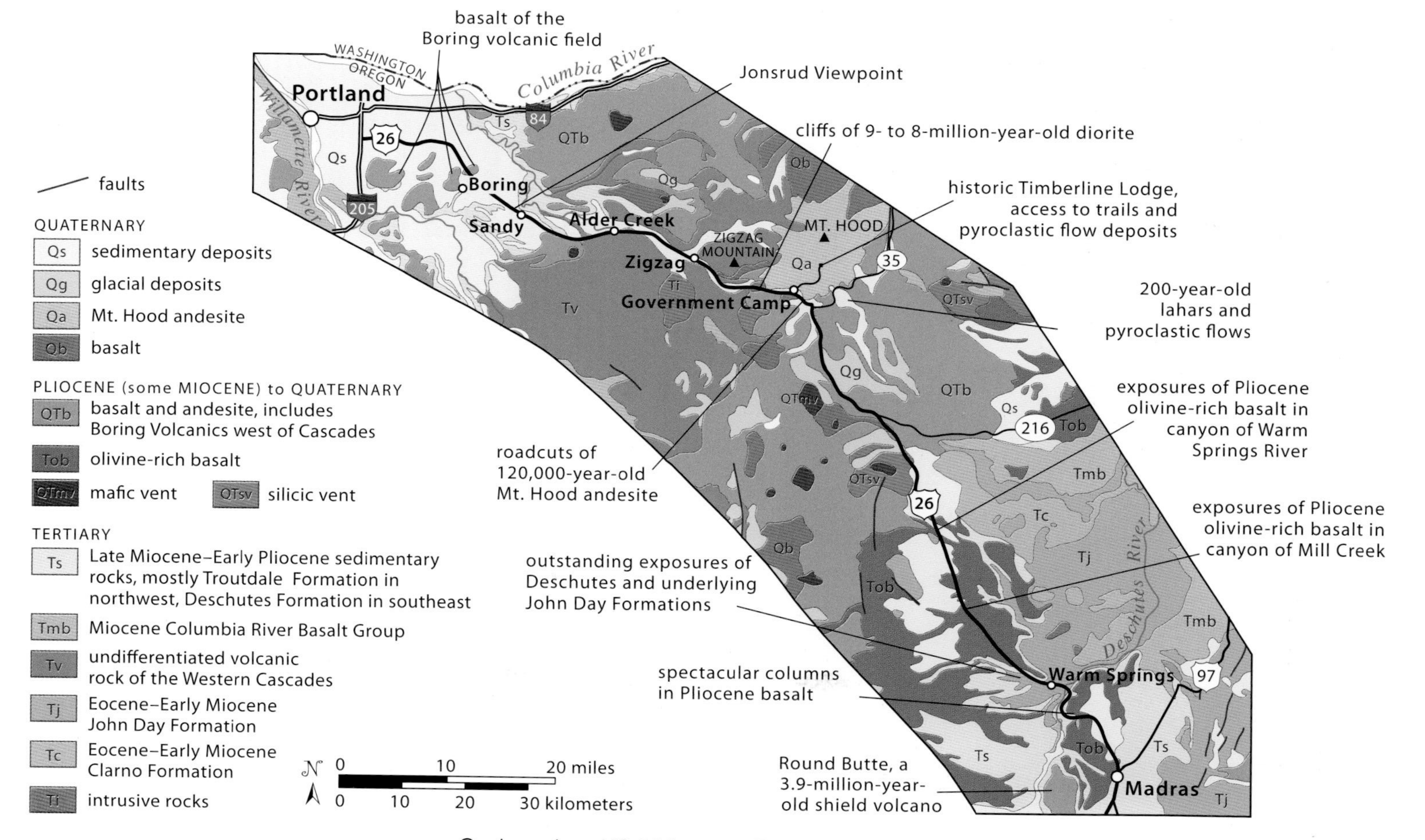

Geology along US 26 between Portland and Madras.

Cliffs of 9- to 8-million-year-old diorite, an intrusive rock of the Western Cascades, line US 26 west of Government Camp.

Mt. Hood

Just east of Government Camp, you can turn northward off US 26 to Mt. Hood on the Timberline Highway. The 5-mile (8 km) road rises some 1,900 feet (580 m) past numerous outcrops of andesite lava and pyroclastic flows to the historic Timberline Lodge. From the lodge, you can look over a fan of pyroclastic flow deposits to the horseshoe-shaped summit crater, which probably formed by a landslide during an eruptive period about 1,500 years ago. To the south, you can see Mt. Jefferson, another stratovolcano of the High Cascades. During winter, Timberline is a popular ski area, so the road is open year-round.

Mt. Hood, Oregon's highest mountain at 11,239 feet (3,425 m), is one of the Cascade's most active volcanoes. Most of the present volcano formed since about 500,000 years ago, but beneath those lavas lie the remains of the Sandy Glacier Volcano, which dates back about 1.5 million years. Mt. Hood's recent eruptions include two major events, one about 1,500 years ago and one between 1781 and the mid-1790s, as well as two minor ones in 1859 and 1865. Today, the volcano hosts a number of steam vents near its summit and occasional earthquake swarms.

Perhaps the most distinctive aspect of Mt. Hood, from a geologic standpoint, is that it contains few deposits of pumice, the hallmark of highly explosive eruptions. The peak consists almost entirely of andesite and dacite lava flows and domes. Crater Rock, a steep-sided lava dome near the mountain's summit, is the youngest dome, marking the vent for the Old Maid eruptive period late in the eighteenth century. Steel Cliffs, just east and slightly lower on the volcano, marks a dome that erupted between about 20,000 and 13,000 years ago.

Mt. Hood's lower reaches are blanketed by pyroclastic flow material, mostly ash-flow deposits with abundant large rock inclusions, and lahars. The collapse of the hot, growing lava domes triggered the pyroclastic flows, whereas many of the lahars formed when pyroclastic flows mixed with melted snow and ice to make fast-moving slurries with a consistency of just-poured concrete. In addition, some of the lahars likely formed from avalanches of cold rock lubricated by groundwater. These avalanches mostly occurred on the steep upper flanks of the mountain, some of which were weakened by groundwater acidified by volcanic gases. Volcanic eruptions or earthquakes likely triggered most of the avalanches.

The upper reaches of Mt. Hood viewed from Timberline Lodge. Note that the bedrock cliffs along the skyline together form a horseshoe shape. Crater Rock, which intruded during the most recent eruptive period, forms the spikelike, reddish peak that doesn't quite reach the skyline in the center of the horseshoe. Steel Cliffs, an older lava dome, forms the prominent whitish cliffs on the right side of the peak. Loose material in the foreground and middle ground consists mostly of pyroclastic flow deposits.

Up to 1 mile (1.6 km) south of the Timberline Highway junction, you pass some roadcuts of 120,000-year-old andesite from Mt. Hood on the left side of US 26. These cuts are the only exposures of andesite from Mt. Hood on US 26; most of Mt. Hood's andesite is closer to the actual mountain or buried beneath younger deposits of lahars or pyroclastic flows. The rocks contain inclusions of different rock, probably blobs of magma that were injected into the andesite just before its eruption. Older andesites show up along the road near milepost 58, just south of the junction with OR 35. These rocks have ages of about 5 or 4 million years.

For the next 20 miles (32 km) or so, US 26 passes through evergreen forest with only scattered rock exposures. Some of the exposures are boulder rich, suggesting they originated either as glacial or volcanic mudflow deposits; others are older andesites. An especially good exposure of andesite (on private property) is on the west side of the road between mileposts 77 and 78.

Beginning about milepost 84, US 26 heads south-southeastward in a straight line, across nearly flat, open terrain developed on flat-lying, olivine-rich basalt flows of Pliocene age. Directly to the north are frequent and awesome views of Mt. Hood, and to the southwest, Mt. Jefferson. US 26 crosses the canyon of Warm Spring River near milepost 85, with exposures of the basalt in the canyon walls, and then 7.5 miles (12.1 km) farther, it crosses the canyon of Mill Creek, with a truly spectacular view of these basalt flows. A small pull-out on the north side of the bridge allows access to some of the rocks, but an even larger pull-out exists on the south side. Look for mounds of silt on the otherwise flat landscape in this area. About 3 feet high (0.9 m) by 15 to 30 feet across (4.6–10 m), they are likely products of frost action during the most recent glacial period.

At about milepost 99, US 26 begins its descent into the Deschutes River canyon and, for the next 12 miles (19.3 km) to the top of the other side, offers probably the best rock exposures of the whole trip. From about milepost 99 to a little past milepost 100, the road descends through well-bedded volcanic and sedimentary rocks of the Deschutes Formation. Below that lies the upper part of the John Day Formation, which makes uniform, nearly white, ashy roadcuts. Most of these John Day exposures are actually part of a larger landslide complex, one of many within the canyon. Some of these landslides temporarily dammed the river, creating lakes that flooded catastrophically when the dams failed. The Columbia River Basalt Group, which lies between these rock units only a few miles to the east, is not exposed here. As this area lies near one of its flow margins, it likely never flowed over this location.

The Deschutes Formation, exposed along the eastern side of the Cascades from here to south of Redmond, records the early history of the High Cascades, from about 7.5 to 4.5 million years ago. These roadcuts, which show just how explosive the early High Cascades volcanoes were, contain several ash-flow, ash-fall, and pumice deposits, as well as sedimentary rocks that are full of volcanic material. Several basalt flows, indicative of quieter eruptions, exist in the Deschutes Formation as well and are exposed as far south as Bend.

As US 26 turns east on the descent into Warm Springs, look for the angular unconformity exposed in the western canyon wall of the Deschutes River. The

Colonnade (columnar joints) in olivine-rich basalt formed by shrinkage during cooling, just south of Warm Springs.

Air-fall ash and pumice deposits, along with volcanic-rich sandstone of the Deschutes Formation, underlie olivine-rich basalt of Pliocene age on the descent into the canyon of the Deschutes River near Warm Springs.

overlying rock consists of flat-lying Deschutes Formation, while the underlying rock consists of gently tilted Columbia River Basalt Group and John Day Formation. East of the river, the road rises back through the John Day Formation, into the Deschutes Formation, and then into the overlying olivine-bearing basalt of Pliocene age. Near milepost 110, toward the top of the grade, a large pull-out allows a closer inspection of the Deschutes Formation and the basalt. The basalt exhibits some beautiful colonnade here.

A similar story is repeated for the descent into Madras but at a much smaller scale. At milepost 117, US 26 descends from the olivine basalt into good exposures of the Deschutes Formation.

US 20
Albany—Bend
121 miles (195 km)

US 20 climbs through the Western Cascades and then over the High Cascades, following the South Santiam River for much of the trip. Perhaps more than any other road that crosses the Western Cascades, this one gives an appreciation for the ruggedness of the province. Since active volcanism gradually shifted eastward to the High Cascades some 8 to 4 million years ago, water erosion has carved deep, steep valleys into the volcanic rocks of the Western Cascades, leaving numerous high cliffs and promontories. The Western Cascades are lower than the High Cascades but far more rugged.

Beginning in Albany, US 20 crosses the eastern side of the Willamette Valley to where it joins the floodplain of the South Santiam River. Near Lebanon, it passes east of Peterson Butte, a large hill cored by a mafic intrusion; a similar intrusion nearby has an age of 31 million years. East of Lebanon, some low tree-covered hills mark the edge of the Western Cascades. The first real rock exposure along this highway, however, isn't until halfway between mileposts 32 and 33, about 5 miles (8 km) east of Sweet Home, where a long roadcut in basalt lies along both sides of the road. Smaller roadcuts expose basalt for the next several miles. These rocks mark the base of the Western Cascades in this area, because their gentle eastward dips and low elevations ensure that rocks to the east lie higher up in the section and are therefore younger.

Volcanic mudflows (lahars) and ash-flow deposits (ash-flow tuffs) are also common along this route but can be difficult to tell apart through the car window. The lahars appear more irregular than ash-flow tuffs, because they consist of a wide variety of material, all thrown together with a muddy matrix. The best exposures of lahars along this route crop out just east of milepost 56. By contrast, the ash-flow tuffs tend to be more uniform, so much so that the more compact ones sometimes look like lava flows. A good example is the ash-flow tuff exposure just west of milepost 52. Ash flows will contain abundant ash with included rock fragments; lava flows will consist of fine crystals. Other good exposures of less-compact ash-flow tuffs exist near milepost 60. These

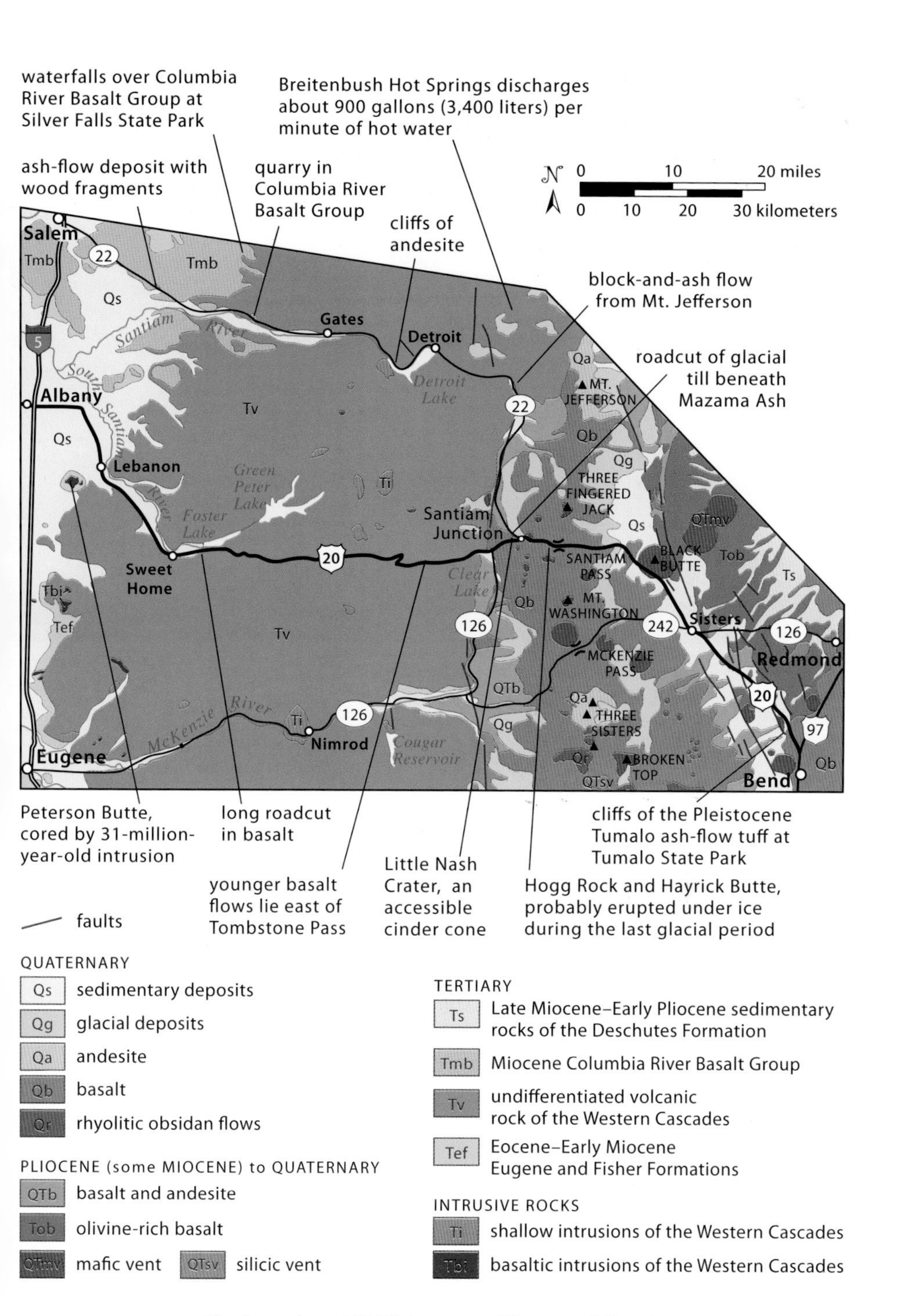

Geology along US 20 between Albany and Bend.

Lahar deposits near milepost 56. Note the lack of stratification and rubbly appearance of the exposure.

rocks have a distinctly green color from the presence of reduced iron disseminated through the rock.

Near Tombstone Pass between mileposts 63 and 64, the road passes abruptly into younger basalt flows, which it follows until its intersection with OR 126. These flows mark the top of the Western Cascades sequence and range from 9 to 4 million years old. Just east of the pass, you can see Three Fingered Jack to the east, a glacially eroded basaltic volcano of the High Cascades.

The intersection of US 20 with OR 126 marks the approximate boundary of the Western Cascades and High Cascades. All of the volcanic features from the intersection east to Bend speak to recent activity. Right at the junction, a 4,000- to 3,000-year-old basalt flow is plainly visible, its highly broken appearance typical of aa flows. It originated from Nash Crater, about 3 miles (4.8 km) to the east. Two other basalt flows of similar ages are also in the immediate vicinity; you can see one less than 0.1 mile (160 m) south on OR 126 and another at Lava Lake, just to the north. US 20 travels over these basalt flows between the junction of OR 126 and OR 22. Little Nash Crater, a source for one of the flows, sits in the triangular area between the two highways 0.5 mile (0.8 km) west of the OR 22 junction. A one-way gravel road accesses a quarry in this crater directly across from the Little Nash Sno-Park. Tall piles of cinders, probably from this quarry, lie stockpiled on the south side of the junction by the

Oregon Department of Transportation for use on snowy roads. Just east of the OR 22 junction, you can look southward to Nash Crater. The side of another cinder cone appears in a roadcut just over 1 mile (1.6 km) east of the junction, at milepost 76.

On the west side of Santiam Pass, US 20 wraps around the south side of Hogg Rock, which provides a display of andesite with thin, variably oriented columns. Much of the andesite is actually glassy, suggesting extremely rapid cooling, and its age of about 90,000 years places it within glacial times. These observations, plus its squarish, flat-topped shape, suggest it might be a small andesitic volcano that erupted beneath glacial ice, with the near-horizontal columns forming where the lava cooled abruptly at the ice margin. Hayrick Butte, immediately to the south, forms a similar-shaped andesitic vent and likely has the same origin. Hoodoo Butte, for which the popular ski area is named, is a large cinder cone.

The descent eastward from Santiam Pass offers numerous roadcuts of basalt as well as views of some of the High Cascades. To the south, you can see Mt. Washington and the North and Middle Sisters volcanoes. Mt. Washington is a basaltic shield volcano with an intrusive plug forming its peak. Magma erupted through this central conduit, then cooled into an intrusive rock. Being more

Quarry inside Little Nash Crater, a cinder cone accessed from US 20 just west of the intersection with OR 22. Note the stratification in the far wall and cinders in the foreground. These cinders are being mined to use as gravel on snowy mountain roads during winter.

resistant than the surrounding layered volcanics, the plug stands out in relief. Recent estimates for the plug's age suggest it is younger than 78,000 years old. North Sister is a steep basaltic-andesite volcano, deeply eroded by glaciers; its activity likely ranged from about 400,000 to 100,000 years ago. Middle Sister was mostly active between about 25,000 and 18,000 years ago and erupted a range of lava types from basalt to dacite. To the east, Black Butte forms an almost perfectly symmetrical cone. Its basaltic lavas erupted about 1.4 million years ago. The Mt. Washington overlook, 0.25 mile (400 m) west of milepost 85, allows a chance to get off the roadway to see some of these features. The overlook also gives a view into the deep blue waters of Blue Lake, which occupies an explosion crater, likely caused by the interaction of shallow magma and groundwater less than 4,000 years ago.

You can also see glacial deposits above the north side of the road across from the Mt. Washington overlook. In other places, glacial deposits lie perched on top of various lava flows, and at milepost 86, you can look down to the south and see a moraine enclosing Suttle Lake. The road to Camp Sherman leads to the head of the Metolius River, a spring that issues from the base of Black Butte.

Sisters rests in the middle of the Sisters Fault Zone, which extends some 40 miles (64 km) to the north and south, part of a normal fault system of the Basin and Range Province. Sisters also lies on the geographic border between the High Cascades and the Lava Plateaus.

Between Sisters and Bend, US 20 has little in the way of rock exposure for about the first 15 miles (24 km), but it does offer some beautiful views back toward the high peaks of the Cascades. In addition to the peaks seen from near Santiam Pass, you can see South Sister and Broken Top. South Sister has been intermittently active from 178,000 years ago almost to the present, with flows as young as 2,000 years. Its flows range in composition from basaltic andesite to rhyolite. Broken Top, a deeply eroded basaltic volcano, hasn't been active for probably 150,000 years.

Sisters to Redmond on OR 126

The 18-mile-long (29 km) stretch of OR 126 between Sisters and Redmond mostly crosses Pliocene-age basalt, exposed in scattered roadcuts and outcrops between Sisters and about milepost 107. On approaching Redmond, however, you'll see the underlying Miocene-Pliocene Deschutes Formation, and Quaternary-age basalt flows derived from Newberry Volcano. The road also offers views to the Three Sisters and Mt. Washington, toward the southwest and west, respectively, and of Newberry Volcano to the southeast.

The Deschutes Formation and overlying basalt are broken by several normal faults. Note that the largest of these faults drops the Deschutes Formation entirely out of view to the left.

An outstanding exposure between mileposts 99 and 100 shows the Deschutes Formation faulted against younger Pliocene basalt. Much of the Deschutes Formation consists of material eroded from the early High Cascades and redeposited as sediments, but it also includes ash-flow tuffs and basalt flows. This exposure contains several well-exposed normal faults, part of the northern extent of the Sisters Fault Zone.

The prominent hills south of the highway near milepost 105 are Cline Buttes, a rhyolitic vent that probably formed during Pliocene time. Near milepost 107, the road descends into the canyon of the Deschutes River, with exposures of the Deschutes Formation. As it crosses the river, the road climbs into exposures of basalt that originated from Newberry Volcano some 30 miles (48 km) to the south. The flat landscape above the canyon is an expression of these flat basalt flows. Cline Falls, on the river north of the highway, drops 20 feet (6 m) through channels in the basalt, but much of the river's water is diverted for irrigation upstream, so the falls are not as voluminous as they might otherwise be.

Low cliffs of Tumalo ash-flow tuff along the Deschutes River at Tumalo State Park.

Between mileposts 15 and 14, the road passes Tumalo State Park, which features 30-foot-high (9 m) cliffs of Pleistocene ash-flow tuff along the Deschutes River. On the south side of the river, the road passes upward through the tuff onto Pleistocene basalt erupted from Newberry Volcano, southeast of Bend.

OR 22

Salem—Santiam Junction

80 miles (129 km)

See map on page 133

OR 22, which follows the North Santiam River east into the Cascade Range, provides one of the more straightforward takes on the Western Cascades—the majority of its rock exposures are of andesite flows erupted between 25 and 17 million years ago. However, it also offers outcrops of the Columbia River Basalt Group, which originated from far outside the Western Cascades and gives some insights into the evolution of the Columbia River system.

For the first several miles east of Salem, OR 22 heads directly southeastward over a flat surface developed on alluvial deposits of the Willamette River. Near milepost 4, it crosses into low hills composed of Columbia River Basalt Group, although little if any rock exposure along the road confirms this.

At milepost 14, at the exit for Sublimity, look for a long roadcut on the north side of the highway. This cut consists of several ash-flow tuffs derived from Late Miocene eruptions in the Western Cascades. A close look at the rock shows that it contains numerous pieces of wood, most of which are charred to various shades of gray and black. The presence of some noncharred wood in the flow indicates it must have cooled significantly by the time it reached this spot. A well-defined paleosol, a soil that formed on the surface of the lower flow, separates two of the flows.

High cliffs to the north of the road between mileposts 19 and 29 consist of Columbia River Basalt Group, and a quarry on the north side of the road halfway between mileposts 28 and 29 shows some beautiful columnar jointing. The Columbia River Basalt Group, the main rock unit of Silver Falls State Park,

Close-up view of a charred wood fragment in the tuff.

Long roadcut of wood-bearing ash-flow tuff near milepost 14. The reddish-brown zone near the middle of the roadcut is the paleosol.

Silver Falls State Park

Silver Falls State Park occupies the confluence area of the North and South Forks of Silver Creek, which drains part of the Western Cascades. These streams, flowing over bedrock of mostly Columbia River Basalt Group, form fourteen waterfalls in the park, ten of which are easily reached by foot trail above the confluence, and six of which drop more than 100 feet (30 m) each.

The bedrock, which in this part of the Western Cascades is covered with thick temperate rain-forest vegetation, is well exposed at the waterfalls. In ascending order, the bedrock consists of relatively flat-lying flows of the Grande Ronde Basalt of the Columbia River Basalt Group; a thin, about 20-foot-thick (6 m) unit of river-deposited sandstone and mudstone; the Wanapum Basalt of the Columbia River Basalt Group; and the Fern Ridge Tuff. The tuff, which erupted explosively from the Western Cascades between about 15 and 7 million years ago, occupies the hilltops on the eastern side of the park. Both members of the Columbia River Basalt Group show beautiful cross sections of complete flows. These cross sections include reddish paleosols at their bases, colonnades and entablatures in their central parts, and tops with abundant vesicles, or gas bubbles. The thin sedimentary unit between them hosts paleosols and molds of tree trunks, to suggest the area was forested before being inundated by lava flows of the Wanapum Basalt. Beneath these rocks, and exposed in the channel of Silver Creek just downstream of the park boundary, lies the Scotts Mills Sandstone, a shallow

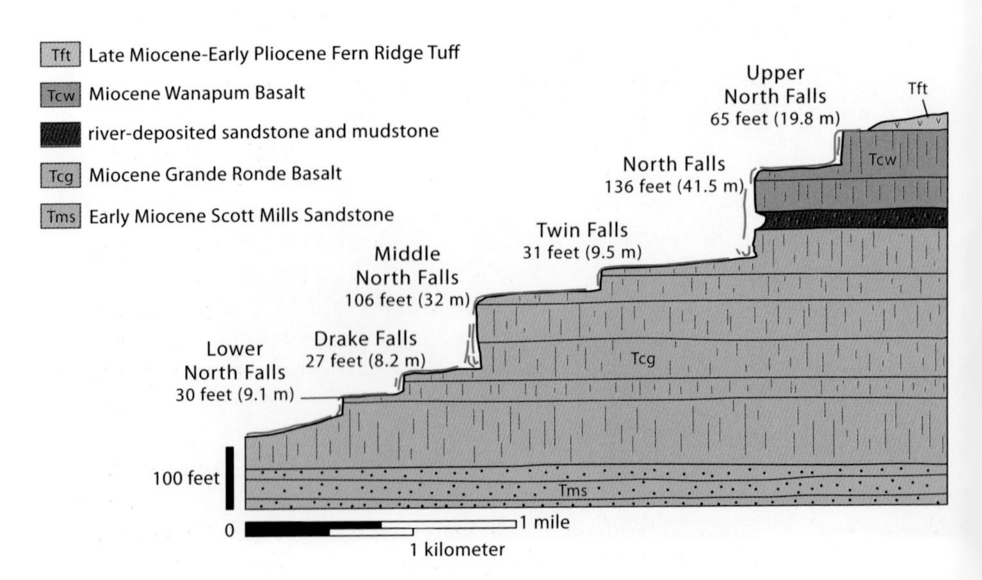

Cross section of Silver Falls State Park along the North Fork of Silver Creek.

marine deposit of Early Miocene age. Visitors can see examples of this unit in the stone walls of the shelter in the South Falls area of the park.

Of these rocks, the basalt is by far the most resistant to erosion and forms the tops of each of the waterfalls. Rapid erosion of the rocks near the plunge pools at the base of the falls undercuts the caprock, which then breaks off. Over time, each waterfall migrates upstream. Probably the best example of this style of erosion is at South Falls, which drops 177 feet (54 m) into its plunge pool. At North Falls, undercutting of the caprock also occurs by rapid erosion of the thin sedimentary unit between the the Grande Ronde and Wanapum Basalts. There, an alcove called the amphitheater has formed directly behind the falls, some 300 feet (90 m) wide. A tree mold appears as a distinct circular hole in the ceiling of this alcove.

---

likely followed an early channel of the Columbia River to reach this area. Subsequent eruptions filled the channel and caused the river to migrate northward toward its present position. The ancient river channel may even have continued westward across the nascent Coast Range to the Pacific Ocean. Similar flows of the Columbia River Basalt Group are exposed south of Lincoln City, but none have been found in today's Coast Range. Perhaps erosion removed any basalt that once flowed over the range before its recent uplift.

Nearly all the bedrock exposed between the towns of Gates and Detroit is andesite flows that erupted between 25 and 17 million years ago. A double roadcut halfway between mileposts 35 and 36 exhibits beautiful columns and weathers nearly black, just like basalt, but a close inspection shows the rock to be andesite. Farther on, the high gray cliff exposures are more typical of andesite, with individual samples containing numerous large white crystals of plagioclase. Some of the best exposures lie along the roadway as it follows the edge of Detroit Lake, a reservoir created in 1953 for hydroelectricity and flood control. The road along Detroit Lake also offers good views of the rugged Western Cascades and, at milepost 48, a glimpse of Mt. Jefferson.

Breitenbush Road, featuring roadcuts of mostly ash-flow tuff, leads northeast out of Detroit toward Breitenbush Hot Springs, probably the best known and most productive hydrothermal area in the Cascades. The springs, which have been developed as a resort, discharge some 900 gallons (3,400 liters) per minute of hot water out of basaltic bedrock.

East of Detroit, OR 22 passes cliffs of ash-flow tuffs that erupted in the Western Cascades between about 25 and 17 million years ago. Some 10 miles (16 km) to the east, the road bends sharply southward and follows the boundary between the Western and High Cascades. The cliff exposures from milepost 60 to milepost 65 are a complicated mix of lahars, pyroclastic flows, and basalt flows, but they all originated in the High Cascades within the last several million years. One pyroclastic flow, exposed on the north side of the road just south of milepost 60, can be traced up Whitewater Creek back to Mt. Jefferson, about 10 miles (16 km) to the east. This flow, called a block-and-ash flow, consists of unusually large rock fragments as well as ash.

OR 22 offers a good view to the northeast of Mt. Jefferson near milepost 68. Similar to Mt. Hood and Middle Sister, Mt. Jefferson consists mostly of andesite, and at an elevation of 10,495 feet (3,199 m), is second only to Mt. Hood in elevation. Although clearly defined dates for most of its eruptions are lacking, the deep glacial erosion on Mt. Jefferson indicates that it hasn't been active since at least 20,000 years ago. It likely experienced two major eruptions in the past 100,000 years, one of which produced the block-and-ash flow near milepost 61. Several glaciers still exist near its summit.

As the road rises into the High Cascades, it passes occasional exposures of glacial till, beginning at about milepost 74. Probably the best exposure lies on the north side of the road at milepost 81, a short distance north of Santiam Junction. There, across from a large pull-out, the till is capped by a thin layer of Mazama Ash, erupted from Mt. Mazama 7,700 years ago. The one-way gravel road behind the pull-out leads out from Little Nash Crater, a 3,000-year-old cinder cone. The road leading into the crater can be accessed from US 20, 0.5 mile (0.8 km) west of Santiam Junction.

Roadcut of glacial till overlain by Mazama Ash at milepost 81. Note the wide range of particle sizes of the till and its lack of layers, a consequence of being deposited directly by a glacier as it melted, without being sorted and stratified by running water.

OR 58
Eugene—US 97
86 miles (138 km)

OR 58 follows the Middle Fork of the Willamette deep into the Cascade Range, but before heading up the road, you should check out a roadcut at the beginning of this trip. The Oligocene Fisher Formation, which underlies the sequence of rocks that make up the Western Cascades, is beautifully exposed along I-5 immediately north of the exit to OR 58. The 50- to 35-million-year-old floodplain and river deposits form prominent thick sandstone and conglomerate beds that overlie more thinly bedded lake deposits, some containing fossil

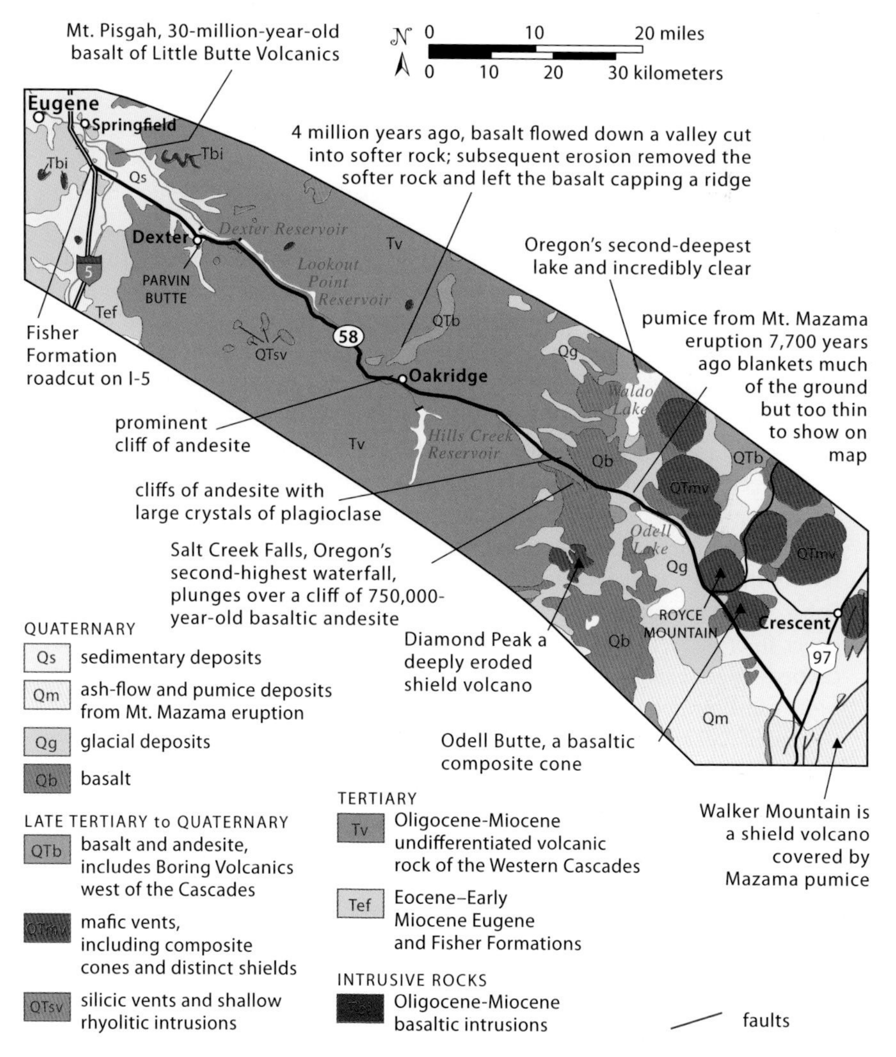

Geology along OR 58 between Eugene and US 97.

leaves. The outcrop is described in more detail in the I-5 road guide. To take this detour, enter I-5 northbound from OR 58, then take the first exit and park on the wide shoulder by the roadcut.

For the 10 miles (16 km) between I-5 and Dexter, OR 58 crosses the floodplain of the Middle Fork of the Willamette River. Mt. Pisgah, made of altered 30-million-year-old basaltic lavas, forms the prominent hill to the north. Spencer Butte, a basaltic intrusion of about the same age, can be seen about 7 miles (11 km) west of Mt. Pisgah. Near milepost 7, good views eastward show the deeply eroded topography of the Western Cascades.

Between mileposts 13 and 25, the road passes Dexter and then Lookout Point Reservoirs, both impounded by dams in 1954 to provide hydroelectric power and control flooding. Views northward across the lakes reveal the eastward dip of the Western Cascades volcanic rock units, specifically basalt flows across Dexter Reservoir and younger andesite flows across most of Lookout Point Reservoir. The bridge to Lowell, at milepost 13, allows access to the north shore of the lake, with beautiful views across the lake as well as good exposures of basaltic rocks just west of the dam outlet. Parvin Butte in Dexter is a basaltic intrusion that is being mined on its south side for road-building purposes. Tragically, the mining continues despite its unprecedented disruption of community life.

At milepost 15, between the two reservoirs, OR 58 wraps around a rhyolitic intrusion expressed as a prominent hill just south of the road. This rock is fairly resistant to erosion and is likely responsible for the localized constriction in the valley that coincides with the Lookout Point Dam. A viewpoint on the north side of the road gives good views of the dam, but the best view of the intrusion is between mileposts 15 and 16, looking westward. Numerous roadcuts for the next several miles consist mostly of east-dipping ash-flow deposits and lahars.

Between mileposts 30 and 31, the ridge on the north side of the river provides a great example of inverted topography, as well as the first inkling of the young High Cascades volcanism. The ridge consists of a young basalt flow of the High Cascades that flowed down a valley about 4 million years ago. Subsequent erosion of the enclosing, older ash-flow tuffs, which are softer and easier to erode, left the lava flow as a nearly flat-topped ridge. Similar basalt-capped ridges extend out from the High Cascades north and south of here. More ash-flow deposits lie along the road just west of Oakridge.

East of Oakridge, OR 58 leaves the Middle Fork of the Willamette and enters the drainage of Salt Creek, one of its main tributaries. County Road 23 follows the Willamette south to Hills Creek Reservoir, impounded by Hills Creek Dam, built by the Army Corps of Engineers in 1961 for hydroelectricity and flood control. Along Salt Creek, OR 58 passes by more ash-flow and lahar deposits, although few show up on the road.

Between mileposts 53 and 54, a double roadcut with ample pull-out space gives access to an andesite flow with large plagioclase crystals. One-quarter mile (400 m) east of milepost 54, basalt is overlain by glacial till. The basalt originated in the High Cascades and shows beautiful cooling fractures just east of the tunnel at milepost 56.

View eastward from the top of Mt. Pisgah toward the eroded peaks of the Western Cascades. OR 58 crosses the fog-enshrouded lowlands.

View across Dexter Reservoir to Parvin Butte in spring 2012, before mining had taken much of a toll on the butte.

Salt Creek Falls spills over a 750,000-year-old intracanyon flow of basaltic andesite. This photo was taken from the overlook on the trail halfway down the north side of the canyon. Note how the plunge pool cuts back into the cliff to create an overhang; this overhang encourages rocks break from its ceiling, which speeds the retreat of the cliff.

Salt Creek Falls, Oregon's second-highest waterfall, is just east of milepost 57. The waterfall plunges more than 280 feet (85 m) over the edge of a basaltic lava flow, erupted from the High Cascades. This flow poured down the canyon within the last 750,000 years to a point about 4 miles (6.4 km) downstream from today's cliff edge. Erosion, much of which proceeds by undercutting at the plunge pool, has caused the cliff to retreat to its present location. A walkway skirts the edge of the cliff and provides dizzying views straight down to the plunge pool. It also gives ample opportunities for viewing columnar jointing in the cliff wall. You can also see a cross-sectional view of the polygon-shaped columns along the steps on the north side of the gorge. In addition, several trails begin from the parking area, one of which descends to an overlook halfway down to the canyon bottom. To reach this trail, follow the rim walkway northward.

East of Salt Creek Falls, OR 58 is entirely within the High Cascades. From the falls to Willamette Pass, the highway passes several exposures of basalt, as well as some overgrown areas of glacial till. In most roadcuts, you can also see a layer of tan-colored ash and pumice on top, erupted from Mt. Mazama about 7,700 years ago during the catastrophic event that formed Crater Lake, about 40 miles (64 km) to the south.

At milepost 59, a road leads about 6 miles (10 km) north to Waldo Lake, Oregon's second-deepest lake. Its average depth is just over 125 feet (38 m) and it has a maximum depth of 420 feet (128 m). Waldo Lake's waters are exceptionally clear because it has an unusually small watershed for its large area of more than 10 square miles (26 km^2). As a result, little sediment is carried into the lake by runoff. The road to the lake passes several exposures of High Cascades basalt overlain by Mazama pumice and ash.

Several viewpoints along OR 58 below Willamette Pass allow good views southward to Diamond Peak, a basaltic shield volcano. The volcano was most likely active since about 100,000 years ago, but glaciers have deeply eroded it, indicating its activity ended before the ice age glaciers melted away. A central cone of pyroclastic material lends some relief and cragginess to its summit area. The spike-shaped peak just to the west marks the remains of another volcanic vent, plugged by magma that cooled into an unusually resistant rock.

Crescent Junction sits between Royce Mountain to the north and Odell Butte to the south, large basaltic composite cones. Crescent Cutoff Road leads eastward to the Cascades Lakes Scenic Byway and the town of Crescent on US 97. It passes by the north side of Odell Butte to a recently erupted basaltic lava flow and a quarried-out cinder cone.

OR 58 continues in a nearly straight line along the south side of Odell Butte to the floodplain of the Little Deschutes River and US 97, just over 13 miles (21 km) away. Walker Mountain, a faulted shield volcano, lies straight ahead on the east side of US 97.

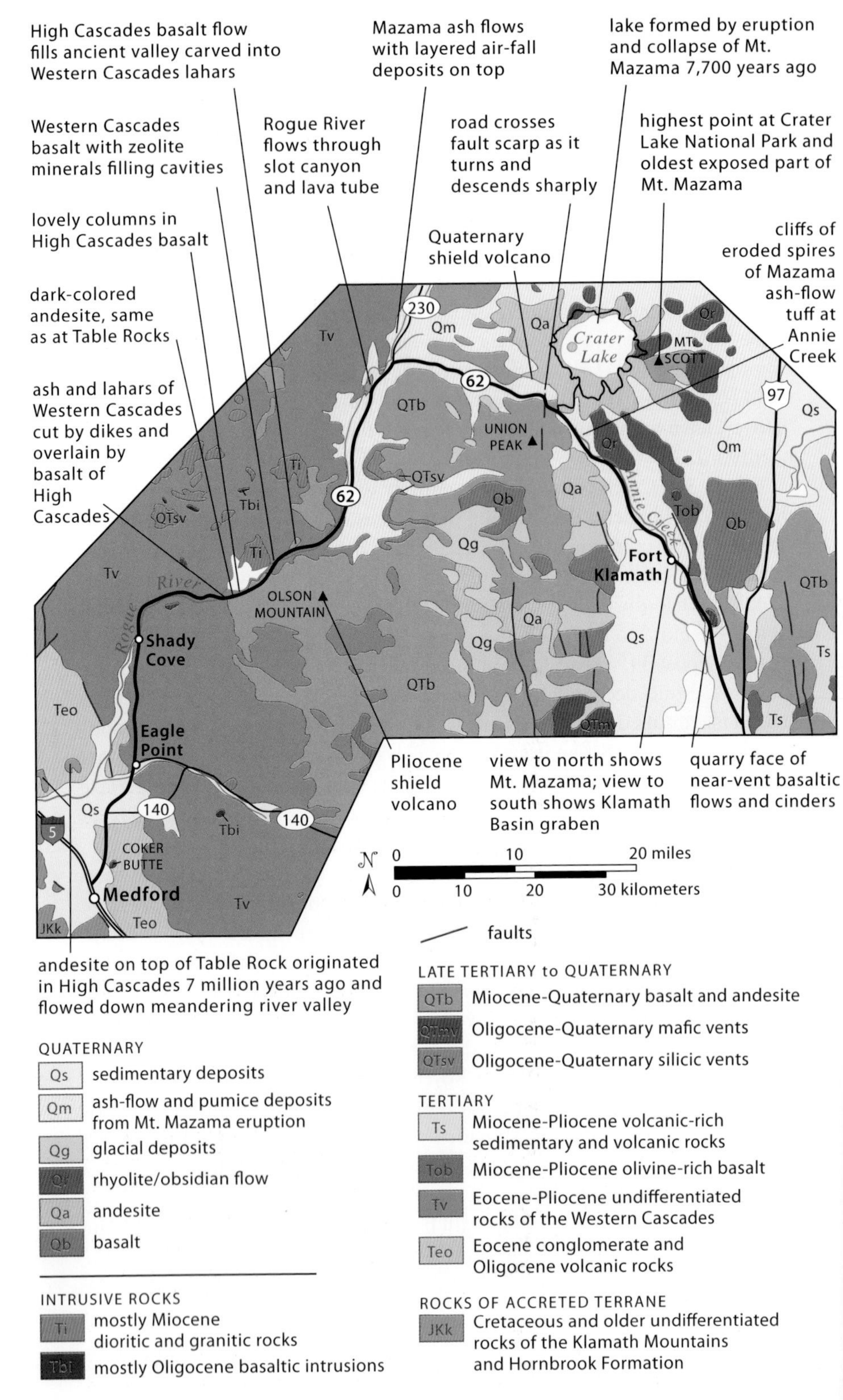

Geology along OR 62 between Medford and US 97.

OR 62
Medford—Crater Lake—US 97
98 miles (158 km)

One of the longer routes across the Cascades, OR 62 follows the Rogue River northeastward, almost as far as its source, but turns eastward up a tributary to the south entrance of Crater Lake National Park. From there, it heads southeastward out of the Cascades and into the Klamath Basin, a fault-bounded valley at the northwestern corner of the Basin and Range Province.

Between Medford and Eagle Point, OR 62 crosses a low relief surface that's part floodplain, part river terrace, and part grass-covered sedimentary bedrock. Coker Butte, a Tertiary cinder cone, rises some 200 feet (60 m) above the east side of the road opposite the airport. East of it, the high cone-shaped Roxy Ann Peak consists of 35- to 25-million-year-old basaltic rocks of the Western Cascades. At Eagle Point, the road crosses into poorly exposed volcanic rock of the Western Cascades. The andesite-capped mesas of Table Rocks lie a few miles to the west.

You can see some basaltic lava flows at Shady Cove, near milepost 19. These flows are between 35 and 25 million years old, which makes them some of the oldest basalts of the Western Cascades. Numerous exposures of these rocks line the highway for the next several miles and include some andesite lavas and ash-flow tuffs as well. Because of the consistent eastward tilt of the rocks, they become younger toward the east.

Just east of milepost 29, the road crosses the Rogue River and climbs past altered ash and lahar deposits. Just west of milepost 30 on the south side of the highway is a roadcut of these rocks crossed by several basaltic dikes and overlain by a basalt flow. The dam of Lost Creek Reservoir, just beyond, uses the basalt flow as its abutments. Pull-outs on the highway offer access to the ash and lahar deposits, but be extremely careful because the traffic here can be unusually fast.

From here to its intersection with OR 230, OR 62 follows the boundary of the Western Cascades and High Cascades. For the next several miles until the bridge over the north end of the reservoir, most exposures along the road came from the High Cascades. North of the reservoir and northeast of the bridge, many of the exposures are part of the Western Cascades. Between mileposts 30 and 31, the double roadcut consists of the same rock as Upper and Lower Table Rocks, more than 20 miles (32 km) away! This distinctive flow looks like a basalt, in that it is dark-colored and even contains the minerals olivine and pyroxene, but the rock contains enough silica to be classified as an andesite. Remnants of this 7-million-year-old flow cap ridges up and down the upper Rogue drainage for more than 30 miles (48 km). At milepost 31, a basalt flow shows a highly irregular base that gives a hint to the topography over which it flowed, as well as beautiful long, narrow columns above.

Across the bridge at the northeast end of Lost Creek Reservoir, you can view a cliff with several basalt flows of the Western Cascades. A large pull-out there allows the safe inspection of the rock, which contains zeolite minerals that filled

View looking south at a roadcut of ash-flow and lahar deposits of the Western Cascades overlain by a younger basalt flow. Note the basaltic dike (brownish, eroding feature to the right of center) that cuts through the lahar and ash-flow tuff but not the overlying lava flow, as well as the eastward tilt of the tuff and lahars on the right side.

gas bubbles. One mile (1.6 km) to the east, a High Cascades lava flow occupies an ancient valley carved into the older rocks. Its eastern side, next to a pull-out on the north, shows the younger basalt cutting progressively downward into east-tilted lahar deposits.

For about 10 miles (16 km) north of Prospect, the road lies within deep forest developed on ash deposits from the Mt. Mazama eruption. Good rock exposures of High Cascades basalt lie in the river channel of the Rogue River at Natural Bridge Campground near milepost 55 and the Rogue River Gorge viewpoint, 1.5 miles (2.4 km) to the north. At Natural Bridge, a basaltic lava tube serves as the main river channel. The river literally goes underground into the tube and then reappears downstream. At the Rogue River Gorge viewpoint, visitors can walk some 500 feet (150 m) along the edge of a slot canyon eroded by the Rogue River through the basalt.

The roadcut at the junction with OR 230 exposes Mazama Ash. During the eruption about 7,700 years ago, ash flows raced down and filled the river valleys that radiated out from Mt. Mazama. Since then, the rivers have reestablished their courses and eroded their modern channels. Only a fraction of the original ash-flow material remains along the margins of the rivers. Much more extensive exposures lie a few miles up OR 230 or on OR 62 just beyond the south entrance to Crater Lake National Park.

Slot canyon carved into High Cascades basalt by the Rogue River at Rogue River Gorge.

OR 62 follows a forested valley developed on ash-flow deposits erupted from Mt. Mazama, the only obvious exposure being a small one between mileposts 59 and 60 on the north side of the road. Less than 0.5 mile (0.8 km) west of the south entrance to Crater Lake National Park, the road bends sharply right and descends some 80 feet (24 m) down a normal fault scarp. Three separate lava flows are exposed in the roadcuts.

OR 230 to Diamond Lake

OR 230, the road to Diamond Lake, follows the Rogue River. You'll see High Cascades basalt just 0.9 miles (1.4 km) from the intersection with OR 62, then the road passes some stunning exposures of Mazama Ash over the next 6 miles (10 km). Just north of milepost 4, you can look westward over a river cut in the ash toward the older rocks of the Western Cascades. The uppermost part of the ash is stratified here, unlike the lower exposures that are homogeneous. The stratification reflects accumulation of ash-fall material, which settled down from the air, on top of the earlier ash-flow deposits, which flowed along the ground. As OR 230 approaches Diamond Lake, it offers a spectacular head-on view of Mt. Thielsen.

River and road exposures of ash-flow tuff erupted from Mt. Mazama, with overlying stratified ash-fall deposits, looking north. The Rogue River (lower left) separates the Western and High Cascades. The small peak on the left (west) is part of the Western Cascades.

Spires of eroded ash-flow tuff, erupted from Mt. Mazama, at Annie Creek. The color change from bottom to top broadly coincides with a decrease in silica content upward.

One mile (1.6 km) east of the road into Crater Lake National Park, OR 62 follows the edge of Annie Creek Canyon for 5 miles (8 km) or so, cut into Mazama Ash. The view from the road shows numerous fluted columns in the ash, some of which have eroded into freestanding pinnacles. Like the Pinnacles, seen from the overlook several miles to the northeast (see the section on Crater Lake at the end of this chapter), these columns remain because they are more resistant to erosion than the surrounding material. Gases escaping upward through the ash flow mineralized the ash, making it more resistant. The deposits also show a color change that corresponds with an upward decrease in silica content. The lower, lighter-colored ash flow erupted first from the top of the magma chamber, which was richer in silica than the lower parts. The lower parts of the magma chamber erupted later, producing the darker-colored ash flow on top.

The road descends steadily in the Annie Creek drainage to Fort Klamath at the northern edge of the Klamath Basin. Looking southward down the almost-flat Klamath Basin, you can see that the valley is bordered by steep range fronts on either side that mark the traces of recently active fault zones. The valley occupies a graben, a block of crust dropped down between two fault zones. Almost imperceptibly, the road passed from the High Cascades into the Basin and Range. The 1993 Klamath Falls earthquake was caused by slip on a similar fault a few miles to the west.

Looking north toward Crater Lake, you can see several low peaks on the skyline. Mt. Scott, an older vent on the side of Mt. Mazama, is the highest, most eastern peak. The others are high points on the caldera rim, parts of the original Mt. Mazama. You can imagine the shape and height of the original volcano by projecting the sides of these peaks upward in the air until they meet.

A wonderful exposure of Pliocene basalt and cinder deposits exists in a quarry face on the east side of the highway about 6 miles (10 km) south of Fort Klamath. The thin layering and preponderance of cinders suggest these rocks originated near a basaltic vent.

OR 66

Ashland—Klamath Falls

57 miles (92 km)

For the first few miles south of Ashland, OR 66 follows over farmland developed on the Hornbrook Formation, the marine deposit of Cretaceous age that overlaps the accreted terranes of the Klamath Mountains. A few poor exposures of Hornbrook shale exist in roadcuts near milepost 3 and just southeast of milepost 6. Emigrant Lake, a reservoir created in 1924 for irrigation and flood control, has a distinctive horseshoe shape, its west arm developed on poorly exposed Hornbrook Formation and its east arm on Eocene river deposits. The Eocene rocks, which are well exposed beginning near milepost 7, consist mostly

Northward view of Eocene sedimentary rocks, dipping eastward beneath volcanic rock of the Western Cascades.

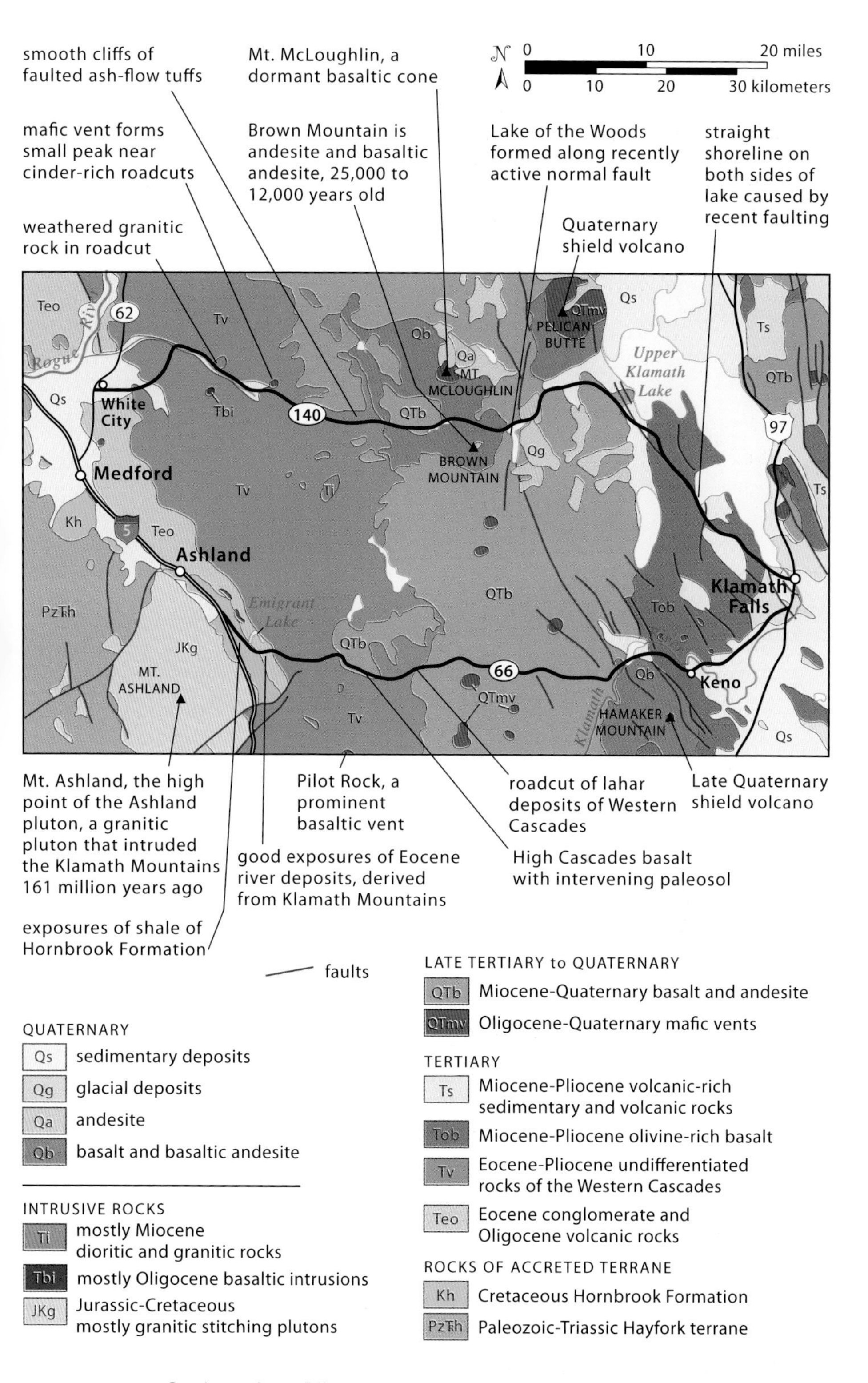

Geology along OR 66 between Ashland and Klamath Falls.

Aerial view northwestward over fog-enshrouded Pilot Rock toward Mt. Ashland, the high point on the left. Pioneers on the Applegate Trail used Pilot Rock as a landmark.

of mica-rich sandstone and conglomerate that was derived from the Klamath Mountains. They dip eastward, similar to the younger volcanic rocks of the Western Cascades.

Near milepost 10, OR 66 passes its first exposures of volcanic rocks of the Western Cascades, which here consist mostly of basaltic lava and lahars. The road also offers good views westward over the Ashland valley and southward to Pilot Rock, an eroded andesitic plug within Oligocene rock of the Western Cascades. Hornbrook Formation lies in the valley, deposited on Jurassic granitic rock that makes up the mountains west of Ashland, including Mt. Ashland.

On the east side of the small Keene Creek Reservoir, near milepost 16, the road encounters younger basalt flows of the High Cascades. These flows poured out over the older rocks of the Western Cascades between 7 and 2 million ago, and so occupy ridgetops in this area. While they don't show up very well along the road, a good exposure near milepost 21 comes complete with a red paleosol between flows. The true edge of the High Cascades approximately coincides with Jenny Creek near milepost 23, although some beautifully exposed lahar deposits just beyond there belong to the older Western Cascades. From there to the Klamath River, the road follows a gentle plateau constructed of the 7- to 2-million-year-old basalt flows. Occasional views to the north and south show several large shield volcanoes of the same age as the basalt.

East of milepost 37, the road crests a low pass and begins a meandering downgrade off the plateau through exposures of High Cascades basalt. Some of the flows are separated by distinct red paleosols. A prominent switchback near milepost 40 marks the trace of a normal fault and the approximate western edge of the Basin and Range Province. The rocks still belong to the High Cascades, having erupted and then flowed into the Basin and Range. The first view of Hamaker Mountain, a Quaternary shield volcano, appears to the east from this switchback. Near milepost 44, the road crosses the John C. Boyle Reservoir, which backs up the Klamath River for hydroelectric purposes, and gains a beautiful view northward to Mt. McLoughlin, Oregon's southernmost High Cascade volcano. Mt. McLoughlin is a basaltic cone that reaches nearly 9,500 feet (2,900 m) in elevation. Some low roadcuts of well-layered pyroclastic deposits line the road near milepost 47.

Between Keno and Klamath Falls, the highway lies wholly within the Basin and Range Province. North of the road, several northwest-trending, elongate ridges coincide with normal faults. South of the highway, the Klamath Basin fills a large graben. The rocks, although mostly basaltic, appear to have originated east of the Cascades. Some Pliocene lakebed deposits make up the faulted hillsides on either side of the intersection with US 97, but these are not exposed. Just 1 mile (1.6 km) south on US 97, you gain a view southward to Mt. Shasta, a Cascade volcano in California.

Roadcut of lahar deposits of the Western Cascades at Jenny Creek.

OR 140
MEDFORD—KLAMATH FALLS
74 miles (119 km)

See map on page 155

Beginning at its intersection with OR 62 just north of Medford at White City, OR 140 crosses the Western and High Cascades to Klamath Falls, at the western edge of the Basin and Range Province. Heading east from White City across a flat floodplain and river terrace, the highway affords a great view toward Mt. McLoughlin about 25 miles (40 km) away. About 6 miles (10 km) east of White City, the highway abruptly enters some hills of eroded basalt flows and well-developed orangish soils. Twin, grass-covered roadcuts just east of milepost 11 contain poor exposures of granitic rock, which intruded the basalt and possibly fed younger eruptions. Between mileposts 13 and 14, you can look northward and sense the rocks are inclined toward the east. The hills tend to have longer and gentler eastern slopes, as well as a subtle layering, and one cliff exposure to the north shows an eastern dip. Driving eastward, therefore, you travel progressively up-section into younger rocks. A good exposure of terrace gravels, deposited by Little Butte Creek but left stranded as it deepened its channel, shows up just west of milepost 13.

Near milepost 18, the road enters landscape more typical of the Western Cascades, with deep, steep narrow canyons and high cliffs. The Western Cascades have experienced a prolonged period of erosion since volcanism migrated eastward to the High Cascades. Numerous good exposures of mostly basalt and lahars exist between mileposts 18 and 26. A mafic vent, with roadcuts of red, tightly consolidated cinders, forms a small peaklike outcrop just west of milepost 20. Beginning about milepost 26, good exposures of 25- to 17-million-year-old rhyolitic ash-flow tuffs, as well as more lahars, show up. Ash-flow tuffs, which tend to form smooth, gray cliffs, are much less common in the southern part of the Western Cascades than the northern part.

At about milepost 28, the road passes east into the less-eroded landscape of the High Cascades, covered mostly by Quaternary basalt flows. Mt. McLoughlin, a high basaltic cone, reaches an elevation of 9,495 feet (2,894 m) to the northeast, but good views of the mountain are hard to come by along this road because of the trees. Most recently active 30,000 to 20,000 years ago during the last major period of glaciation, the volcano is heavily eroded by glaciers on its north side. The south side remains relatively smooth and uneroded. Fish Lake sits at its base. Brown Mountain, a remarkably circular andesitic volcano from 25,000 to 12,000 years old, lies south of Fish Lake and is plainly visible to the southwest from about milepost 37.

East of Fish Lake, OR 140 rises over a low pass and gradually descends to Lake of the Woods, an elongate lake stretched out along the base of a cliff defined by a normal fault. This fault marks the western edge of the Basin and Range Province. East of Lake of the Woods, the highway gradually descends to the shores of Upper Klamath Lake, passing scattered exposures of basalt along the way. If you drive partway down the West Side Road toward Rocky Point,

View of Mt. McLoughlin, a basaltic cone of the High Cascades, from its east side off the road to Rocky Point.

Along Klamath Lake, much of OR 140 follows the base of a recently active normal fault. The steep hillside (the fault scarp) has moved up relative to the lake basin. View toward the north.

you can see Mt. McLoughlin and Pelican Butte, yet another imposing basaltic cone, between 120,000 and 25,000 years old.

One of the great things about driving the shoreline of Upper Klamath Lake is that the landscape so clearly reflects the faulting. To the west are steep ridges uplifted by a series of normal faults. Across the lake to the east is another ridge, also uplifted by faults. In between, the Klamath Basin has dropped down along the faults as a graben. Moreover, the ridges are steep and straight, indicating the faulting has occurred recently enough that erosion has not had enough time to soften the edges. You can soak in views of this faulting at the pull out near milepost 57, on the trace of one of these faults on the edge of Howard Bay. For the next 12 miles (19 km) to its intersection with OR 66 in Klamath Falls, the road passes scattered exposures of Pliocene basalt. Approaching the intersection, make sure to notice the view southward into the southern part of the Klamath Basin, another graben formed by Basin and Range extension.

OR 126

Eugene—US 20

75 miles (121 km)

OR 126 follows the McKenzie River from the Willamette Valley all the way to its source at Clear Lake on the western edge of the High Cascades. A few miles north of Clear Lake, OR 126 joins US 20. Use the road guide for US 20 for that section of road. An alternate, highly recommended route to Sisters is OR 242 later in this chapter, the Old McKenzie Highway, which splits off from OR 126 near McKenzie Bridge. Open only during the summer months, this road crosses recently erupted basalt flows of the High Cascades. See the road guide for OR 242 later in this chapter for that route.

In Springfield you can see the river-deposited sandstone and gravel of the Fisher Formation in a roadcut less than 0.5 miles (0.8 km) south of Business 126 on S. Second Street. A small fault offsets the sandstone on the south side of the roadcut.

Between Springfield and Leaburg Dam, near milepost 24, OR 126 passes over the McKenzie River floodplain and into the Western Cascades. The only rock exposures are some basalt on the north side of the road between mileposts 17 and 18, 1.5 miles (2.4 km) east of Deerhorn. At Leaburg Dam, several cliffs of andesite lie on the north side of the road. Leaburg Dam was constructed in 1929 to provide hydropower for Eugene. A canal, which runs next to the highway for several miles below the dam, diverts water to a small power plant downstream.

Scattered exposures of basalt, andesite, and ash-flow tuffs show up for the next several miles as the road follows the McKenzie River's north bank. Between milepost 32 and just past the community of Nimrod, cliffs on the road consist of granitic rock called the Nimrod Granite, which intruded the surrounding andesitic rocks about 16 million years ago. This intrusion is one of many similar intrusions in the Western Cascades, but only a few are well exposed

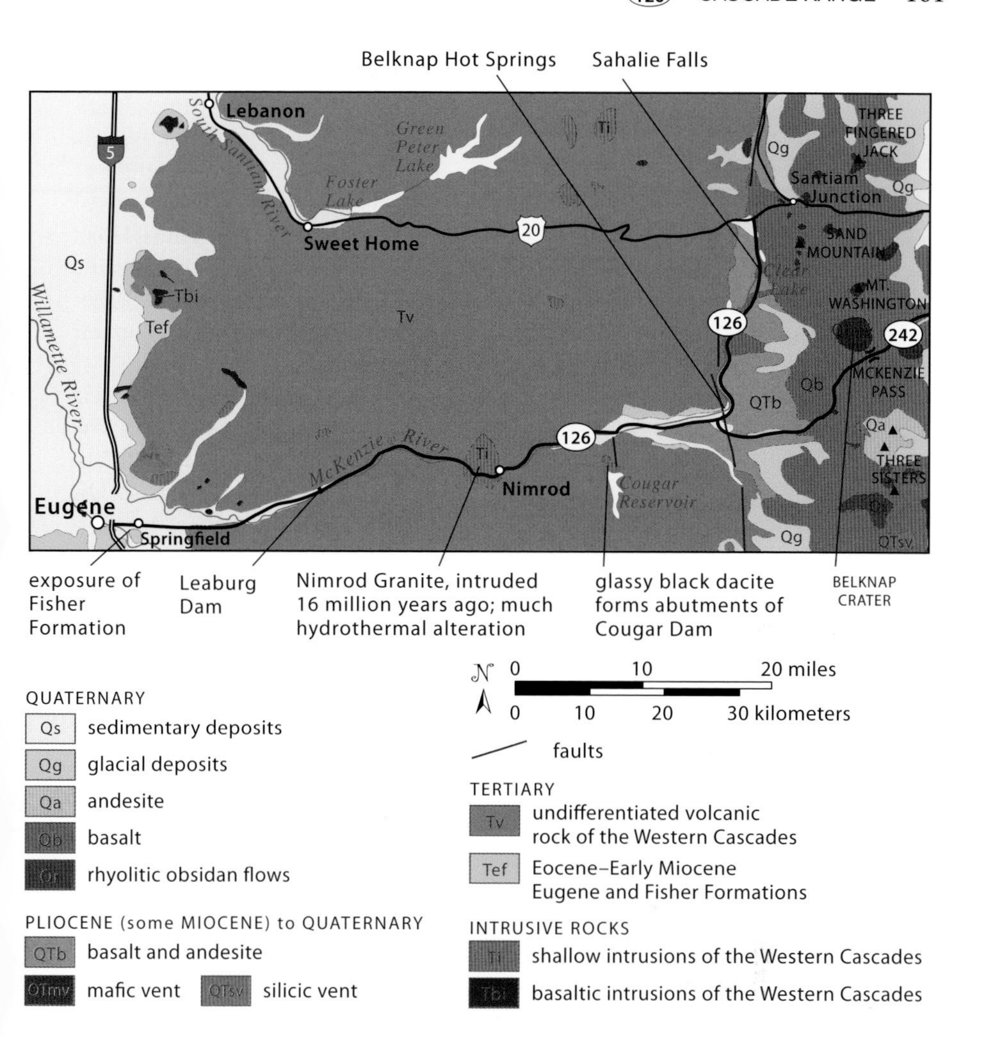

Geology along OR 126 between Eugene and US 20.

along a major highway. They provide a firsthand look at some of the more shallow magma chambers that may have fed the Western Cascades volcanoes. In addition, fluids from the intruding magmas caused hydrothermal mineralization in the surrounding rock, as well as in the granite. A small quarry about 0.3 mile (480 m) east of milepost 33 shows some of this alteration as yellowish, oxidized rock with numerous pyrite crystals.

East of Nimrod, the road passes some gravels deposited by the McKenzie River about 0.25 mile (400 m) east of milepost 36 and then travels back into andesitic rocks and ash-flow tuffs. Many of the tuffs contain fragments of preexisting volcanic rocks that were included in the flow. Outstanding exposures of these rocks show up between here and the community of Rainbow, at milepost 47. Many of them show some degree of alteration, which may be related to deep, unexposed intrusions similar to the Nimrod Granite. The prominent peak south of the highway is Castle Rock, a 9-million-year-old plug of andesite.

Forest Road 19 leads 3 miles (4.8 km) south to Cougar Reservoir, passing outcrops of ash-flow tuffs and lahars. At the dam, you can see cliffs of 16-million-year-old dark gray dacites that form the east and west abutments of the dam. A dacite is a volcanic rock with more silica than andesite but less than rhyolite. A close inspection of the rocks shows that they are dark gray because they contain dark-colored glass. These dacites likely intruded at shallow levels, cooling quickly near the surface. The elongate valley filled by Cougar Reservoir is eroded along a large down-to-the-east normal fault.

Just east of milepost 54, OR 126 curves northward, parallel to the McKenzie River. Poor exposures of glacial till show up sporadically along the road. The Old McKenzie Highway (OR 242), described in a separate road guide, continues roughly eastward over McKenzie Pass to Sisters. Belknap Hot Springs on OR 126, 1 mile (1.6 km) past the turnoff, discharges hot water at some 80 gallons (300 liters) per minute along one of the faults that bound the west side of the High Cascades graben. The area is a commercially developed resort that is open year-round.

North of Belknap Hot Springs, the now north-trending OR 126 follows the boundary between the Western Cascades on the west and the High Cascades on the east. From Belknap to about milepost 12 (this stretch of road has a new milepost numbering system that counts down to the intersection with US 20),

Exposure of dark glassy dacite forms the abutment of Cougar Dam. This dacite intruded at shallow levels about 16 million years ago.

the boundary is irregular and obscured by forest. As a general rule though, older basalt lavas of the Western Cascades lie at low elevations on both sides of the river, and those of the High Cascades lie at higher elevations. Beginning at Trail Bridge Reservoir, which has been ponded by a hydroelectric dam since 1963, the road passes into the younger High Cascades basalts. An unusual roadcut just north of the reservoir contains basaltic ash-flow tuffs, as well as a south-dipping fault.

Near milepost 7, the road passes by a 1,500-year-old lava flow, its surface marked by sparse vegetation. This flow originated at Belknap Crater, about 8 miles (13 km) to the southeast near McKenzie Pass, and shows the hackly broken surface typical of aa flows. Older lava flows that date back more than 1 million years form the tree-covered cliffs just beyond.

Koosah and Sahalie Falls, between mileposts 6 and 5, spill over the fronts of two lobes of a basaltic lava flow that flowed down the valley about 2,900 years ago. The flow originated from a series of cinder cones that form a north-south chain, including Sand Mountain, about 3 miles (4.8 km) east of the road. Koosah Falls spills over the lower lobe; Sahalie Falls spills over the upper one.

Clear Lake, the headwaters of the McKenzie River, is fed by unusually cold springs, which prevents the proliferation of algae and keeps the water remarkably clear. The lake formed because the northern margin of the same lava flow that created Sahalie and Koosah Falls dammed the McKenzie drainage. The backed-up lake waters inundated a stand of Douglas fir, parts of which are

Sahalie Falls on the McKenzie River spills over a 2,900-year-old lava flow.

preserved and still visible in the cold lake water. Near its south side, the lake reaches a depth of 178 feet (54 m).

Just north of Fish Lake, which is dry during dry seasons, the highway passes lava flows that erupted from the Sand Mountain chain between 4,000 and 3,000 years ago.

OR 138
Roseburg—US 97
100 miles (161 km)

Quite possibly the most geologically instructive route across the Cascades, OR 138 follows the North Umpqua River and one of its tributaries to the crest of the range, where it meets the north entrance to Crater Lake National Park. Along the way, it samples basalt of Siletzia, a wide variety of volcanic rock of the Western Cascades, and recent lava flows of the High Cascades, including ash flows from the eruption of Mt. Mazama, about 7,700 years ago. This stretch of highway is also known for its waterfalls, especially those near the boundary of the High and Western Cascades. Numerous pull-outs line the highway to ensure easy access to the rocks as well as the river.

East of Roseburg, OR 138 passes scattered exposures of basalt of the Siletz terrane for the first 7 miles (11 km) or so. East of the basalt, the road passes through open valleys with little in the way of outcrops; more easily eroded sandstone and shale of the Umpqua Group lie underneath. The Umpqua Group accumulated in a marine setting during the accretion of Siletzia in Eocene time.

Just east of milepost 16 where the road crosses the Little River, however, the geology starts to get really interesting. A roadcut west of the bridge shows east-dipping sedimentary rock of the Umpqua Group. East of the bridge, the Colliding Rivers Viewpoint allows visitors to stop and contemplate the turbulent confluence of the Little and Umpqua Rivers, which run into each other nearly head-on and then flow together peaceably downriver. Several basaltic sills in the Umpqua Group cause this unusual channel configuration because they prevent the channels from migrating to a gentler confluence. One of the sills shows up clearly on both sides of the road immediately west of the bridge. From the overlook, you can follow a foot trail underneath the bridge to some fossil-rich outcrops of the sandstone along the Little River.

For the next several miles east of the confluence, more sandstone lies along the Umpqua's course, but beginning at Idleyld Park, the river exposes Western Cascades volcanic rocks, formed between about 35 and 10 million years ago and everywhere younger than the Umpqua Group. They also dip toward the east and so tend to get younger in that direction. These rocks on the far western side are therefore some of the oldest rocks of the Western Cascades. They are part of a sequence of lahars (volcanic mudflows), ash flows, andesite flows, and some basalt flows called the Little Butte Volcanics. Between Idleyld Park and several miles east of Steamboat, they reach nearly 15,000 feet (4,600 m) in thickness. A steep, narrow trail (watch for poison oak!) from the Narrows Wayside leads to the river, where you can examine numerous water-polished outcrops of andesite and lahars—as well as some potholes formed by the grinding action of cobbles trapped in rock recesses during high water.

The road passes many cliffs, roadcuts, and small outcrops of the Little Butte Volcanics for the next 35 miles (56 km) or so before it reaches the first young

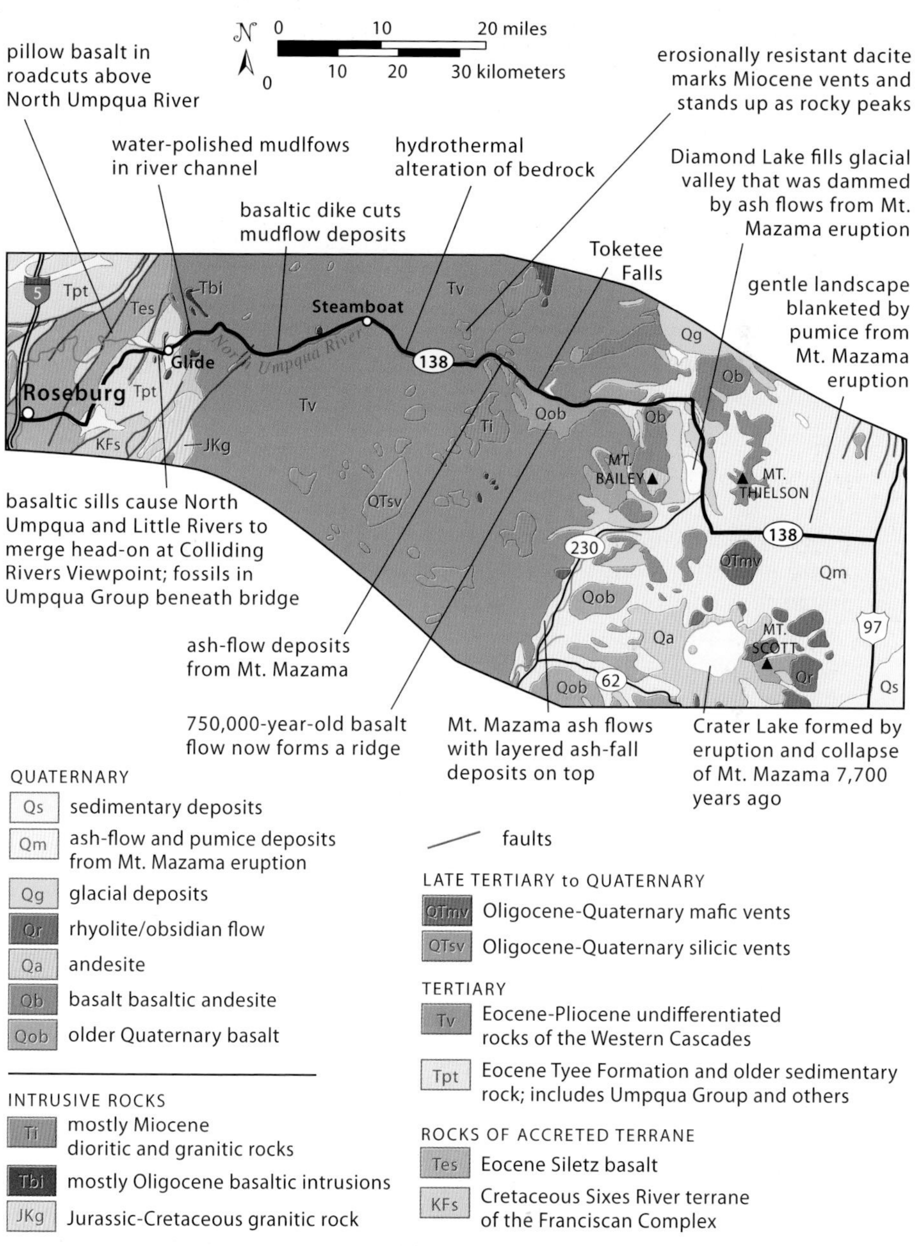

Geology along OR 138 between Roseburg and US 97.

Water-polished lahar deposits of the early Western Cascades in a channel of the Umpqua River. Lahars are volcanic mudflows that incorporate pieces of rock as they rush downslope.

basalt flows of the High Cascades at Toketee Falls. While it's impossible to detail all the interesting features, the following few are especially worth mentioning. Just west of milepost 27, a basaltic dike that intrudes deeply weathered lahar deposits is exposed in roadcuts on both sides of the road; look for the pull-out 0.1 mile (160 m) to the west of the roadcut. A pull-out between mileposts 31 and 32 allows inspection of a weathered ash-flow tuff beneath a basalt flow with well-developed entablature and colonnade. Fine-grained sedimentary rocks just east of milepost 34 were deposited in lakes that resided alongside the Western Cascade volcanoes. Just east and west of the intersection with the Steamboat Creek Road, at milepost 39, andesite flows display some beautiful columns. At Horseshoe Bend, between mileposts 46 and 47, the rocks show a wide variety of yellow colors, indicating hydrothermal alteration, driven by hot groundwater as it passed through the rock. Some of these weak, altered rocks are especially prone to landsliding, as can be seen in the numerous scars in the hillsides. At milepost 50, the road crosses to the south bank of the North Umpqua River and offers views to several rocky summits including Eagle Rock; most of these are erosionally resistant dacites between 25 and 17 million years old.

Beginning about milepost 53, you can see ash-flow deposits from the eruption of Mt. Mazama, which is about 30 miles (48 km) to the southeast. The flows traveled existing river valleys out from the volcano in all directions; to get here, they flowed north and then west, approximately along the route of OR 138, a distance of more than 40 miles (64 km). Imagine this lush valley being incinerated almost instantly as it filled with the hot ash flow only 7,700 years ago. Since then the river has reclaimed its channel and erosion has removed most of the deposits.

Just west of the turnoff to Toketee Falls at milepost 59, basaltic rocks, which form a high ridge to the south, mark the western margin of the High Cascades. These flows erupted about 750,000 years ago from vents at the edge of the High Cascades and flowed down the Umpqua and Clearwater valleys. The flows filled the existing valleys to depths of several hundred feet. Subsequent erosion reestablished narrow river channels along the edges of the flows and left behind a somewhat flat-topped ridge of basalt where the valley used to be.

Younger basalt flows, originating from Mt. Bailey within the last 25,000 years, flowed down the narrow channel of the Clearwater River and into the Umpqua to a few miles past where Toketee Falls is today. Since then, the river has cut its channel once again and spills out of a notch and over a cliff of columnar joints in this younger basalt. A 0.5 mile (0.8 km) hike leads from the parking lot to a lovely view of Toketee Falls. Another waterfall, Watson Falls, pours off the 750,000-year-old basalt near milepost 61.

The road turns southward near milepost 73 and travels over tree-covered ash-flow deposits for the next 10 miles (16 km). From the road, you can get

Toketee Falls, as seen from the overlook. Here, the Umpqua River flows through a notch it eroded through a 25,000-year-old basaltic andesite flow. Note the column-shaped cooling fractures.

a few views toward Mt. Bailey, an andesitic volcano, and the remnants of Mt. Mazama, which erupted about 7,700 years ago to produce Crater Lake. Just north of milepost 79, a road leads to Diamond Lake, which fills most of the valley bottom for more than 3 miles (4.8 km). According to David Sherrod, the US Geological Survey geologist who mapped much of the High Cascades, this lake likely formed after the Mazama eruption, when the area became inundated with ash-flow deposits. These ash flows disrupted the existing drainage, and along with the bedrock spur at the north end of the lake, created a natural dam.

On the east side of the highway halfway between mileposts 79 and 80, the Diamond Lake Viewpoint offers exceptionally informative displays as well as good views of Diamond Lake and Mt. Bailey to the west and 9,183-foot (2,799 m) Mt. Thielsen to the southeast. Deeply eroded by long-gone glaciers, Mt. Thielsen was an active basaltic cone some 300,000 years ago. Its summit spire, which attracts numerous lightning strikes, is held up by the more resistant intrusive rock of its main magma conduit. One tiny glacier, the Lathrop Glacier, remains in a north-facing cirque on the peak.

Between mileposts 82 and 83, OR 138 intersects OR 230, which passes by the west side of Crater Lake National Park. The Mt. Thielsen Viewpoint, only 0.25 mile (400 m) down the road, is well worth the detour, because it gives an unobstructed and altogether different view of the mountain.

OR 138 passes over a low-relief landscape blanketed with pumice from the Mazama eruption for another 3 miles (4.8 km) to the Crater Lake North

Mt. Bailey and Diamond Lake from Diamond Lake Viewpoint. Mt. Bailey consists of an andesitic cone atop a shield of basaltic andesite. It erupted the lava flow over which Toketee Falls spills.

Entrance Road. Turning eastward, OR 138 continues another 12 miles (19 km) over pumice deposits to US 97. Between mileposts 99 and 100, an unpaved road leads north 0.3 mile (500 m) into the core of a cinder cone blanketed by pumice from Mt. Mazama.

OR 242
McKenzie Bridge—Sisters
37 miles (60 km)

Open only in summer

The Old McKenzie Highway, a 37-mile-long (60 km) "shortcut" to Sisters, takes over an hour, partly because the road is windy and slow, but also because the views from McKenzie Pass are positively distracting. The road follows a heavily forested, deep glacial valley for its first few miles to about milepost 64, the trailhead for Proxy Falls. The road then climbs steeply to an elevation of about 4,500 feet (1,370 m) as it switchbacks across a lava flow that poured down the valley between about 20,000 and 7,600 years ago. The lava originated as a series of flows from Sims Butte, perched nearly 1,000 feet (300 m) above the road just

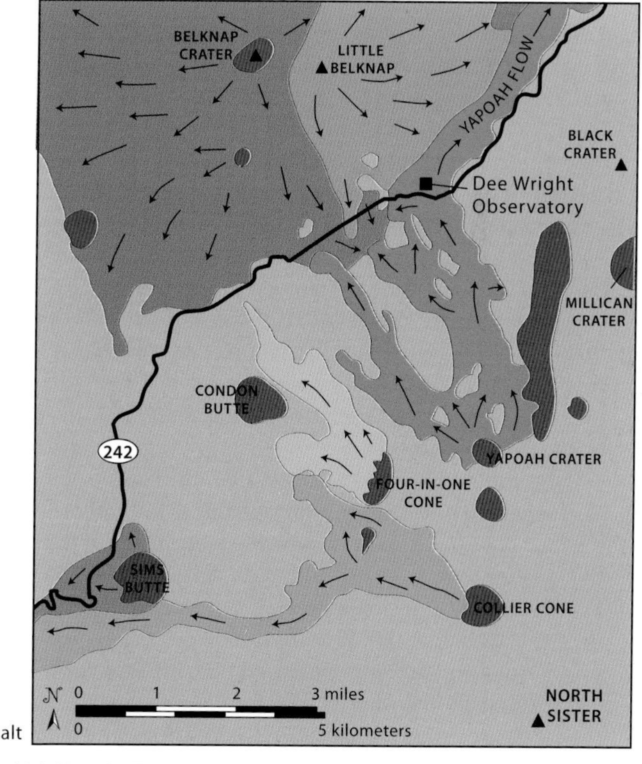

Geology of McKenzie Pass area. –Modified from Taylor and Smith, 1987

east of the uppermost switchback. Occasional views from the upper switchbacks, near milepost 69, give a sense of the deep, U-shaped nature of the valley below, a sign of glaciation.

Near milepost 70, the road levels off and passes by the west side of Sims Butte, occasionally visible through the trees. Just north of milepost 74, a pull-out allows views and access to a flow from Belknap Crater, the central vent of the Belknap shield volcano. The volcano and its satellite cone, Little Belknap Crater, lie behind the flow on the horizon to the north. The road follows the margin of this flow and the one from Little Belknap almost to the pass.

The Dee Wright Observatory at McKenzie Pass gives arguably the best view of Cascade volcanoes and recent lava flows of any roadside stop in Oregon. Completed in 1935 by a crew of the Civilian Conservation Corps, the observatory consists of an open-air viewing platform and stone shelter, built from basaltic rocks from the surrounding lava flows. It was named for Dee Wright, the foreman of the construction crew, who died the year before the observatory was completed.

You can see in all directions from the observatory, and a circular bronze monument helps locate features. To the south, you look across basaltic lava flows to the North and Middle Sisters volcanoes. Between them, they host the Collier Glacier, which flows toward the west and northwest. The most recent lava flows are treeless, having erupted only about 2,000 years ago from Yapoah Crater, which can be seen directly in front of the face of North Sister. Collier Cone, which also erupted about 2,000 years ago, is visible just to the right on North Sister's west flank. Collier Cone produced a lava flow about 1,500 years ago that reaches OR 242 near Proxy Falls.

Views to the north include the north edge of the Yapoah flow, on which the observatory is built. Mt. Washington, a deeply eroded basaltic shield volcano, and Mt. Jefferson, an andesitic stratovolcano, form the highest peaks. Belknap Crater and its subsidiary cone, Little Belknap, form broad basaltic shield volcanoes that occupy much of the foreground. They produced most of the lava flows on the north side of the road during the past 3,000 years. Two kipukas, a Hawaiian term for islands of land surrounded by lava flows, are visible within these flows. On clear days, Mt. Hood is even visible. A 0.5 mile (0.8 km) paved trail leads from the observatory through the Yapoah flow and offers glimpses into a collapsed lava tube, cooling fractures, and many beautiful examples of young, fresh basalt.

Just below the pass on its east side, the road descends through the Yapoah flow, giving a seldom-seen glimpse of a flow interior. It consists of several distinct flow units, each of which has a relatively thin, dense interior of lava that grades outward into rubble. Windy Point, slightly less than 3 miles (4.8 km) below the pass, offers a view of Belknap shield volcano and Mt. Washington with the Yapoah flow in the foreground. The prominent outcrop at the pull-out is from an older flow that came from nearby Black Crater. The road crosses an exposure of glacial till 2.8 miles (4.5 km) north of Windy Point, and then travels on poorly exposed older lava flows and glacial outwash, descending gently through a ponderosa pine forest to Sisters.

View of Belknap shield volcano from Windy Point. The Yapoah lava flow, which erupted about 2,000 years ago from Yapoah Crater, just south of the Dee Wright Observatory, occupies the foreground.

Crater Lake National Park

No other spot in Oregon even comes close to Crater Lake in combining scenery with geology. The extraordinarily blue lake occupies a caldera in Mt. Mazama, a volcano that erupted so catastrophically about 7,700 years ago that it collapsed during the eruption into its rapidly emptying magma chamber. Visitors can travel around the caldera edge, peer into the lake water, and see natural cross-sections of the insides of this volcano on the walls. Crater Lake became the country's seventh national park in 1902 and receives about a half million visitors each year. During the summer months, the National Park Service leads boat tours around the lake to offer an inside glimpse of the volcano.

Mt. Mazama was a typical large Cascade volcano. It had steep slopes and probably reached an elevation between 10,000 and 11,000 feet (3,000 and 3,350 m). Howel Williams, one of the first geologists to study the area in great detail, published this conclusion in 1942 after comparing Mazama's topographic profile to that of Mt. Rainier, Mt. Adams, and Mt. Shasta. Deep glacial valleys, which later provided channels for the ash flows of the catastrophic eruption, cut into the mountain. The U-shaped profiles of these valleys show up beautifully as U-shaped notches in the east rim of the caldera.

In the caldera walls, you can see multiple layers of volcanic rock, formed through a long history of eruptions. These layers consist mostly of andesite and dacite lava flows, rather than pyroclastic deposits from explosive eruptions,

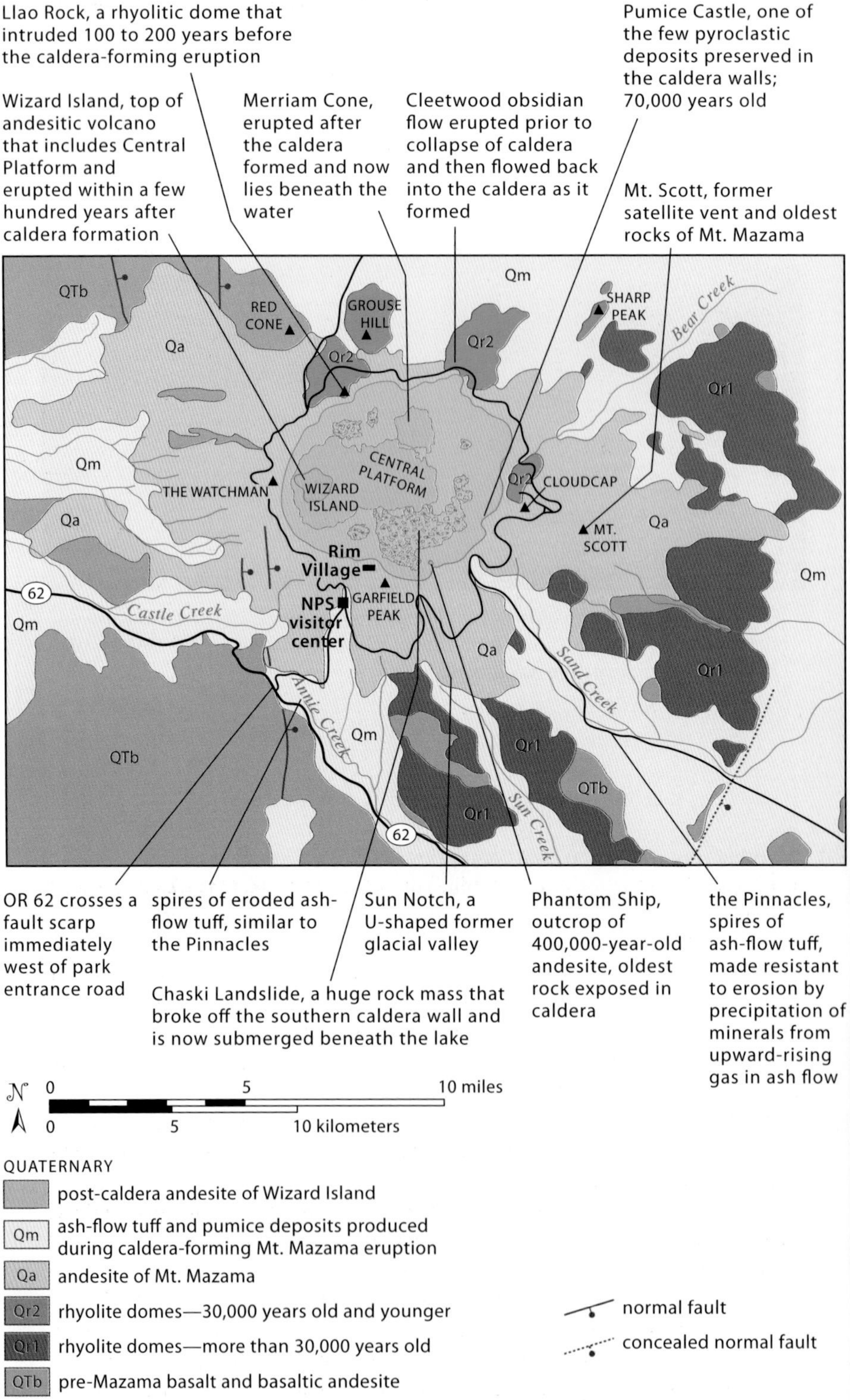

Llao Rock, a rhyolitic dome that intruded 100 to 200 years before the caldera-forming eruption

Wizard Island, top of andesitic volcano that includes Central Platform and erupted within a few hundred years after caldera formation

Merriam Cone, erupted after the caldera formed and now lies beneath the water

Cleetwood obsidian flow erupted prior to collapse of caldera and then flowed back into the caldera as it formed

Pumice Castle, one of the few pyroclastic deposits preserved in the caldera walls; 70,000 years old

Mt. Scott, former satellite vent and oldest rocks of Mt. Mazama

QTb
RED CONE
GROUSE HILL
Qr2
Qm
SHARP PEAK
Bear Creek
Qa
Qr2
Qr1
Qm
CENTRAL PLATFORM
Qr2
CLOUDCAP
THE WATCHMAN
WIZARD ISLAND
Qa
MT. SCOTT
Qa
Rim Village
Qm
62
NPS visitor center
GARFIELD PEAK
Castle Creek
Qm
Qa
Sand Creek
Qr1
Annie Creek
Qm
QTb
Qr1
QTb
Qr1
Sun Creek
62

OR 62 crosses a fault scarp immediately west of park entrance road

spires of eroded ash-flow tuff, similar to the Pinnacles

Sun Notch, a U-shaped former glacial valley

Phantom Ship, outcrop of 400,000-year-old andesite, oldest rock exposed in caldera

the Pinnacles, spires of ash-flow tuff, made resistant to erosion by precipitation of minerals from upward-rising gas in ash flow

Chaski Landslide, a huge rock mass that broke off the southern caldera wall and is now submerged beneath the lake

N
0 5 10 miles
0 5 10 kilometers

QUATERNARY

post-caldera andesite of Wizard Island

Qm ash-flow tuff and pumice deposits produced during caldera-forming Mt. Mazama eruption

Qa andesite of Mt. Mazama

Qr2 rhyolite domes—30,000 years old and younger

Qr1 rhyolite domes—more than 30,000 years old

QTb pre-Mazama basalt and basaltic andesite

normal fault

concealed normal fault

Geology of Crater Lake National Park.

Crater Lake and Wizard Island from the Watchman. Wizard Island is a cinder cone at the top of a submerged andesitic volcano that erupted a few hundred years after the caldera-forming eruption. The U-shaped notches in the far side of the caldera rim are the edges of glacial valleys that lost their upper reaches during the eruption. The high peak in the center background is Mt. Scott, a satellite cone and the oldest exposed part of Mt. Mazama.

suggesting its early history was less explosive than many that of other Cascade volcanoes. In places, you can see dikes that cut through these layers to feed younger flows above.

As a distinct volcano, Mazama has likely been active since before 400,000 years ago, the approximate age of its oldest rocks. Phantom Ship, the small jagged island along the lake's southeast edge, consists of 400,000-year-old andesite, while Mt. Scott, the small peak that rises just east of the caldera, consists of dacitic flows that are about 420,000 years old. Younger lavas overlie these rocks, indicating intermittent but frequent periods of activity throughout its 400,000-year history. Several dacitic domes and lavas and associated pumice falls erupted on the north and east sides of the mountain between about 30,000 and 8,000 years ago. Their chemistries suggest they may have been precursors to the caldera-forming eruption. They now form topographically high points, such as Grouse Hill and Sharp Peak. The second youngest one, Llao Rock, formed just 100 to 200 years prior to the eruption.

We have long understood the basic idea behind the formation of the Crater Lake caldera: the eruption 7,700 years ago was so large that it emptied its magma chamber and the mountain collapsed into the void. More recently, largely because of detailed mapping by Charles Bacon of the US Geological Survey, we've come to learn some of the details of that process. It seems that the

Phantom Ship, formed of 400,000-year-old andesite, reaches 170 feet (52 m) above the water surface.

caldera-forming eruption proceeded in two steps, first from a single vent, and second from fractures that formed around the caldera as it began to collapse. A towering eruption column rose from the initial vent to a height of 30 miles (48 km) and spread mostly northward by the prevailing winds. This phase of the eruption ended when the column collapsed and fed ash flows that traveled down valleys on the mountain's northern and eastern flanks. Ash flows from the second, caldera-collapse phase, poured down the mountain in all directions, overfilled the valleys, and blanketed the mountain. Remnants of these flows today are still found in valleys as far away as 40 miles (64 km) from the volcano. All told, between 12 and 15 cubic miles (50–60 km^3) of magma was ejected from the volcano. By comparison, the Mt. St. Helens eruption in 1980 produced less than 0.75 cubic mile (3 km^3).

The eruption appears to have proceeded quickly. Some of the best evidence of the time frame lies along the road near Cleetwood Cove. There, the Cleetwood lava flow, which erupted just prior to the caldera-forming eruption, was still hot as the caldera collapsed. It oozed back into the newly formed caldera and baked the pumice that fell on its surface during the caldera-collapse phase.

The most obvious expression of volcanic activity after the main caldera-forming event is Wizard Island, the andesitic cinder cone that rises some 500 feet (150 m) above the water. Because it is within the caldera, we know it formed after the caldera collapse. Studies of the lake bottom, however, show that Wizard Island reflects only a small fraction of the post-caldera activity and is the top of a large drowned volcano. Nearly all of this activity was andesitic and happened within only a few hundred years after the caldera collapsed. Two prominent features beneath the water are Merriam Cone and the central platform east of Wizard Island.

Crater Lake's incredible blue color derives from clarity and depth. With a maximum depth of 1,932 feet (589 m) and a clarity that often surpasses a depth of 100 feet (30 m), it is at once the deepest and possibly clearest lake in the United States. High clarity means deep light penetration, which in turn causes more absorption of all but the short, blue and purple wavelengths of light. The reason the lake is so clear is because the lake water carries little suspended sediment; the lake has no watershed except for the crater walls.

With the exception of some input of thermal waters into the lake bottom, Crater Lake today is virtually inactive as a volcano. The caldera-forming eruption and post-caldera activity depleted the magma supply to the extent that it will likely take thousands of years to replenish. While that does not rule out minor eruptions in the near future, those seem unlikely, simply because the volcano has been so quiet for the last 5,000 years. Even so, Mt. Mazama and Crater Lake reside in the High Cascades, an active volcanic arc fueled by magma from below.

Guide to the Rim Road and Entrance Roads

Two entrances access Crater Lake National Park. The north entrance lies on a road that leads south from OR 138 and is closed during the winter. The south entrance, which is open year-round, lies just north of OR 62. The Rim Road, which is closed during the winter, circles the lake and connects the entrances. Given space limitations, this guide only discusses some of the major features. For more detailed road guides, see those by Charles Bacon, published in 1987 and 1989. The ages for individual rock units come from Bacon, 2008. Please remember that rock collecting is prohibited in the national park.

The south entrance road passes the campground and after a few miles reaches the visitor center at park headquarters and begins its steep climb to the rim. It passes roadcuts of highly broken-up dacite blocks—avalanche deposits from a lava dome that formed about 35,000 years ago high on Mazama. Just over 1 mile (1.6 km) beyond, Rim Village offers overlooks and visitor amenities. From the overlooks, note the Devils Backbone, a dark-colored dike that intrudes the walls of the caldera behind Wizard Island to the left. Also note Llao Rock, behind Wizard Island to the right, made of dacite that plugged a vent just prior to the eruption. The dacite now forms a sheer cliff some 1,200 feet (370 m) high and is exposed along the rim road.

The trailhead for Garfield Peak lies at the east edge of the Crater Lake Lodge parking lot at Rim Village. At an elevation of 8,054 feet (2,455 m), Garfield Peak is one of the highest points along the rim. The 1.7-mile (2.7 km) trail climbs nearly 1,000 feet (300 m) past exposures of 350,000- to 225,000-year-old andesite and dacite lava flows and breccias.

Between Rim Village and the parking lot for the Watchman Trail, the Rim Road passes several overlooks between imposing cliffs of andesite and dacite.

Devils Backbone and older lava flows as viewed from lake level. This dike intruded Mt. Mazama and fed lava flows that were later removed in the caldera-forming eruption.

The views southward, away from the lake, show a landscape dotted with low, conical hills, many of which are cinder cones. Union Peak, in the middle distance, caps a broad shield volcano that is about 160,000 years old. In the far distance, Mt. McLoughlin and Mt. Shasta rise some 40 and 100 miles (64 and 160 km) to the south, respectively. The Watchman consists of 50,000-year-old dacite and hosts a fire lookout at its top. A trail from the parking lot leads 0.7 mile (1.1 km) up to the fire lookout and more wonderful views. At the parking lot, you can look directly down on Wizard Island. Across the lake, note the U-shaped Sun and Kerr notches, the cutoff upper reaches of Sun and Kerr valleys. Glaciers eroded the U-shaped valleys during the Pleistocene Epoch.

The North Entrance Road descends the volcano between Red Cone to the west and Grouse Hill to the east. Red Cone marks a basaltic vent, some 35,000 years old. Grouse Hill is a dacitic dome between 30,000 and 25,000 years old. Some 5 miles (8 km) from the rim, the road flattens and crosses a barren landscape called the Pumice Desert. Shrubs and trees do not thrive there because the pumice deposits are as much as 200 feet (60 m) thick, inhibiting typical soil-forming processes. Water drains quickly through the pumice to depths that are out of reach of many plants. Several trees around the edge of the desert, however, indicate that through time, the forest of lodgepole pine will gradually reclaim the land.

North of the North Entrance Road junction, the Rim Road passes north of Llao Rock. Good exposures of glassy rhyolite crop out here, many of which contain small crystal-lined cavities. A large pull-out on the west side of the road exists just over 0.25 mile (400 m) north of the junction with the North Entrance Road.

Just under a quarter mile east of the large parking lot for boat tours, a pull-out on the south allows parking to view the Cleetwood lava flow and overlying pumice deposits. On the south side of the road, the Cleetwood flow, which erupted prior to the catastrophic eruption, consists largely of glassy rhyolite with flow-produced layering. Note that the layering is vertical here but is horizontal deeper within the lava flow. Below, a tongue of this lava descends into the caldera as far as the lakeshore. On the north side of the road, you can see the contact where pumice from the eruption was deposited on the flow. Near the contact, the pumice becomes increasingly oxidized and welded. Both of these observations indicate that the catastrophic eruption proceeded quickly, before the Cleetwood flow had cooled.

Numerous exposures of Mazama andesite and dacite line the road between Cleetwood Cove and Mt. Scott. These flows formed between about 160,000 and

Air-fall pumice on top of the Cleetwood lava flow at Cleetwood Cove. The Cleetwood flow, which erupted just prior to the caldera-forming event, is the dark red solid rock at the bottom of this photo; the pumice, which erupted during the caldera-forming event, forms a dark red ledge just above the flow and becomes lighter-colored and less welded farther up the slope. The red colors are from oxidation, and the compactness is from welding as the pumice landed on the still-hot Cleetwood flow.

The Pinnacles

The overlook at the Pinnacles is well worth the 6-mile (10 km) drive away from the caldera. The road follows the canyon of Sand Creek, a glacial valley that was inundated by pyroclastic flows during the caldera-forming eruption. Since then, the stream has reestablished its channel, leaving remnants of the flows along the edge. The overlook at the end of the road gives a close-up view of what the flows look like: a random assortment of all sizes of blocks and pumice in a matrix of ash. You can follow a short trail beyond the parking lot for more views.

The Pinnacles form solitary spires because they are resistant to weathering and erosion. They consist of mineralized pipes, much harder and resistant than the surrounding material. After the ash flow had settled, gases naturally rose to the surface, particularly where the flows covered streams and bogs. These gases concentrated into conduits and, as they rose, precipitated the minerals. In addition, the ash-flow deposit is chemically zoned, with lighter-colored, more silicic material at the bottom, and darker-colored, less silicic material near the top. The first pyroclastic flows, from the top of the magma chamber, were richer in silica than the later flows, which originated from deeper in the chamber.

Ash-flow tuff of one of the spires at the Pinnacles.

27,000 years ago. Mt. Scott consists of dacitic rock but formed a separate cone on the side of Mt. Mazama. At 420,000 years old, Mt. Scott is the oldest known part of Mt. Mazama. Glaciation carved the western side of Mt. Scott into a cirque with a steep face above it.

For the 6 miles (10 km) between Mt. Scott and the Pinnacles turnoff, the Rim Road passes several worthwhile overlooks, particularly Cloudcap and the one for Pumice Castle. Pumice Castle highlights a thick, 70,000-year-old pyroclastic deposit of Mt. Mazama. Because pyroclastic deposits provide the most direct evidence of explosive eruptions, this deposit indicates a major explosive event about 70,000 years ago.

Pumice Castle is one of only a few old pyroclastic deposits preserved on Mazama. This observation might suggest that Mazama had a relatively quiet history overall and that its eruption 7,700 years ago was an aberration. However, pyroclastic deposits tend to be fairly loose and are easily removed by erosion, especially during glacial periods. It's quite possible that direct evidence from many explosive eruptions is simply missing! Given the number of dacitic domes and flows that erupted on Mazama in the last 70,000 years, it seems reasonable that the deposits of Pumice Castle mark a change from a less explosive to more explosive history.

For the 8 miles (13 km) between the Pinnacles Road and the park visitor center, the Rim Road winds in and out of three glacial valleys. It passes exposures of andesite of the older part of Mt. Mazama the entire way. You can see southward into the Klamath Basin, down-dropped as a graben between two fault systems. On clear days, Medicine Lake Volcano in California is visible south of Klamath Falls, at the south end of the graben. The 0.25 mile (400 m) walk to Sun Notch, at the top of the middle valley, gives a spectacular view of the lake and Phantom Ship, nearly 1,000 feet (300 m) below.

Colorful pumice fragments on the ground near Cleetwood Cove.

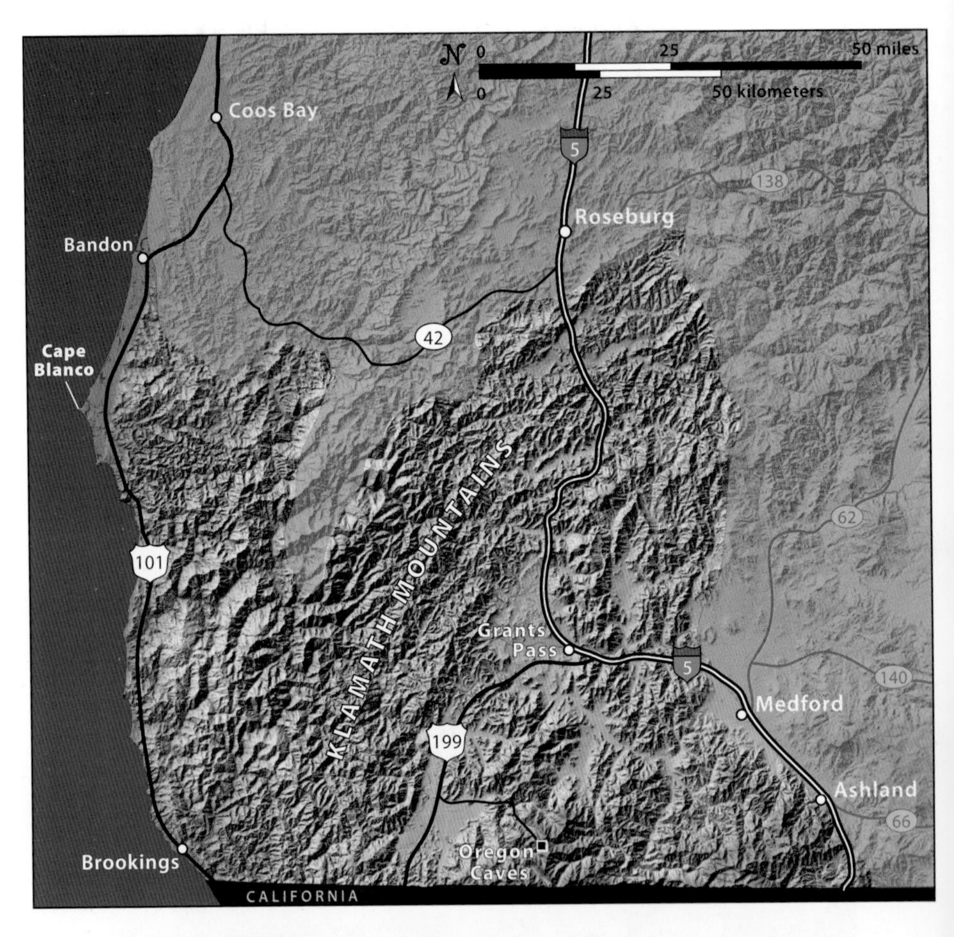

Physiographic map of the Klamath Mountains in Oregon. –Base image from US Geological Survey National Elevation Data Set Shaded Relief of Oregon

Klamath Mountains

The Klamath Mountains Province, which includes some of the most rugged and inaccessible country of Oregon, forms the southwestern corner of the state and extends southward into California nearly as far as Red Bluff. Steep river valleys are separated by ridgelines and peaks exceeding 5,000 feet (1,500 m) in elevation. The Illinois and Rogue Rivers traverse the Oregon Klamaths, merging some 20 miles (32 km) from the coastline. The Chetco and Coquille Rivers drain parts of the southern and northern Klamaths in Oregon, respectively. This book includes the adjacent coastline as part of the province, although many researchers consider the coastline part of the Coast Range and not the Klamaths.

Like the Coast Range to the north, the noncoastal part of the Klamaths tends to be heavily forested, so roadcuts provide a large portion of the accessible bedrock. Unlike the Coast Range, however, the geology exposed in the roadcuts offers comparatively little consistency or predictability, making the Klamaths' geology some of the most interesting and complicated of the state.

A geologic map of the Klamath Mountains resembles a patchwork quilt in which each patch is completely different from the next. These patches are fault-bounded terranes, which consist mostly of oceanic rock that was accreted to North America at various times throughout Mesozoic time. Within some patches and across some boundaries are igneous stitching plutons that invaded the terranes throughout the accretionary process. To further complicate the picture, the rocks are highly deformed, having experienced multiple accretion events.

Individual terranes of the Klamaths are grouped into six composite terranes, or belts. The age of the rocks in the terranes and the timing of accretion decrease toward the coast. The two oldest composite terranes, called the Eastern Klamath terrane and the Central Metamorphic Belt, contain rocks as old as Ordovician age but lie entirely within California. Progressively younger belts, which also exist in Oregon, are the Condrey Mountain Schist, the Paleozoic-Triassic Belt, the Western Klamath Belt, and the Franciscan Complex.

The Blue Mountains also consist of accreted oceanic terranes intruded by stitching plutons. Terrane accretion in both the Klamaths and Blue Mountains occurred along the same north-trending coastline, prior to 70 degrees of clockwise rotation during Cenozoic time that moved the Blue Mountains east relative to the Klamaths. Unlike the Blue Mountains, which display accreted basement rock as scattered exposures in between younger sedimentary or volcanic rock, the Klamath Mountains are almost entirely basement rock, with little younger

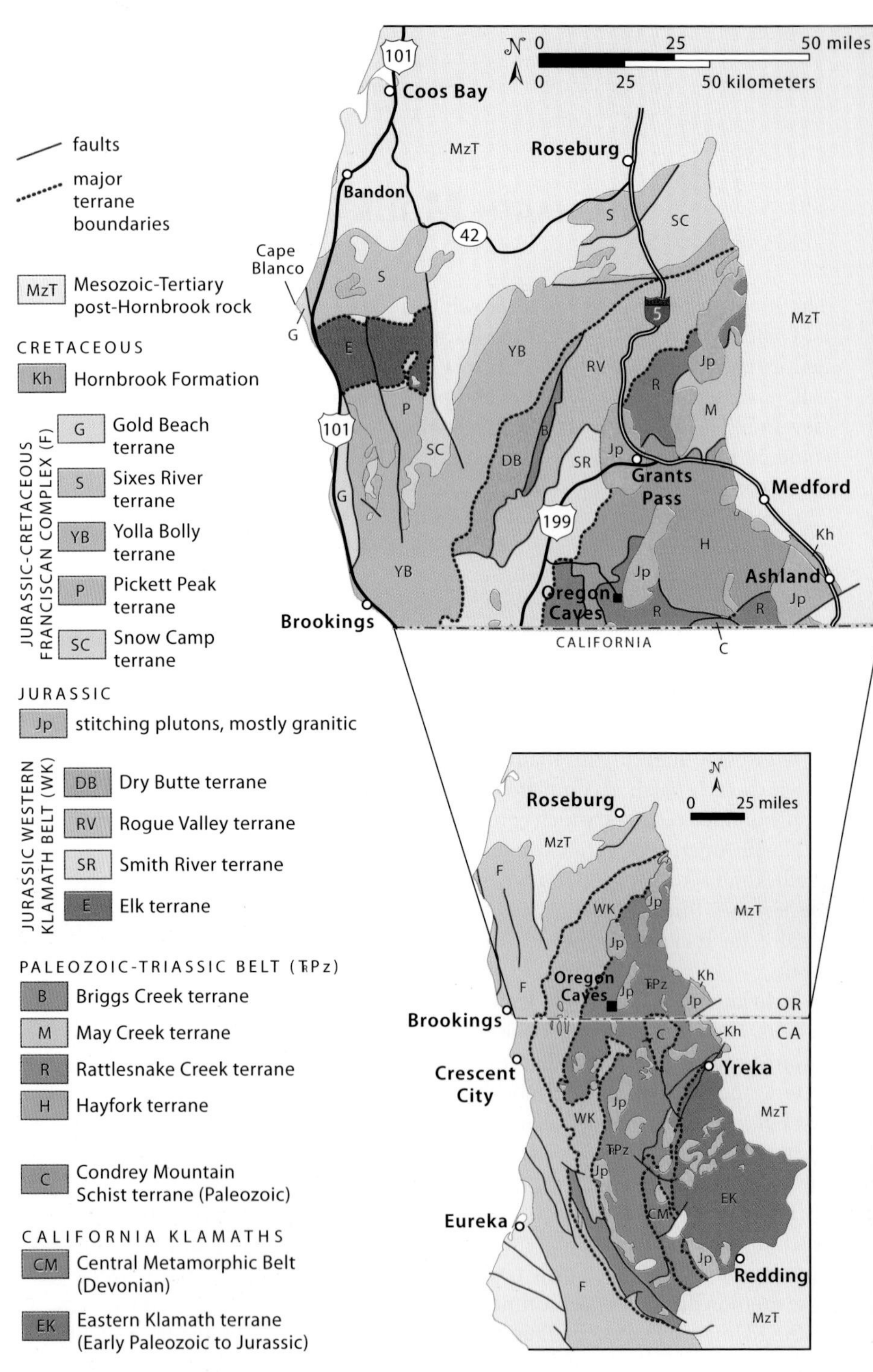

Map of accreted terranes of the Klamath Mountains. Map on the bottom includes the California Klamaths; the enlarged map on the top depicts only the Oregon Klamaths.

–Modified from Snoke and Barnes, 2006; and Blake and others, 1985

material. The best known, post-accretion rock unit of the Oregon Klamaths is the Cretaceous Hornbrook Formation, which lies on top of the older accreted terranes in a narrow belt between Medford, Oregon, and Yreka, California.

The faults that separate the different terranes of the Klamaths are exposed in only a few places, but geologists have determined that they are eastward-dipping thrust faults that place older rock over younger rock. Accretion at

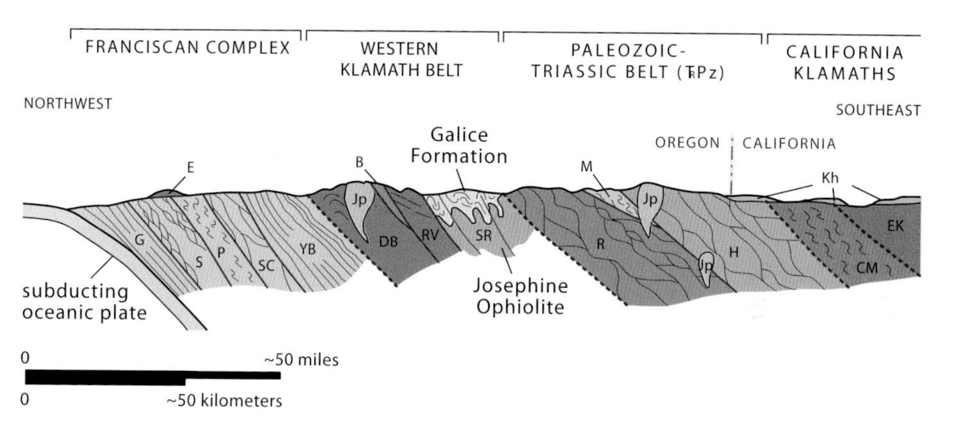

Schematic cross section across the Klamath Mountains of Oregon and California, showing the major belts and their terranes.

Outcrop of serpentinite in the Franciscan Complex near Myers Creek on US 101. Serpentinite bodies mark many of the fault zones along the coast. Cape Sebastian lies in the background.

an eastward-dipping subduction zone would have produced these eastward-dipping thrust faults, supporting the idea that a subduction zone existed to the west of the growing continent.

Plutons intruded the Klamath Mountains at a variety of times, beginning prior to 400 million years ago in the Eastern Klamath terrane near Mt. Shasta, California. Most of the plutons, however, are Jurassic in age. In Oregon, the oldest Klamath pluton is the 174-million-year-old Squaw Mountain pluton and the youngest is the 139-million-year-old Grants Pass pluton. Like many of the volcanic rocks of the Klamaths, these plutons likely formed in volcanic arc settings, cooling at depths from the magma chambers that fueled the volcanism.

Some stitching plutons cut across terrane boundaries, seeming to stitch the terranes together. Their ages provide a younger limit to the time those terranes came together. The Grants Pass pluton, for example, intrudes the boundary of the Western Klamath Belt with the Paleozoic-Triassic Belt, indicating they were joined prior to 139 million years ago. The 161-million-year-old Ashland pluton and the 160-million-year-old Grayback pluton intrude across the boundary of the Rattlesnake Creek and Hayfork terranes of the Paleozoic-Triassic Belt, indicating those terranes joined before the age of those plutons.

With the accretion of so many terranes as well as multiple intrusions of plutons, the Klamath Mountains experienced numerous episodes of mountain building, known in the geologic world as orogenies. The best-known of these events was the Nevadan Orogeny, originally named in 1914 to collectively describe Jurassic-age mountain building along the entire length of the west coast from Alaska to Mexico. Its usage now seems restricted mostly to the Klamath Mountains and the foothills of the western Sierra Nevada in California. In the Klamath Mountains, the Nevadan Orogeny occurred between about 152 and 150 million years ago when the Western Klamath Belt rode up against the already-accreted Paleozoic-Triassic Belt. An older but less well-understood event was the more localized Siskiyou Orogeny, which took place between about 169 and 160 million years ago with accretion of the Paleozoic-Triassic Belt. Other smaller episodes of mountain building likely occurred throughout the Cretaceous Period as the terranes of the Franciscan Complex were accreted.

Accreted Terranes

Paleozoic-Triassic Belt

In Oregon, the Paleozoic-Triassic Belt contains the Hayfork and Rattlesnake Creek terranes, and possibly the May Creek and Briggs Creek terranes, although the affinities of the latter two are not clear. The Hayfork terrane consists of a mélange of oceanic rocks formed during the Late Permian to Triassic Periods, as well as Early Jurassic–age volcanic arc rocks. Most of the mélange rocks are sandstone, chert, limestone, and basalt. Some of the Permian fossils found in the mélange are from animals that lived in the ancient Tethys Sea, in the vicinity of today's Mediterranean. It appears that the Hayfork terrane was accreted during two episodes: the eastern Hayfork episode, about 180 million years ago, and the western Hayfork episode, about 177 million years ago, although in Oregon, this distinction is not clear.

The Rattlesnake Creek terrane contains abundant mélange of serpentinite, ultramafic to intermediate intrusive rock, volcanic rock, and a wide range of sedimentary rock, including limestone and chert. These rocks formed at various times between Permian and Middle Jurassic time. The Rattlesnake Creek terrane also hosts some plutons that intruded prior to accretion, at about 200 million years ago. The terrane was likely a volcanic arc during the latter part of its formation, and these plutons were the magma chambers that fed the volcanoes. The May Creek terrane consists principally of schist and may be a metamorphosed part of the Rattlesnake Creek terrane.

Western Klamath Belt

The Western Klamath Belt, sometimes called the Western Jurassic Belt, contains three principal terranes, each of which exists in Oregon: the Rogue Valley, Dry Butte, and Smith River terranes. A fourth terrane, called the Briggs Creek terrane, is enigmatic because it lies within the Western Klamath Belt but seems more related to the Paleozoic-Triassic Belt. When viewed together, these terranes reflect an island arc landscape that formed during the Middle to Late Jurassic Period, between about 164 and 157 million years ago. The island arc had a subduction zone on its west side and a smaller marine basin on the east.

The Rogue Valley terrane forms the cliffs along I-5 south of Canyonville and from there extends southwestward almost as far as Cave Junction. Consisting of mostly basaltic and andesitic lava flows, tuff, and volcanic-rich sedimentary rocks, the Rogue Valley terrane represents the island arc's lava flows. Many of these rocks are now greenstones because low-grade metamorphism caused innumerable tiny flecks of the green mineral chlorite to grow in the rock and impart a green color.

The roots of the island arc and parts of the volcanoes appear to belong to the Dry Butte terrane, which borders the Rogue Valley terrane to the west. Rocks of this terrane consist mostly of mafic to intermediate intrusive rocks (the roots), as well as scraps of metamorphosed volcanic rock that are between about 160 and 155 million years old. They are are usually referred to as the Chetco Complex.

The Smith River terrane, immediately southeast of the arc terranes, consists mostly of the Josephine Ophiolite, a fragment of oceanic lithosphere. It consists of a sequence of ultramafic to mafic intrusive and volcanic rock types, as well as overlying deep marine sedimentary rocks. Most researchers agree that the Josephine Ophiolite was the floor of a small ocean basin that formed between the arc and the continent as the result of extension. Called a back-arc basin, this setting likely resembled the modern-day Sea of Japan, located between the Japanese Islands (the arc) and the Asian continent. The back-arc basin of the Western Klamath Belt persisted until accretion of the three terranes at about 152 to 150 million years ago.

The Galice Formation was deposited over the top of each of the Western Klamath terranes. Initially, it accumulated in the deep water of the back-arc

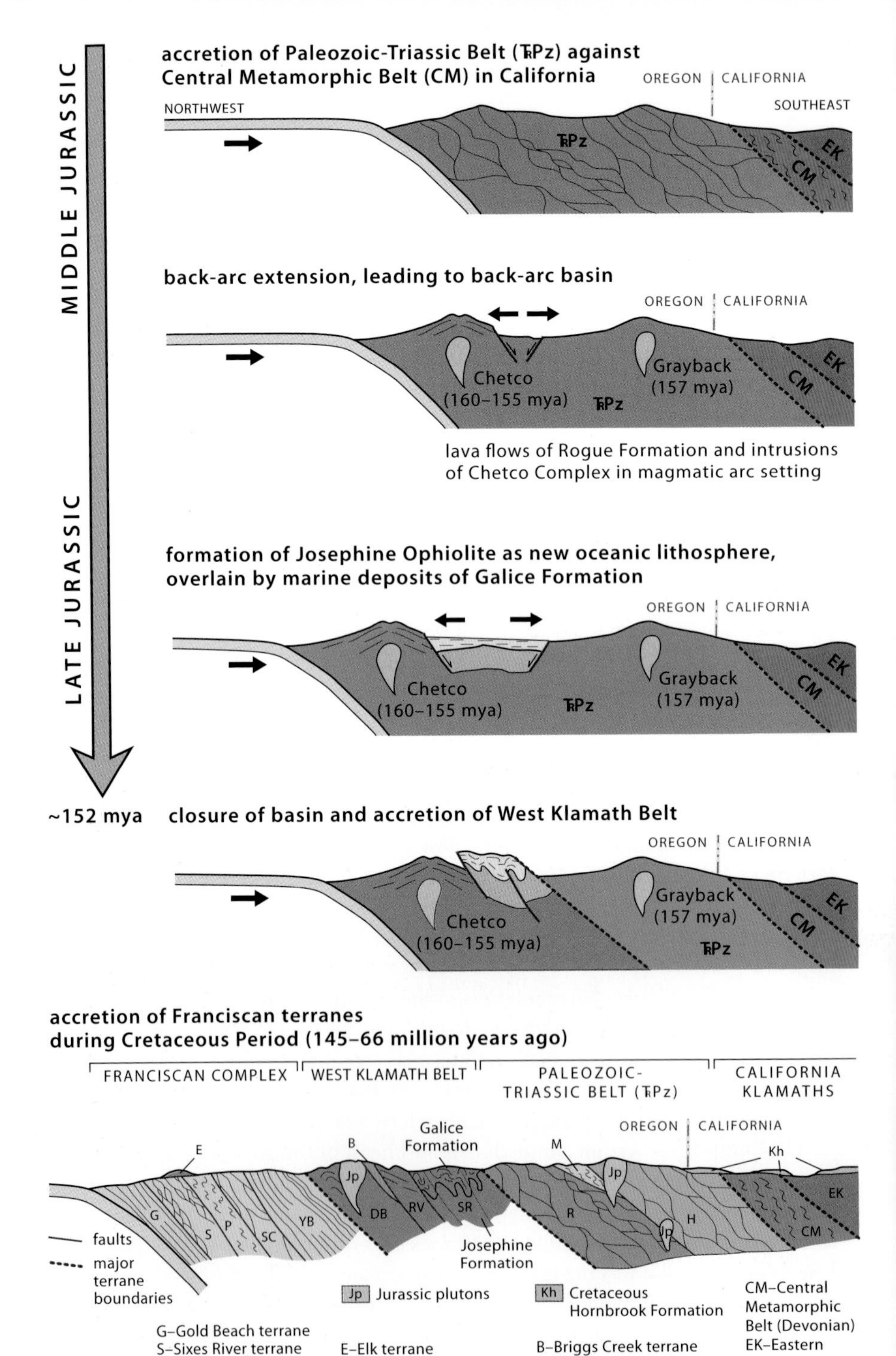

Sequence of accretionary events that formed the Western Klamath Belt during the Middle and Late Jurassic Period. Top cross section shows the starting point, after accretion of the Paleozoic-Triassic Belt; bottom cross section depicts the Klamaths after accretion of the Franciscan Complex.

basin above the Josephine Ophiolite, but it gradually covered the arc, too, as it began to subside. During accretion, the entire arc complex, including the Galice, became metamorphosed and strongly deformed.

Franciscan Complex

Perhaps the most accessible and possibly the most complicated terrane belt is the Franciscan Complex, sometimes referred to as the Franciscan-Dothan Belt. In Oregon, it consists of five different terranes. The Gold Beach, Sixes River, Pickett Peak, and Yolla Bolly terranes lie along the coast, and the Snow Camp terrane lies slightly to the east. The Elk terrane, which also lies along the coast, is part of the Western Klamath Belt, thrust westward over the Franciscan Complex.

While individual terranes differ in details and ages, the Franciscan Complex as a whole is the youngest of the Klamath belts, having formed from Late Jurassic through Middle Cretaceous time. Most interpretations describe the Oregon Franciscan as an assemblage of rock that came together in a subduction zone. Much of this rock originated in the accretionary wedge next to the continent, but other parts originated as island arc or oceanic crust fragments.

Folded radiolarian chert, a deep-sea deposit, of the Yolla Bolly terrane of the Franciscan Complex at Rainbow Rock near Brookings.

Mélange zone within the Otter Point sandstone of the Gold Beach terrane of the Franciscan Complex, exposed at Bandon Beach.

The Franciscan Complex consists mostly of marine sedimentary rock, minor volcanic flows, mélange, and scraps of ophiolite. Some of the mélange bodies contain blocks of blueschist, which indicates deep burial under relatively low-temperature conditions, a characteristic of subduction zones. The Pickett Peak terrane, which consists mostly of schist, was metamorphosed from what was originally marine shale. Studies of the Gold Beach terrane suggest that it originally formed far to the south, possibly at the latitude of southern California, and was transported northward by strike-slip faulting.

Klamaths after Accretion

A number of sedimentary rocks were deposited over the top of the terrane bedrock in Cretaceous time, both during and after accretion. On the east side of the Klamaths, the approximately 90-million-year-old Hornbrook Formation of mostly marine sandstone and shale was deposited in a marine basin. It overlies several of the Klamath terranes, indicating deposition after accretion. Among its fossils, the Hornbrook contains scattered ammonites and pelecypods, including a partly uncoiled variety of ammonite.

On the coast, a number of sedimentary units overlie the strongly deformed Franciscan rocks, indicating they postdate that deformation, which was probably related to the initial accretion of the Franciscan. However, the younger sedimentary rocks are also deformed and restricted to individual terranes. The Humbug Mountain Conglomerate, for example, only exists on the Elk terrane, whereas the approximately 80-million-year-old Cape Sebastian Sandstone and 75-million-year-old Hunters Cove Formation only exist on the Gold Beach terrane. These relations suggest the accretion process continued after deposition of those sedimentary rocks. The terranes may have been moved there by strike-slip faults parallel to the coast. Some researchers argue that the conglomerates were deposited on the accreted Franciscan rocks as far south as southern California.

The oldest sedimentary rocks that overlie multiple terranes of the Franciscan Complex are Eocene in age. They include sedimentary rocks of the Umpqua Group, which were deposited in marine settings during and after accretion of Siletzia, Oregon's most recently added terrane and now the basement of the Coast Range.

After all this accretion and mountain building, the Klamaths were eroded back to sea level and submerged. We know this erosion and submergence took place because we can find outcrops in the California portion of the Klamaths of various Late Miocene and Pliocene sedimentary rocks that were deposited, at least in part, in marine environments. The younger sedimentary rocks appear to have been deposited over a fairly level, low-lying surface eroded in the older rocks. In Oregon, scattered remnants of this erosion surface now reach elevations as high as 4,000 feet (1,200 m)

The modern Klamaths are young and have been uplifted and erosionally sculpted since deposition of those younger rocks. Like uplift of the Coast Range to the north, uplift of the Klamaths results from the ongoing plate convergence and subduction taking place immediately to the west.

GUIDES TO THE KLAMATH MOUNTAINS

INTERSTATE 5
ROSEBURG—CALIFORNIA BORDER

127 miles (204 km)

The interstate route between Roseburg and Ashland follows sections of the South Umpqua and Rogue Rivers and their tributaries and crosses a number of mountain passes in between. Because the structural grain of the Klamaths trends northeast-southwest, the interstate cuts across most of the main components of the province.

Although Roseburg is situated on bedrock of Siletzia, the interstate crosses a fault and into older accreted terranes of the Jurassic to Cretaceous Franciscan Complex a few miles to the south, near exit 119. Good exposures of these rocks

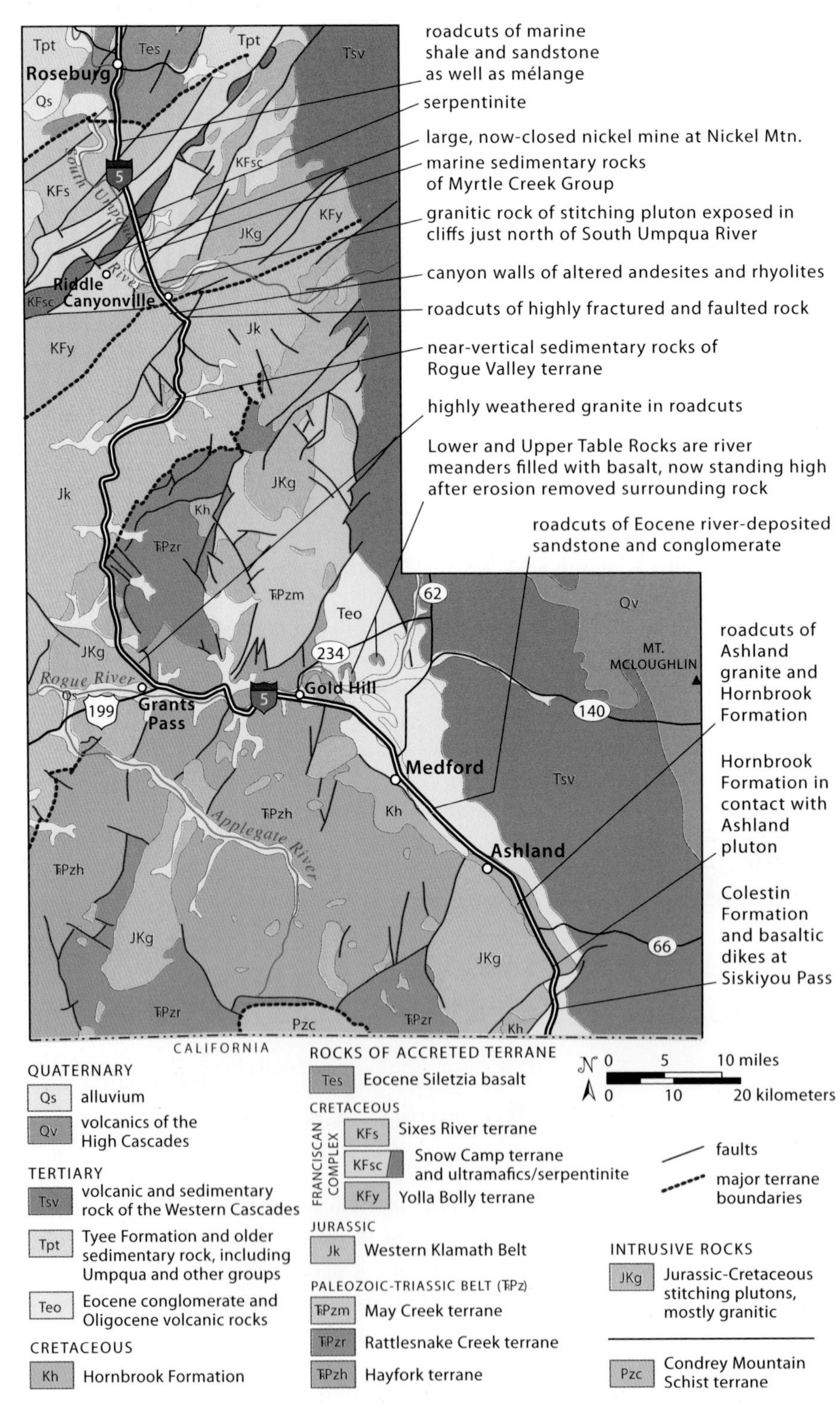

Geology along I-5 between Roseburg and the California border.

show up sporadically on both sides of the highway all the way to Canyonville and demonstrate the Franciscan's variability—a mishmash of material accumulated in and near a subduction zone. Between exit 119 and milepost 110, the Franciscan consists of marine sandstone and shale of the Sixes River terrane. In some places, this rock is well bedded and clearly sedimentary, but in others, it consists of fine-grained, sometimes weakly foliated material without recognizable bedding. These latter exposures, characteristic of this part of the Franciscan, are often described as mudstone mélange and exemplify the high degree of deformation that occurs in weak, mud-rich sediments near a subduction zone. The mudstone forms the matrix that surrounds a jumble of other rock types.

Just south of milepost 110, the road crosses a thrust fault and passes a long roadcut of ultramafic rock, a slice of Jurassic-age oceanic lithosphere. This ultramafic rock, which has less silica than basalt, has been metamorphosed, or highly altered, so that it is now serpentinite. At Nickel Mountain, near the town of Riddle, deeply weathered zones of the ultramafic rock contain rich deposits of nickel. Until the mine closed in the late 1990s, this area was the only producer of nickel in the United States. South of another fault, marine sedimentary rock of the Jurassic and Cretaceous Myrtle Creek Group appears. These rocks belong to the Franciscan's Snow Camp terrane and are generally much less deformed than the Franciscan farther north.

Outcrop of mudstone mélange of the Sixes River terrane between mileposts 117 and 116 on the east side of the interstate.

Canyonville lies in a valley cut by the South Umpqua River. The underlying bedrock is a Cretaceous granitic pluton that intruded after this part of the Franciscan was accreted. Roadcuts of the pluton lie on the east side of the highway between mileposts 103 and 102, just north of the river crossing.

Immediately south of Canyonville, I-5 enters the upper reaches of the South Umpqua River system and heads toward Canyon Creek Pass some 9 miles (14 km) to the south. The canyon walls consist of highly altered rock of the Jurassic-age Rogue Valley terrane of the Western Klamath Belt. Originating as lava flows of an island arc, these rocks consist mostly of andesitic and rhyolitic tuffs and breccias that have since been metamorphosed and highly altered. Pull-outs allow close inspection. Between mileposts 93 and 92, numerous quartz veins and faults, many with zones of scaly, striated rock, cut through the extremely deformed rock.

Volcanic-rich sedimentary rocks appear in roadcuts at Canyon Creek Pass and continue to just south of Stage Road Pass. These rocks, eroded from the lava flows, also belong to the Rogue Valley terrane. Immediately north of milepost 87, a beautiful exposure of these rocks shows them in a near-vertical orientation, dipping steeply to the north.

Grants Pass lies on the Rogue River at the eastern edge of the Grants Pass pluton, a granitic stitching pluton that intruded along the boundary of the Western Klamath and Paleozoic-Triassic Belts about 139 million years ago. Its presence indicates these two belts of the Klamaths had come together by then. The pluton is deeply eroded through here and offers little in the way of outcrop. Only some weathered exposures on the east side of the highway near milepost 60 and some nondescript roadcuts near the edge of town give any indication of the intrusion. You can see a large quarry exposure of the granite by taking exit 61 for Merlin and following Highland Avenue south for 1.4 miles (2.3 km) on the east side of I-5. Just southeast of Grants Pass, the prominent roadcut on the northeast side of the road at milepost 54 is altered volcanic rock of the Paleozoic-Triassic Belt.

Between Grants Pass and Gold Hill, the interstate follows the floodplain of the Rogue River, which forms a nearly flat, meandering valley across the grain of the Klamaths. Only a few roadcuts show up along this stretch, although a couple of quarries on the north side between mileposts 45 and 44 and mileposts 43 and 42 expose volcanic rock of the Paleozoic-Triassic Belt. It is not immediately clear why the river has such a low gradient through this stretch, as it flows at a much steeper gradient to the east in the Western Cascades and to the west in the Klamaths. Several other rivers along this route decrease their gradients as well, only to increase them again after cutting partway into the Klamaths or Coast Range. Most likely, ongoing uplift of the Coast Range to the west has decreased the gradient on its east and increased the gradient on its west.

Gold Hill lies near the eastern edge of one of Oregon's largest gold mining areas. Gold mining began in Oregon in the early 1850s and peaked several times until the end of World War II, when it practically ended. Most of the gold in the Klamaths originated from hot fluids emanating from the stitching plutons as they cooled. Lode deposits, in which the gold is in its original host bedrock, and placer deposits, in which the gold has been eroded and redeposited, were mined extensively in the Klamaths. Although most of the gold production came from placer deposits, some spectacular finds came from lodes. Gold Hill, for example, was named for a deposit that yielded some 30,000 ounces (850 kg) of gold from a cut only 15 feet (4.6 m) deep!

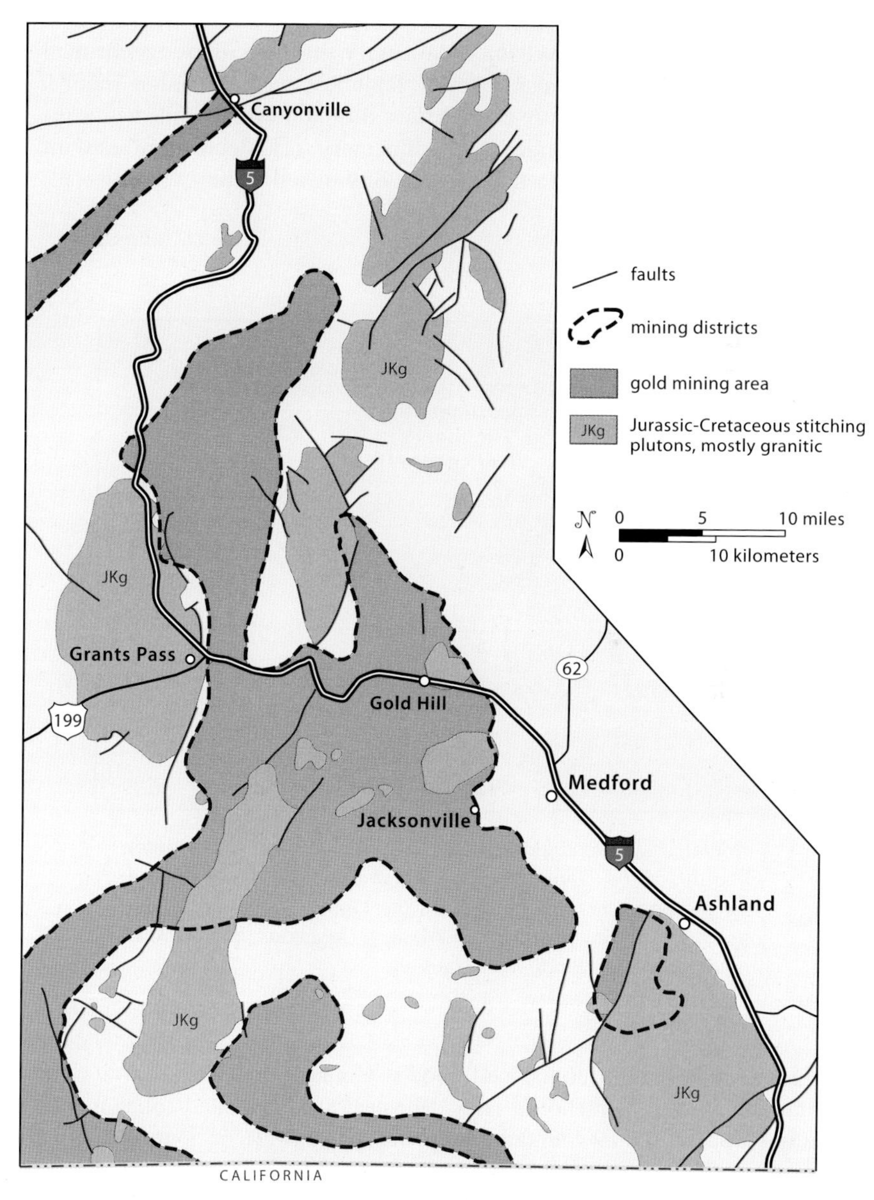

Stitching plutons and gold mining areas along I-5 between Canyonville and the California border. –Mining area boundaries from Ferns and Huber, 1984

At milepost 38, you can see the pyramidal peak of Mt. McLoughlin to the east and Table Rocks to the north. Mt. McLoughlin is the southernmost stratovolcano of the High Cascades in Oregon. It consists of basaltic andesite that last erupted some 30,000 to 20,000 years ago. Named for John McLoughlin of the Hudson's Bay Company, the peak reaches an elevation of 9,495 feet (2,894 m). Evidence for ice age glaciation appears restricted to a cirque near its summit and a moraine near its base on the north side.

Table Rocks are flat mesas capped by a 7-million-year-old andesite flow of the Western Cascades, overlying sedimentary rock of Eocene age. In map view, you can see that Upper and Lower Table Rocks are horseshoe-shaped, like meanders of a stream. Sure enough, lava flowed down and filled a meandering canyon cut into the underlying sedimentary rock. Being more resistant to erosion, the lava remained while the surrounding sedimentary canyon walls eroded away.

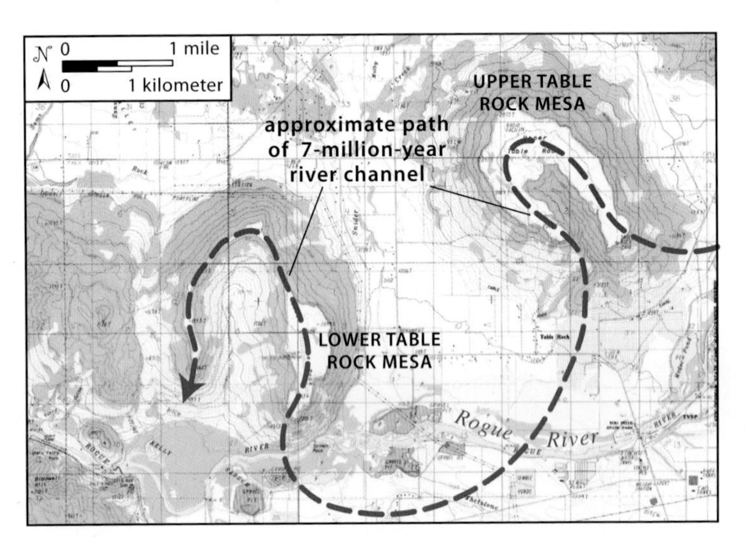

Annotated US Geological Survey topographic map (Sams Valley, Oregon) of the Upper and Lower Table Rocks. The 7-million-year-old andesite, which occupies the tops of these mesas and originated from a volcano in the Western Cascades, flowed down an ancient, meandering river channel.

From about Medford to Ashland, good roadcuts of the Eocene Payne Cliffs Formation reveal a sequence of river-deposited sandstone and conglomerate that was eroded from the Klamath Mountains. In some places, however, marine sandstone of the underlying Cretaceous Hornbrook Formation is exposed. Near Jacksonville, just west of Medford, one of Oregon's first placer gold deposits was discovered in 1851.

Ashland occupies a beautiful valley eroded into the Hornbrook Formation, which was deposited on top of accreted terranes of the Klamaths as well as granite of the 161-million-year-old Ashland pluton, one of the stitching plutons. It therefore provides the oldest unambiguous tie between the accreted terranes and North America. To the west, granitic rock of the Ashland pluton rises abruptly over the valley, while to the east a more gradual ascent leads up through the Eocene rock and into the Western Cascades.

Between Ashland and Siskiyou Pass, the interstate passes scattered outcrops of the Hornbrook Formation and Ashland pluton. The Hornbrook appears as east-tilted sandstone, while the granite is more evenly textured, although it contains dikes and zones of hydrothermal alteration in some places. Between mileposts 10 and 9, exposures of biotite schist and gneiss reveal some of the original terrane rock into which the pluton intruded. Just north of the off-ramp to Mt. Ashland, the Hornbrook Formation lies in depositional contact on top the granite, which has weathered into grus, small fragments of rock that look like the original granite but readily disintegrate into smaller fragments.

A side trip, 4 miles (6.4 km) up the road toward Mt. Ashland Ski Area at exit 6, allows inspection of the Ashland pluton in relative calm, away from the busy interstate. Outcrops of Oligocene-age, volcanic-rich sedimentary rock appear

Granitic rock and zone of inclusions of the 161-million-year-old Ashland pluton, exposed along the road to Mt. Ashland Ski Area. The inclusions came from rock of the Paleozoic-Triassic Belt into which the pluton intruded.

along the road for the first 2 miles (3.2 km), but at milepost 4, you can see elongated inclusions of the Paleozoic-Triassic Belt within the granite.

Siskiyou Pass offers an enormous roadcut with gigantic pull-outs. You can see the same rock on both sides of the highway, so do not try to cross this intensely busy interstate. The roadcuts show a cross section of volcanic-rich sandstone, conglomerate, and ash-flow tuffs of the Oligocene-age Colestin Formation. Some of the conglomerate appears to have originated as volcanic mudflows, or lahars, and some beds contain abundant coal-like wood fragments. You can see younger basaltic dikes on both sides of the highway toward the south end of the roadcut. The Colestin Formation is considered to mark early stages of the Western Cascades. Its coarse-grained nature suggests its source area was nearby.

I-5 descends for the next 5 miles (8 km) to the California border through more Colestin Formation. Dark-colored cliffs to the south consist of 30-million-year-old basalt, which appears to have flowed down a canyon that has since eroded away. Turbidite deposits of the Hornbrook Formation are exposed in a roadcut on the west side of the highway right at the state line.

The double roadcut at Siskiyou Pass displays lahars, stream deposits, and tuffs of the Colestin Formation of Oligocene age. Note the large basaltic dike in the roadcut near its south side, slanting down to the left (center left of photo). Just left of the dike is a well-preserved channel of conglomerate that cuts down into a green-colored unit. View of the west roadcut.

US 101
Bandon—California Border
88 miles (142 km)

Between Bandon and the California border, US 101 traverses the Oregon Coast's oldest rocks—those accreted to the west coast of North America at various times toward the end of Mesozoic time. Most of the rocks along this route belong to the Gold Beach and Yolla Bolly terranes of the Franciscan Complex, although one stretch of the route crosses part of the Western Klamath Belt. The most prevalent and dramatic rock exposures, however, are unquestionably of the Gold Beach terrane, which consists of a basement of Jurassic-age Otter Point Formation, overlain by a variety of Cretaceous sedimentary rocks. The Otter Point Formation contains a great deal of homogeneous-looking sandstone as well as some conglomerate, mudstone, zones of interbedded sandstone and shale, and scattered blocks of chert, greenstone, and blueschist. The formation is best described as a mélange because the various rock types appear to be thrown together in an almost random fashion. The erosion-resistant sandstone forms many of the sea stacks and headlands along the southern Oregon coast. The chert, greenstone, and blueschist are also resistant but consist of smaller bodies and tend to form scattered blocks along the beaches. The blueschist is especially interesting because it is a metamorphic rock that formed in the comparatively low temperatures and high pressures characteristic of subduction zones. Many of the finer-grained zones of the Otter Point Formation are highly deformed by shear zones that run parallel to bedding, whereas the coarser-grained rocks are broken by myriad fractures that run in all directions.

Old Town Bandon lies at sea level along the Coquille River, about 100 feet (30 m) below the Whisky Run terrace, the lowest marine terrace along the coast. It formed at sea level about 80,000 years ago and was uplifted to its present position. The south jetty, where the river empties into the ocean, is constructed of huge boulders of blueschist, quarried from the Otter Point Formation. Walking south along the beach from the jetty, you can see numerous sea stacks, and at low to moderate tides, you can walk around Coquille Point to Bandon Beach, also accessed from crossroads off US 101 on top of the terrace just south of town.

Bandon Beach hosts more than a dozen picturesque sea stacks of coarse-grained sandstone of the Otter Point Formation, many of which are easily reached during low tides. It also features large blocks of greenstone and blueschist and even some smaller ones of red chert. Finer-grained parts of the Otter Point Formation have mostly eroded away but are still present in the hillside beneath the terrace. At Coquille Point, you can find striking examples of shear zones within shales and mudstones. A small sea arch in one of the sea stacks can also be seen there.

Between Bandon and Port Orford, US 101 follows an uplifted marine terrace. Acres of cranberry bogs are grown on the terrace because its wet, acidic, sandy

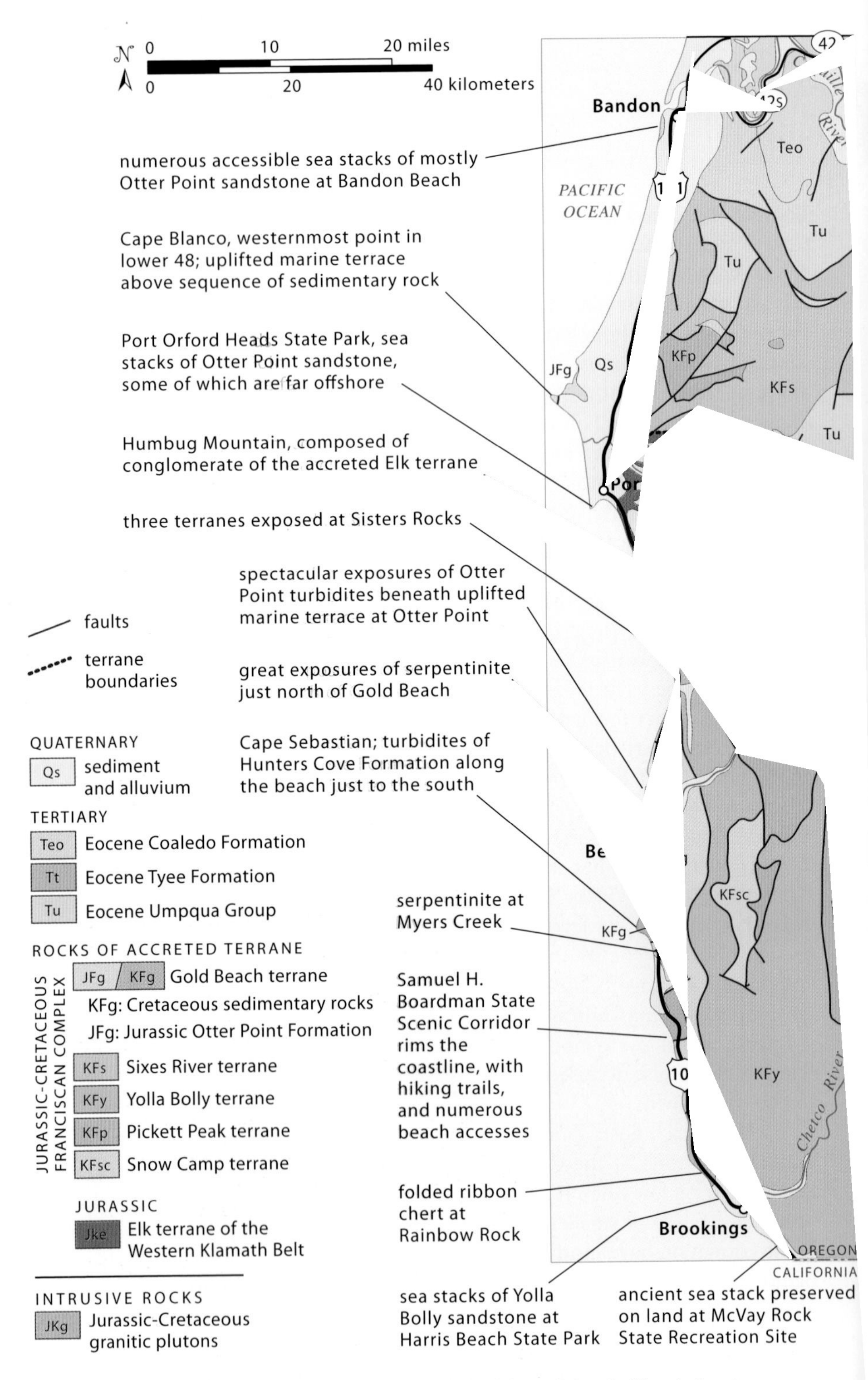

Geology along US 101 between Bandon and the California border.

Sea stacks on Bandon Beach consist mostly of sandstone of the Jurassic-age Otter Point Formation. Notice how the flat tops of the large sea stacks in the background are at the same level as the mainland on the right. This surface is the Whisky Run terrace.

soil provides the right conditions for this kind of agriculture. Oregon's cranberries are of an especially dark color because the moderate climate allows for a longer growing season. Occasional small roadcuts show deposits of sand over the terrace. Between mileposts 296 and 297, a side road accesses Cape Blanco, the westernmost point in the conterminous US.

Four miles (6.4 km) south of Cape Blanco is Port Orford Heads State Park, which was a Coast Guard lifeboat station from 1934 to 1970. It offers hiking trails and views of some sea stacks of Otter Point Formation at the coastline. Other sea stacks of Otter Point Formation, some 3 miles (4.8 km) off the coast, mark an especially shallow part of the seafloor just east of where it slopes into a submarine canyon that cuts through the continental shelf.

Just south of Port Orford, the highway crosses over highly oxidized, reddish marine sands of a marine terrace. Beneath these deposits lies a fault that separates the Gold Beach terrane of the Franciscan Complex from the Elk terrane of the Western Klamath Belt. For the next 10 miles (16 km), to milepost 314, the road passes outcrops of sedimentary rocks of the Elk terrane, which tend to be poorly bedded and not easily recognizable. They include sandstone of the Rocky Point Formation and the Humbug Mountain Conglomerate, which makes up Humbug Mountain.

Sisters Rocks, just north of milepost 315, marks an unusually complex area, which has been mapped in various ways by different people. It's most likely, however, that the area consists of three terranes separated by thrust faults. The sea stacks consist of Otter Point Formation of the Gold Beach terrane. Just

Cape Blanco

Cape Blanco, the westernmost point of the conterminous United States, protrudes 0.5 mile (0.8 km) into the sea as a windswept, nearly treeless plateau, some 200 feet (60 m) above the waves. It ends abruptly as a cliff, flanked by long empty beaches that stretch in either direction. The cape hosts Oregon's oldest and highest continuously operating lighthouse, built in 1870 and now open to the public during spring and summer months.

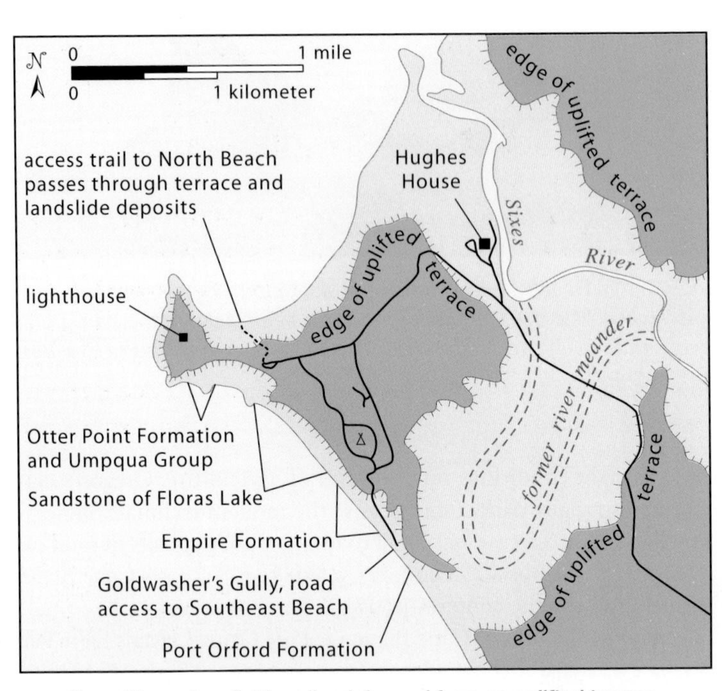

Cape Blanco is a flat headland, formed from an uplifted terrace.

Perhaps the most striking thing about Cape Blanco is its apparent flatness, formed from an uplifted marine terrace, called the Cape Blanco terrace. Coastal sand and gravel deposits, which compose the terrace, are well exposed on the beach trail north of the lighthouse parking lot. These deposits are estimated to be 80,000 years old, so the Cape Blanco terrace is probably equivalent to the Whisky Run terrace, which is the lowest terrace at Cape Arago, north of Bandon.

Just inland of the Cape Blanco terrace and at a slightly higher elevation lies the older and much more extensive Pioneer terrace, estimated to be about 105,000 years old. It is also the second-youngest terrace at Cape Arago. Because of its great extent, we can see that it is gently folded into an anticline. At Cape Blanco, the terrace slopes ever so slightly toward the south, joining sea level some 5 miles (8 km) away; to the north, the terrace also reaches sea level in about 5 miles (8 km). Deformation at the subduction zone, less than 50 miles (80 km) to the west, has formed this young fold.

Beneath the terrace lies bedrock, much of which is beautifully exposed in the sea cliffs. You can see these cliffs from the beach along the southwest side of the cape, reached by a road at the far end of the campground. Most of the rocks dip toward the southeast, so they get younger in that direction. The oldest rock, the Jurassic-age Otter Point Formation, consists of dark-colored mudstone and sandstone with scattered blocks of greenstone and even blueschist. The Otter Point is a mélange that formed elsewhere but was accreted to North America sometime at the end of the Mesozoic Era. Rocks of the Eocene-age Umpqua Group, part of a fault slice here, consist of gray shale that was deposited in deep water during and following the accretion. The Miocene-age Sandstone of Floras Lake, named for outcrops near Floras Lake north of Cape Blanco, consists mostly of sandstone and conglomerate, whereas the Miocene-age Empire Formation is largely tan-colored sandstone and siltstone. Needle Rock, the prominent sea stack below the parking lot on the south side of the cape consists of the lower part of the Sandstone of Floras Lake. The Pleistocene-age Port Orford Formation consists mostly of marine-deposited, darker colored sandstone and thin beds of conglomerate that overlies a 30-foot-thick (9 m) base of river-deposited conglomerate and sandstone.

Aerial view of Cape Blanco toward the north.

Numerous landslides are eroding the narrow peninsula between the mainland and the lighthouse area. Much of the trail to the beach on the north side winds down through one of these slides. Immediately south of the parking lot, you can see small faults forming in the gravel deposits as they pull away from the terrace edge at the top of one of these slides. Numerous other slides affect the southeast edge of the cape.

North of Cape Blanco, the Sixes River meanders over its floodplain. An example of how its channel changes position through time is the large cutoff meander that occupies the floodplain about 2 miles (3.2 km) from the river's mouth. The road to the cape crosses this area where it descends almost to sea level before rising up to the terrace, about 1 mile (1.6 km) east of the parking lot.

---

uphill from the sea stacks, a fault separates them from similar-looking rocks of the Dothan Formation of the Yolla Bolly terrane; above the road, another fault separates the Dothan from the Colebrook Schist of the Pickett Peak terrane!

Scattered exposures of Dothan Formation occur for the next 4 miles (6.4 km) to Ophir, where the road crosses another fault and back into the Otter Point Formation, although few rocks are actually exposed along the road until just north of the town of Gold Beach, near milepost 328. Beautiful dunes lie along the coast near Ophir Wayside State Park.

Otter Point State Recreation Site offers exposures of vertically bedded Otter Point Formation and overlying terrace deposits. Otter Point is the type locality for the Otter Point Formation, the first place it was described in detail and named. To reach this state park, turn west on Old Coast Road some 300 feet (90 m) south of milepost 324 (it intersects US 101 in several places, so this milepost is important), turn left after another 300 feet (90 m), then turn right at the sign after a quarter mile or so. You can follow the trail out to the peninsula. Exercise extreme caution, because the terrace sands are undercut in some places and could give way if someone walks too close to the edge.

The town of Gold Beach lies at the mouth of the Rogue River, which rises in the High Cascades near Crater Lake. The Rogue drains more than 5,000 square miles (12,950 km^2) over a distance of some 215 miles (346 km), the largest area and longest distance of any Oregon river that flows into the Pacific save for the Columbia. Still, its estuary is smaller than many of those farther north in the Coast Range, because it drains much more resistant rock of the Klamaths. Because more resistant rock tends to maintain a steeper gradient, the distance inland to which tides can influence the river is greatly restricted.

At Gold Beach, the roadside geologist can get a great look at serpentinite, a greenish scaly rock. It forms through low-temperature metamorphism of

Thin-bedded shale and sandstone of the Otter Point Formation at Otter Point occupies the foreground, while thick-bedded sandstone, also of the Otter Point Formation, forms the sea stack in the background.

silica-poor, iron- and magnesium-rich rocks, such as those that make up the oceanic lithosphere. Exposures of serpentinite often mark fault zones, because its scaly nature allows it to be somewhat mobile under high pressures. It can practically squirt its way into available openings. Individual pieces of the rock are often quite beautiful and sometimes mistaken for jade, but jade is made of altogether different minerals. Exposures of serpentinite as well as sandstone-mudstone mélange of the Otter Point Formation show up along the roadside between milepost 327 and the Rogue River Bridge. Much larger exposures of serpentinite lie only a quarter mile up the road along the Rogue River's North Bank Road (Co. Hwy 545), which splits off US 101 just north of the bridge.

South of Gold Beach, US 101 continues past a tall outcrop of Otter Point Formation of the Gold Beach terrane near milepost 331 and a series of coastal dunes. As it heads uphill, the road encounters the overlying 80-million-year-old Cape Sebastian Sandstone, which is best exposed on the east side of the road

Turbidite deposits of the Hunters Cove Formation as seen on the beach between mileposts 336 and 337. Coarser grains that settled first form the beds of orangish sandstone, and finer sediments form the intervening dark shaly layers. Many of the sandstone beds are graded, with coarser grain-size sand at their bottoms and finer sand at their tops.

just south of milepost 334. The sandstone also makes up most of Cape Sebastian, which offers wonderful views of the coastline. A road to the overlooks and trails of Cape Sebastian intersects US 101 just north of milepost 335. Excellent exposures of turbidites of the 75-million-year-old Hunters Cove Formation, which overlies the Cape Sebastian Sandstone, exist in sea cliffs about 0.5 mile (0.8 km) north along the beach from pull-outs between mileposts 336 and 337. Some of these rocks are highly deformed. Turbidites, layered rocks with graded beds, are deposited on the seafloor by sediment-laden currents.

A third rock unit, the Houstenaden Creek Formation, lies beneath the Cape Sebastian Sandstone. It doesn't crop out along the road but is an important part of the story, worked out by Joanne Bourgeois and Robert Dott in 1985. Taken together, the three rock units—the Houstenaden Creek, Cape Sebastian, and Hunters Cove—consist of thickly bedded and crossbedded sandstone, some conglomerate beds, and turbidite deposits, which were deposited in a marine setting that was actively being faulted. Their studies of the conglomerate clasts of the Houstenaden Creek Formation indicate that they probably originated near the latitude of southern California. Strike-slip faults, older than but resembling the San Andreas Fault in California, must have transported the terrane to Oregon.

Just north of the bridge across Myers Creek, at milepost 337, US 101 passes a fault zone, which appears along the road as some beautiful green serpentinite exposures. The hillside east of the highway has a hummocky appearance, characteristic of landsliding, likely promoted by the presence of the serpentinite. A faulted sliver of the Otter Point Formation makes up the sea stacks at Myers Creek Beach between mileposts 337 and 338. Just south of the turnoff to the community of Pistol River, the road passes the estuarine part the Pistol River. A prominent spit separates the river from the ocean.

Between Pistol River and Brookings, US 101 passes through Samuel H. Boardman State Scenic Corridor, which offers numerous views, points of access to the coast, and hiking trails, but little in the way of rock exposures along the road. In the park, Indian Sands requires a short hike but offers interesting shapes eroded into Quaternary sand deposits, which overlie exposures of conglomeratic Otter Point Formation. A sea arch is visible along the coast. Indian Sands is an important archeological site, yielding artifacts that are upward of 12,000 years old. Please remember that collecting artifacts is prohibited. Whaleshead Beach lies at the fault between the Gold Beach terrane to the north and Yolla Bolly terrane to the south. Lone Ranch Beach offers an exposure of serpentinite and sea stacks of the Yolla Bolly terrane.

Rainbow Rock, a headland just north of milepost 354, consists of folded green, white, and red ribbon chert of the Yolla Bolly terrane. Ribbon chert is a deep-sea deposit made of tiny organisms called radiolaria that produce small intricate skeletons of silica. Radiolaria are sometimes preserved as thin layers of chert, referred to informally as ribbons. A small sea cave cuts through the west side of the headland. The beach can be reached via a rough trail at the south end of the long pull-out.

Harris Beach State Park north of Brookings offers numerous accessible sandstone sea stacks of the Yolla Bolly terrane. Some narrow mélange zones also lie within the bedrock.

Brookings sits on top of a marine terrace on the west side of the Chetco River, Oregon's southernmost estuary and small port. The port suffered more than $7 million in damages during the March 11, 2011, tsunami caused by the magnitude 9 Tohoku earthquake in Japan. Steeply dipping sandstone beds of the Yolla Bolly terrane lie at the southeast edge of Macklyn Cove. Two and half miles (4 km) south of the Chetco River, an exposure of sandstone of the Yolla Bolly terrane lies on the east side of US 101.

South of Brookings, US 101 follows the top of a marine terrace. Rocks along the coast 2 miles (3.2 km) to the southwest are steeply dipping and overturned Dothan Formation. McVay Rock, at the entrance to McVay Rock State Recreation Site, about halfway between Brookings and the California line, is an ancient sea stack that sticks up through the terrace.

Rainbow Rock and nearby sea stacks, as seen from the large pull-out on US 101. Note the tight folds in the rock, which consists of ribbon chert of the Yolla Bolly terrane.

US 199
Grants Pass—California Border
43 miles (69 km)

This lovely road begins near the center of a Jurassic-age stitching pluton in Grants Pass, and then follows the border between two belts of accreted terranes of the Klamath Mountains. To the east lies the Paleozoic-Triassic Belt, and to the west, including the highway, lies the Western Klamath Belt.

After crossing to the south side of the Rogue River, US 199 follows a low terrace on the edge of the floodplain. Beneath the alluvial deposits, as well as in the hills to the south, lies Jurassic-age granitic rock of the 139-million-year-old Grants Pass pluton. Because this stitching pluton intrudes the Paleozoic-Triassic Belt and the Western Klamath Belt, we know the two terranes had joined by 139 million years ago. Deeply weathered granite shows up in the hillsides on the south side of the road between mileposts 4 and 5.

Just west of milepost 7, the road crosses the Applegate River and enters a more rugged landscape made of slates of the Galice Formation, part of the Western Klamath Belt. The shale and siltstone of the Galice Formation were deposited in a marine basin over the tops of the other terranes of the Western Klamath Belt during Jurassic time. The Galice Formation and the Western Klamath terranes became metamorphosed between 152 and 150 million years ago when the belt was accreted to the western edge of the continent. At the time, this western edge consisted of the Paleozoic-Triassic Belt, which makes up the mountains to the south of US 199. The Galice Formation crops out along the highway for the duration of this route but is best exposed between the Applegate River and about milepost 23.

Probably the easiest and most instructive pull-out to inspect the rocks lies on the north side of the highway along the wide shoulder immediately west of the Applegate River. The rocks here are best described as phyllites instead of slates, because they take on a distinctive sheen from the growth of white mica, or muscovite, during metamorphism. Because the rocks lose this sheen with distance from the Grants Pass pluton, it is likely that they grew as a result of a later, local metamorphism driven by heat from the Grants Pass pluton when it intruded and cooled. Look also for pinhead-sized clots of the mineral cordierite within the rocks, another indicator of this local metamorphism. In addition, notice that the foliation here is nearly vertical. Elsewhere along the route, the foliation lies at different orientations, hinting at the complexity of the deformation that metamorphosed these rocks. Numerous side roads, such as Waters Creek Road just north of milepost 13, give the chance to inspect the Galice off the busy highway.

As you approach Selma from the north, you can see Eight Dollar Mountain, the near-conical peak rising more than 2,500 feet (760 m) above the valley to the southwest. The peak consists mostly of peridotite, the low-silica igneous rock that forms most of Earth's mantle, including the deep oceanic lithosphere. Because peridotite also makes up most of the Smith River terrane, another

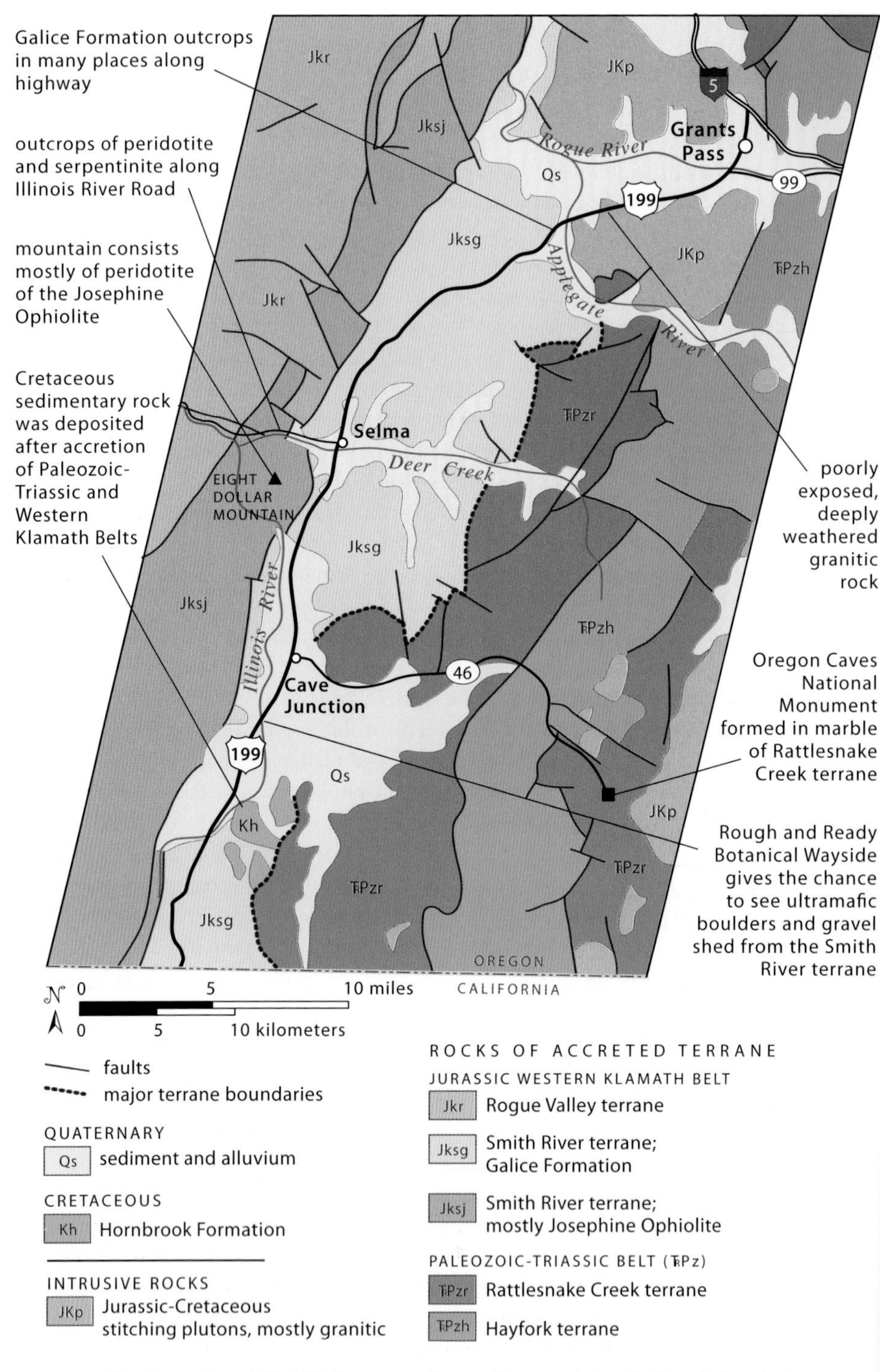

Geology along US 199 between Grants Pass and the California border.

Just west of the Applegate River, the layering in the Galice Formation, called foliation, dips steeply to the southwest (left). Farther south on US 199, the orientation of this foliation changes.

part of the Western Klamath Belt, the terrane probably originated as oceanic lithosphere. The Illinois River Road, which heads west out of Selma, offers a worthwhile side trip to inspect the peridotite and serpentinite, which crop out along the roadway only 2 miles (3.2 km) to the west, just past a large pull-out on the left. Better exposures, as well as some small pull-outs, exist another mile up the road.

The peridotite of the Smith River terrane has long been known to contain low-grade nickel deposits. The deposits are concentrated in laterites, extremely deep-weathered soils in which most other elements have been leached away. These laterites now reside high on the mountains but likely formed on the Klamath erosion surface some 5 million years ago before the Klamaths were uplifted to their present elevations.

Cave Junction occupies the floodplain where two forks of the Illinois River come together. OR 46 to Oregon Caves National Monument leads eastward through exposures of the Hayfork and Rattlesnake Creek terranes, both belonging to the Paleozoic-Triassic Belt.

Five miles (8 km) south of Cave Junction, US 199 passes Rough and Ready Botanical Wayside, which offers access to a large alluvial fan deposit shed from the mountains to the west. The fan spreads out to the north and south along

Oregon Caves National Monument

Resting high in the Siskiyou Mountains of the Klamath Province, Oregon Caves National Monument offers an unprecedented amount of geology in a relatively small area. The cave is formed in marble of the Rattlesnake Creek terrane of the Paleozoic-Triassic Belt. It's the only easily accessed limestone or marble cave system in Oregon, and it serves up a satisfying fare of cave formations, called speleothems. Its lower reaches contain an underground stream, which issues from the cave entrance as a gushing spring. The National Park Service runs daily tours from April through November. A historic hotel, built by the Civilian Conservation Corps in 1937, is purported to be haunted.

Marble is metamorphosed limestone. Caves form in limestone and marble because both rock types consist of the mineral calcite, which slowly dissolves in slightly acidic groundwater. This dissolution creates the passageways and large rooms. Several long narrow passageways in the caves follow fractures or faults in the rock, places where increased water flow accelerated the dissolution.

The process is more complex than simple dissolution, however, because precipitation forms the speleothems. Precipitation, or the formation of minerals from water, will occur if the water becomes less acidic, which happens when it loses carbon dioxide to the air. For this reason, most caves form in two stages: the rooms and passageways form by large-scale dissolution when the bedrock is below the water table, and the speleothems form later, by precipitation after the water table drops. Falling water tables affected many parts of the western United States as the Pleistocene Epoch ended and the climate dried. Oregon Caves contain a variety of speleothems, including stalactites, which grow downward from the ceiling, and stalagmites, which grow upward from the ground. They also include columns, where a stalactite and stalagmite join, and countless examples of flowstone, sheets of newly precipitated calcite covering the bedrock.

The marble bedrock that forms the caves originated as limestone, probably deposited as a reef behind a volcanic arc during Triassic time. The actual extent of the marble is limited to the general area around the caves. To the north, east, and west, the marble runs into sandstone, shale, or lava flows of approximately the same age, and to the south, it is intruded by a granitic pluton. The marble can best be described as an isolated block within a mélange of other rock types, all of which are part of the Rattlesnake Creek terrane. The mélange likely formed as a submarine landslide before the sediment hardened into rock, or by faulting at or near a subduction zone during accretion.

Inside a cave at Oregon Caves National Monument, a stalactite growing down from the ceiling has merged with a stalagmite growing up from the floor, producing a column.

The best place to see the marble is along the park road between the parking lot and the cave entrance. There, the marble contains thin layers of chert, which are strongly folded with the limestone. The folding is an artifact of the terrane accretion.

The granitic body is the 160-million-year-old Grayback pluton, one of the stitching plutons that intruded in Late Jurassic time after the terrane had been accreted. When the pluton intruded, the heat from its crystallizing magma baked the original limestone into marble. Mafic dikes and sills from this pluton are visible within some of the cave rooms. Because they are insoluble, they stand out from the marble in relief like narrow, jagged draperies. Some of the dikes are offset along small faults. The main body of the pluton, being insoluble, forms an abrupt boundary to the cave system.

Many of the cave rooms also contain Quaternary-age sediments, which have yielded a host of fossils, including lungless salamanders, mice, shrews, and beavers. The fossils also include a 50,000-year-old grizzly bear and a 38,600-year-old North American jaguar. The grizzly bear is one of the oldest known in North America, and the jaguar is both the northernmost known example and the most complete skeleton west of the Mississippi River.

the mountain front and consists of three different surfaces, each formed at a different time. At the wayside, the fan surface is the youngest, probably deposited in just the past 2,000 years. It contains active braided channels, poor soil development, and scattered Jeffrey pine. By contrast, older surfaces, which sit at slightly higher elevations to the south, exhibit more soil development and host Douglas fir. A brief inspection of the gravels on the fan show that they exhibit a wide variety of brownish coloration but consist almost entirely of peridotite, which makes sense given that they were derived from the Smith River terrane. The wayside exists to highlight and protect a variety of rare plant species that grow only in the iron- and magnesium-rich soils derived from these ultramafic rocks.

For a wider selection of gravel and boulders, but without the vegetation, stop at the active channel of Rough and Ready Creek. It is easily accessible from where US 199 crosses it, only 0.1 mile (160 m) south of the wayside.

The view southward along US 199 toward California shows a clear demarcation of the accreted terrane belts. The rocks of the Paleozoic-Triassic Belt to the east form a higher, more rugged range of mountains than those of the Western Klamath Belt to the west. Perhaps even more striking is the difference in vegetation. The Western Klamath Belt, which is underlain mostly by peridotite in this region, forms poor soils and hosts much sparser vegetation.

Rough and Ready Creek, near where US 199 crosses it, flows over brown boulders of peridotite, a typical color derived from weathering of ultramafic rock. View is northwestward. Fog obscures the mountains of the Western Klamath Belt, only about 2 miles (3.2 km) away.

LAVA PLATEAUS

The Lava Plateaus hold some of the Oregon's loneliest landscapes, with vast stretches of nearly flat or rolling, sagebrush-covered land, interrupted by the occasional mountain range or canyon. Cut deep into the rock, the canyons expose a seemingly endless succession of lava flows, all erupted since Early Miocene time. The lava covers up the underlying basement of accreted terranes.

The Lava Plateaus are divided into three subprovinces based on location and the age and types of lava flows present. The Columbia Plateau, which lies north of the Blue Mountains and reaches north into Washington, contains the Columbia River Basalt Group. The High Lava Plains, immediately south of the Blue Mountains, and the Owyhee Upland, in the far southeastern corner of the state, consist predominantly of younger basalt as well as rhyolitic rocks. The Owyhee Upland is distinguished from the High Lava Plains, because its rhyolitic rocks are several million years older.

Columbia Plateau and the Columbia River Basalt Group

The Columbia Plateau, which straddles the Oregon-Washington border, is dominated primarily by lava flows of the Columbia River Basalt Group. It also contains some interbedded lakebed deposits and, in northernmost Oregon, Pliocene sedimentary rock and a relatively thin covering of windblown silt and fine sand called loess. The lakebed deposits accumulated during quiet times between major lava eruptions, when water filled up drainages that had been blocked by the lava. The loess was ultimately derived from abrasion of rock high in the mountains by glaciers during the Pleistocene Epoch. As the glaciers melted, the finely ground up rock, or silt, was easily blown from the barren, unvegetated land and redeposited elsewhere.

Most of the Columbia River Basalt Group originated from fissure eruptions in eastern Oregon and Washington between 17 and 6 million years ago, although some 96 percent of its volume had erupted by 14.5 million years ago. The lava flows are called flood basalts because they literally flooded the landscape. Including the Steens Basalt, which is now considered to be a part of the group, the volume of the Columbia River Basalt Group exceeds 52,800 cubic miles (220,000 km^3), spread over an area of more than 77,220 square miles (200,000 km^2). The lava flows extend well beyond the Columbia Plateau, covering much of the Blue Mountains, as well as parts of the High Lava Plains, Owyhee Upland, and Basin and Range to the south.

At first glance, individual flows of the Columbia River Basalt Group look identical. Geologists who study these rocks distinguish them based on a combination of things, including the types of minerals visible to the naked eye, the rock chemistry, the rock magnetic properties, and the flow's position in a series of exposed flows, such as in a cliff face. These researchers divide the Columbia River Basalt Group into seven principal units, which from oldest to youngest are the Steens, Imnaha, Grande Ronde, Prineville, Picture Gorge, Wanapum, and Saddle Mountains Basalts. The Prineville erupted at the same time as the Grand Ronde, and some researchers consider it part of the Grand Ronde. Each of these seven units contains many individual flows, some of which were

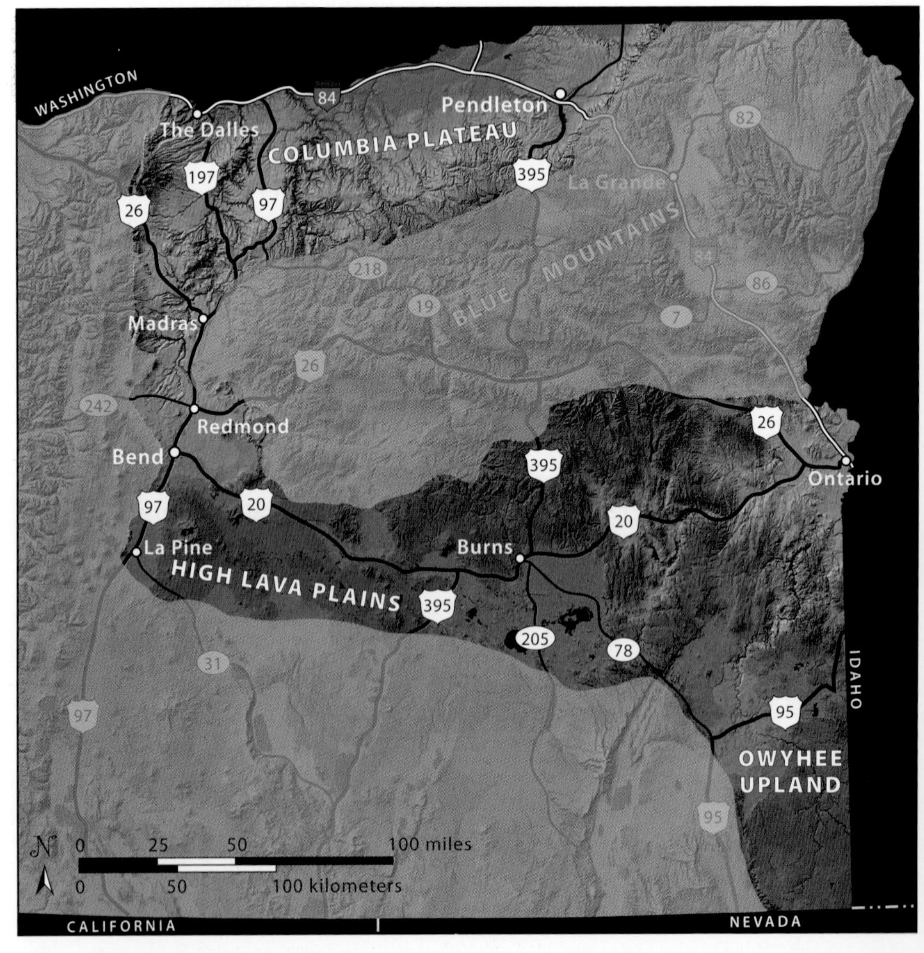

Physiographic map of the Oregon Lava Plateaus and its three main subprovinces. The Columbia Plateau lies to the north and west of the Blue Mountains. The High Lava Plains region lies south of the Blue Mountains. The Owyhee Upland occupies the southeastern corner of the state. –Base image from US Geological Survey, National Elevation Data Set Shaded Relief of Oregon

enormous, with volumes greater than 480 cubic miles (2,000 km^3). Some flows even reached the Pacific Ocean. By comparison, the 1783–84 Laki eruption in Iceland, which was the largest known basalt flow in recorded human history, reached a volume of 3.6 cubic miles (15 km^3).

Many basaltic lava flows exhibit a distinctive feature, called columnar jointing, in which the flow is cracked into tall, multiple-sided columns. The columns form as the lava contracts and breaks during cooling. Looking down on the upper surface of the columns, you can see that they are roughly hexagonal. They resemble mud cracks, which form by shrinkage brought on by drying rather than cooling.

River canyons cut into the basalt reveal beautiful cross sections of individual lava flows. Some of the more dramatic examples include the Columbia Gorge, the Deschutes River canyon, the Cove Palisades State Park, and deep canyons of the Blue Mountains in northeasternmost Oregon. In the most general sense, these flows consist of a bottom, a dense interior (the middle), and a top. The columns, which collectively form a zone called the colonnade, tend to be part

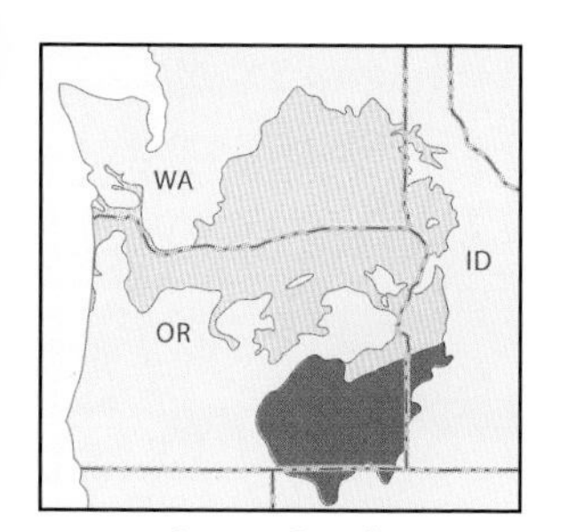

Steens Basalt
(16.8 to 16.6 million years)

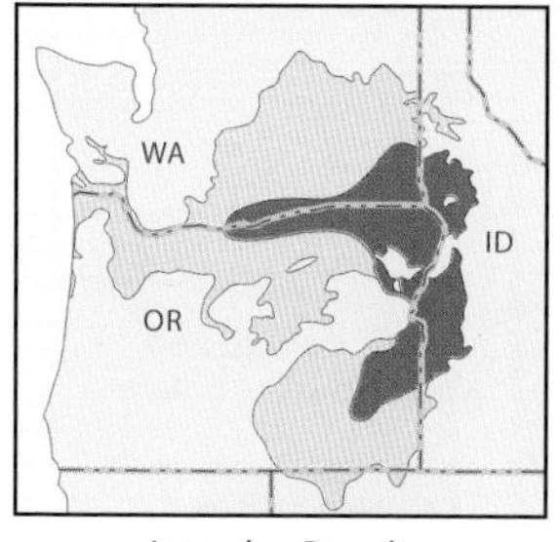

Imnaha Basalt
(16.7 to 16.0 million years)

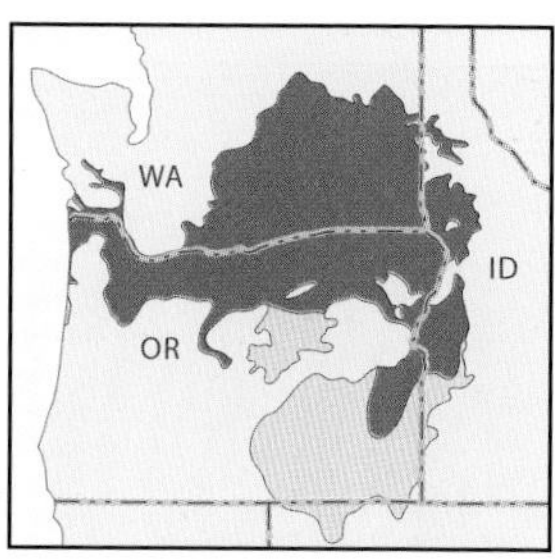

Grande Ronde Basalt
(16.0 to 15.6 million years)

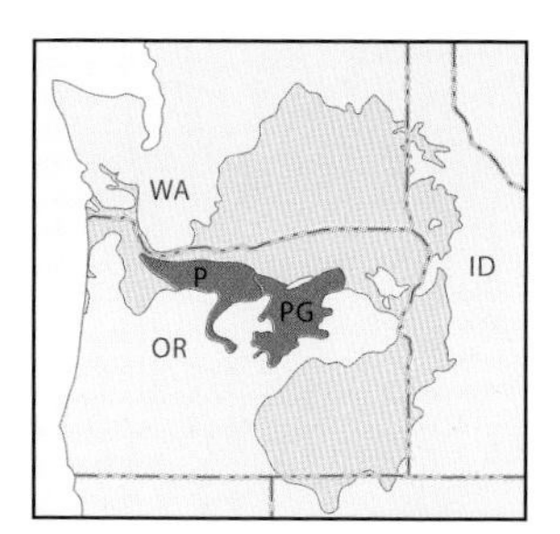

Prineville (P) and
Picture Gorge (PG) Basalt
(16.4 to 15.2 million years)

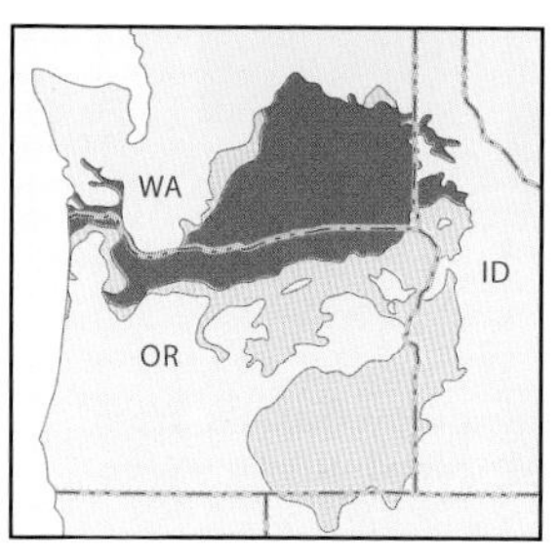

Wanapum Basalt
(15.6 to 14.5 million years)

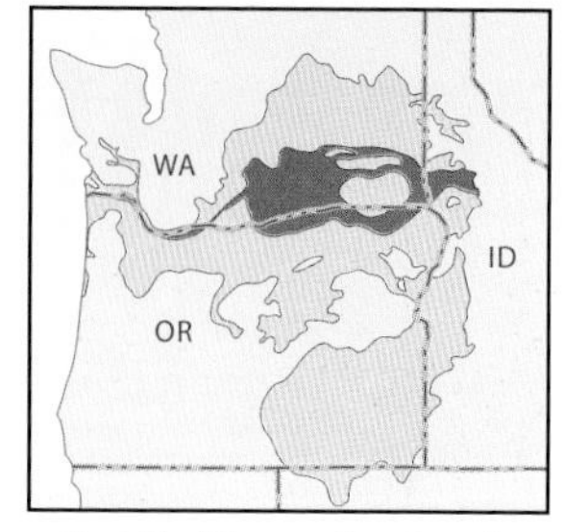

Saddle Mountains Basalt
(about 14 to 6 million years)

Distribution of major units of the Columbia River Basalt Group. The light brown region shows the limits of the Columbia River Basalt Group as a whole; the darker brown signifies the extent of an individual unit.

–Modified from Tolan and others, 1989; Riedel and others, 2013

of the dense interior. Upward, the colonnade may change to a zone called the entablature, which contains many closely spaced and irregular cooling fractures. Above the entablature, the flow top usually consists of basalt that is either full of bubbles, from gases rising through the lava, or broken into innumerable angular blocks.

The flow bases tend to be the most variable part of the flow, because they depend on the nature of the ground over which the lava flowed. If the ground was dry for example, the base will likely be a thin, less than 3-foot-thick (0.9m)

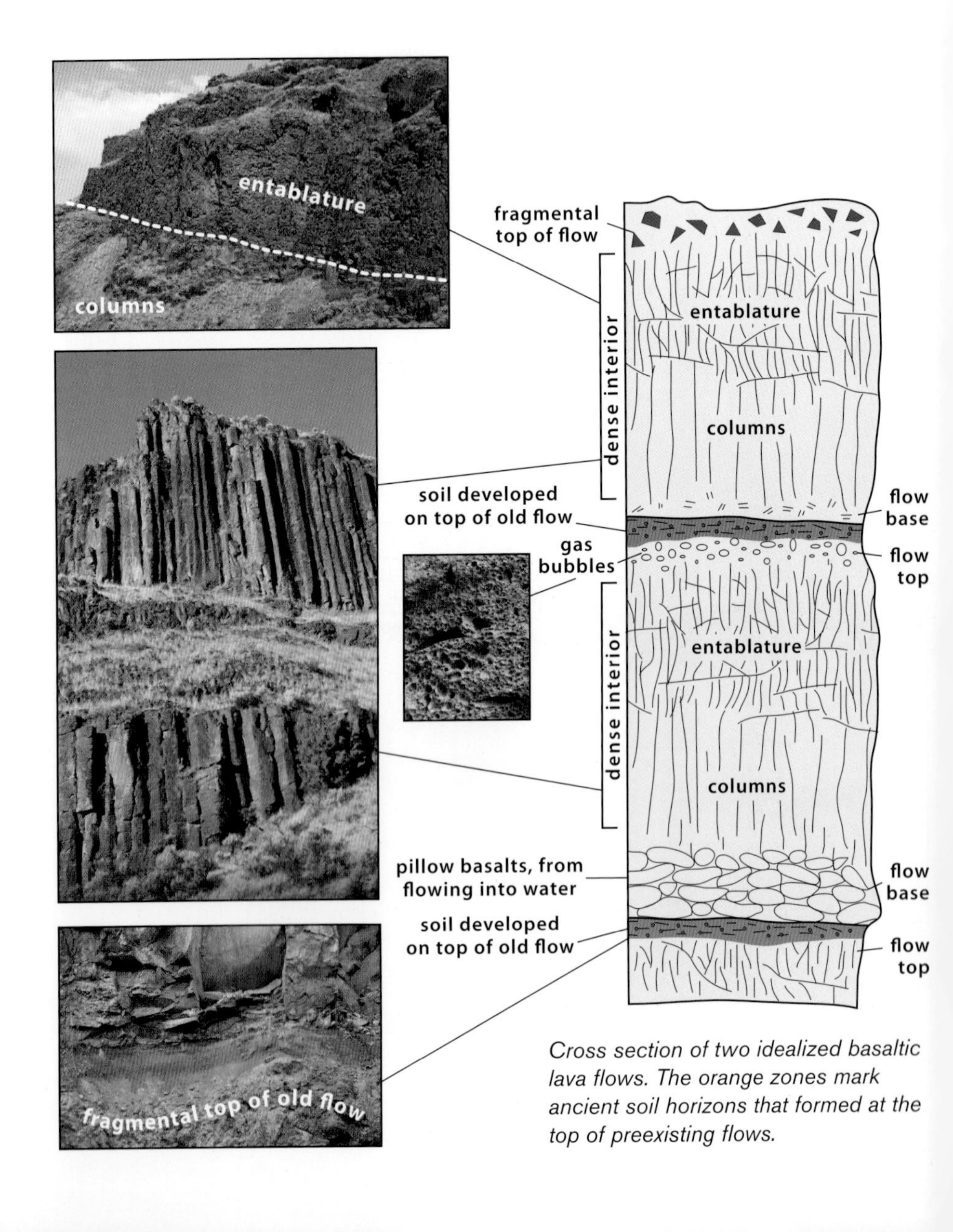

Cross section of two idealized basaltic lava flows. The orange zones mark ancient soil horizons that formed at the top of preexisting flows.

zone of fine-grained or even glassy rock because of fast cooling against the cool ground. If the flow poured into a lake, its base may consist of a much thicker zone of pillows, rounded blobs that form when basalt cools underwater. Many flow bases lie directly on top of a relatively thin, bright red, crumbly zone of rock up to 3 feet (0.9 m) thick. These zones mark ancient soil horizons, or paleosols, that formed on top the older flows before eruption of the overlying flow, which baked the soil to create its distinctive red color.

High Lava Plains

The High Lava Plains consist almost entirely of basaltic lava flows, rhyolitic ash-flow tuffs, and scattered rhyolite domes, a combination of lava types known to geologists as bimodal volcanism. Basalt, with low silica, is present with rhyolite, with high silica, but there is little of the intermediate andesitic types of lava. Bimodal volcanism often forms in regions undergoing crustal extension. In these areas, basaltic magma can rise easily into the upper crust and partially melt some of it to create rhyolitic magma.

Because the lava and ash-flow tuffs of the High Lava Plains are flat-lying, the semiarid landscape tends to be fairly flat, with occasional small mountains, mesas, and ridges breaking the skyline. The mountains mark locations of rhyolite domes, rounded accumulations of rhyolitic lava surrounding volcanic vents. The mesas consist of resistant basalt or ash flows capping more easily eroded material. Many ridges mark locations of normal faults that have displaced the lava and ash-flow tuffs up or down. Topographically and geologically, the High Lava Plains form a transition from the north-trending basins and ranges to the south to the east-northeast-trending mountains and valleys of the Blue Mountains in the north.

One of the hallmarks of the High Lava Plains is its abundance of ash-flow tuff, a rock solidified from hot ash and other pyroclastic material that explodes from a volcano and flows at incredible speeds across the land. During Late Miocene time, three major ash-flow tuffs explosively erupted somewhere in the vicinity of Burns. These huge eruptions blanketed enormous areas that extend north into the Blue Mountains and south into the Basin and Range. The tuffs are the 9.7-million-year-old Devine Canyon Tuff, the 8.5-million-year-old Prater Creek Tuff, and the 7-million-year-old Rattlesnake Ash-Flow Tuff, each of which is well exposed along US 395 north of Burns. The Rattlesnake Ash-Flow Tuff covers nearly a tenth of the state of Oregon!

The sixty or so rhyolite domes on the High Lava Plains show a general decrease in age from southeast to northwest. The domes range from about 10 million years old near Malheur Lake to less than a half million years old for Paulina Peak at Newberry Volcano. Newberry's Big Obsidian Flow, which is rhyolite, erupted only 1,300 years ago. There are exceptions to this trend, such as Wagontire Mountain, along US 395 southwest of Burns, which has an age of 14.7 million years. By contrast, the basaltic volcanism does not show a well-defined trend. Numerous flows, erupted between about 8 and 7 million

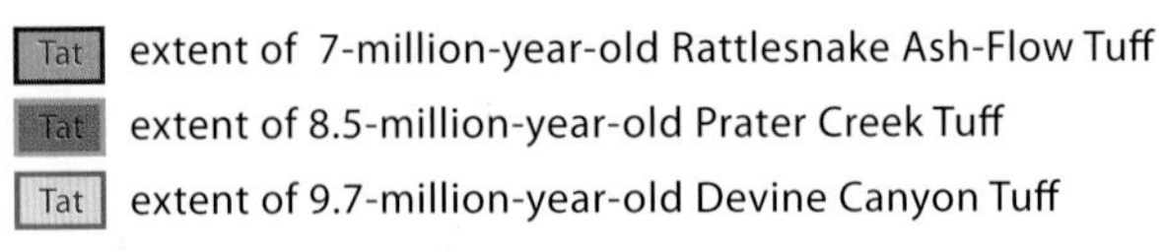

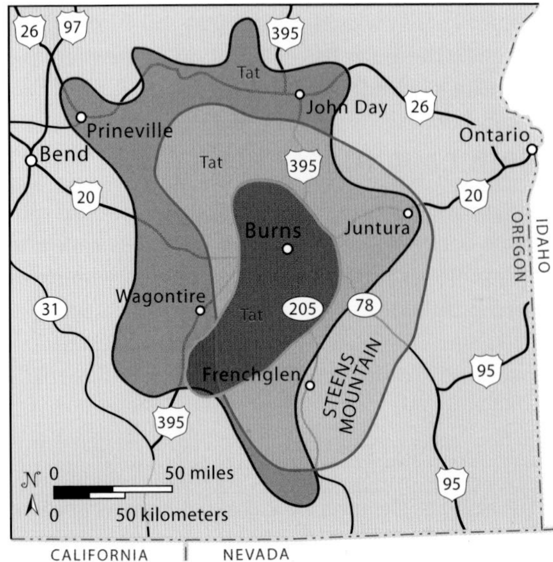

Map showing the extent of the principal Late Miocene ash-flow tuffs centered on Burns. The symbol Tat *stands for Tertiary ash-flow tuff and is used throughout the road guides to designate these units.* –Modified from Jordan and others, 2002

years ago, cover much of the length of the High Lava Plains, and younger flows exist on both the east and west sides. However, most of the young features are on the western side. These features include countless cinder cones and fresh, uneroded lava flows, as well as some interesting landmarks, such as Fort Rock and Hole-in-the-Ground. Fort Rock marks the location of an eruption through a Pleistocene lake, whereas Hole-in-the-Ground is a crater that marks the site of an eruption where magma encountered shallow groundwater.

The age progression of rhyolitic eruptions across the High Lava Plains forms a mirror image of the Yellowstone–Snake River Plain eruptions, which march across Idaho and become younger toward the northeast. It is not altogether clear why this trend occurs, or if it is somehow related to the Yellowstone hot spot. Some researchers argue that the heat source for the Yellowstone eruptions was partially deflected toward the west, because Oregon's subsurface is so complex. Oregon consists mostly of accreted terranes and has a thin lithosphere, but it resides next to thick continental lithosphere to the east. Most researchers agree that the trend is somehow related to the steepening of the oceanic plate subducting offshore of Oregon and Washington.

A seemingly uncountable number of northwest-trending normal faults cut the High Lava Plains. Compared to many of the range-bounding normal faults to the south in the Basin and Range Province, these faults are relatively small, with lengths of up to only a few miles and vertical displacements of tens to possibly hundreds of feet. Because these faults are recently or currently active, their vertical displacements break the landscape into numerous small northwest-trending ridges.

In the northern part of the High Lava Plains, these faults collectively define the Brothers Fault Zone, which extends in a west-northwest direction from south of Burns near Harney Lake to almost as far as Bend. As a whole, the zone appears to show an oblique motion, with elements of mostly normal and minor right-lateral slip. Its origin, although not entirely clear, is likely related to the effects of the ongoing clockwise rotation of Oregon and extension to the south in the Basin and Range. The Brothers Fault Zone marks where part of Oregon's crust is tearing to accommodate these two processes.

Owyhee Upland

The Owyhee Upland occupies the most remote part of Oregon and possibly the most remote region in the Lower Forty-Eight. The Owyhee River drains much of the region, flowing north toward the Snake through deep canyons cut in lava flows. Similar to the High Lava Plains, the Owyhee Upland is characterized by two types of lava—basalt and rhyolite. Most rhyolitic eruptions in the Owyhee Upland took place between 16 and 14 million years ago, some 4 million years before the earliest rhyolites of the High Lava Plains. These eruptions produced large accumulations of ash-flow tuff, now exposed over much of the upland. The basaltic eruptions on the Owyhee Upland began about the same time as the rhyolites but continued until nearly the present time at Jordan Craters.

GUIDES TO THE LAVA PLATEAUS

INTERSTATE 84
The Dalles—Pendleton
125 miles (201 km)

At The Dalles, eastbound travelers emerge from the constricted Columbia River Gorge into a wider valley, still framed by cliffs of the Columbia River Basalt Group, for the most part flows of the Wanapum Basalt. This section of I-84 traverses the length of what was Lake Condon, which repeatedly formed during the Missoula Floods between 18,000 and 15,000 years ago. The Columbia River Gorge west of The Dalles constricts the flow of the modern river, and during the Missoula Floods, the narrowing gorge created a hydraulic dam that backed up the floodwaters, inundating everything behind it. Many of the floods caused lake depths to rise more than 800 feet (240 m), which back flooded the many smaller drainages in the area. The lakes were short-lived, lasting only a week or

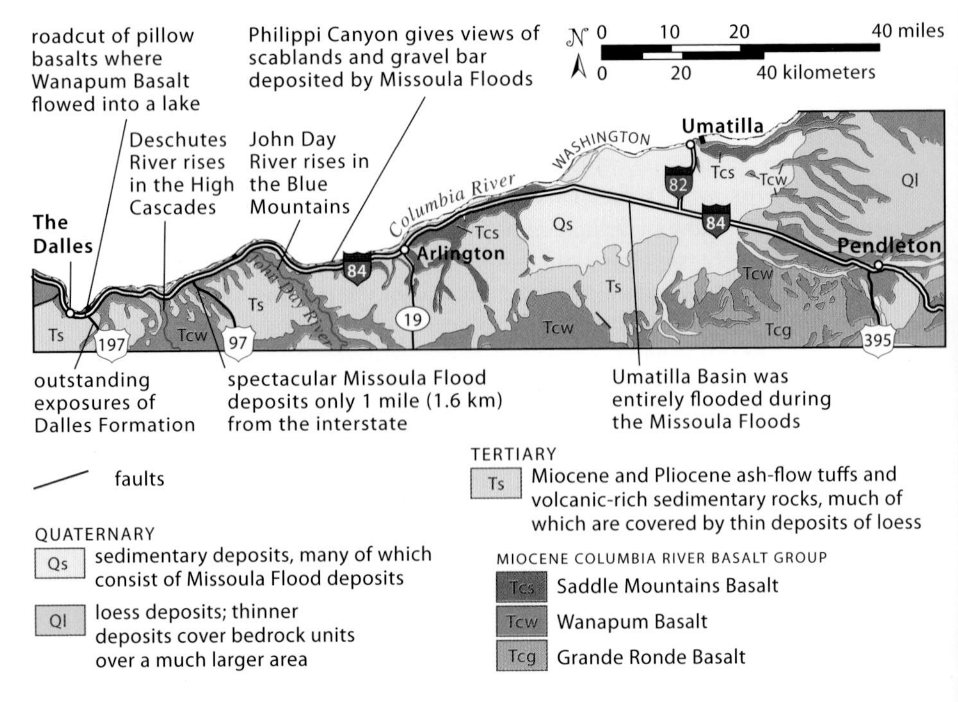

Geology along I-5 between The Dalles and Pendleton.

two before completely draining through the gorge, but they left strandlines, or shorelines, eroded into the landscape. You can also find erratics throughout the region at an elevation of about 1,000 feet (300 m), rafted there by ice chunks floating on the lake surface. During the 3,000 years of the Missoula Floods, Lake Condon, when it was full, had a surface area of about 1,500 square miles (3,900 km^2). The erosional features along this stretch consist mostly of scablands, odd-shaped outcrops of bare basalt that show up frequently along the highway near the river.

Just off the interstate at exit 87, a beautiful roadcut of pillow basalt and palagonite marks a location where Wanapum Basalt flowed into either an older channel of the Columbia River or a lake. Take exit 87 for US 197 and follow it about 0.25 mile (400 m) to its intersection with US 30. The roadcut lies on the east side of the road across from a large pull-out. See the road guide for US 197 later in this chapter for a photo and more detail.

Just east of exit 87, you can see a small area of scablands near the river. Fifteenmile Creek enters the Columbia near milepost 88, just downstream from the Dalles Dam. This drainage backed up considerably during the Missoula Floods but also was flooded from above, as the floodwaters spilled through gaps in a narrow divide. You can see one such gap just south of the highway near milepost 94. Flood deposits, including large bars and ripples, lie on the slopes behind the divide. For the next 10 miles (16 km) or so, you can see strandlines on the Washington side of the Columbia.

The Deschutes and John Day Rivers empty into the Columbia River at milepost 100 and 114, respectively. The Deschutes, which originates at Little Lava Lake near Bend, drains much of the east side of the Cascades. It backed up some 50 miles (80 km) during the Missoula Floods, almost as far as Maupin. The John Day River, which drains much of central Oregon, backed up about 35 miles (56 km) during the floods. Between these rivers US 97 follows a minor drainage that contains flood deposits on the west side of the highway less than 1 mile (1.6 km) from the interstate.

Across from the mouth of the Deschutes River, Miller Island divides the Columbia into two channels. Most of the island consists of basaltic bedrock, but its downstream edge, near milepost 99, displays a gravel bar deposited during the Missoula Floods on the downstream side of the bedrock. Between mileposts 111 and 114, cliffs of Wanapum Basalt come right down to the highway, and at about milepost 120 they form some impressive scablands. Views across the Columbia River in many places show triangular-shaped facets on the hillsides, eroded during the floods.

Along this stretch of road is a seemingly endless array of windmills. The year-round climate difference from western to eastern Oregon creates a consistent difference in air pressure, generating wind that flows through the natural wind tunnel of the Columbia River Gorge.

Philippi Canyon, at exit 123, provides views to some spectacular scablands as well as the upper reaches of the narrow divide between the Columbia and the John Day Rivers. Follow the steep road into the canyon east of the exit for just over 1 mile (1.6 km) and turn right at the first intersection; within 0.25 mile (400 m), the road passes alongside the scablands. Because the scablands lie on private property, they should be viewed from the road. As with other canyons nearby, the floods overtopped the divide and flooded the John Day River from above, as well as backing up from below. You can see a large gravel bar, which appears as an unusually smooth, steep slope, near the mouth of the canyon on its east side. Another short stretch of scablands shows up between mileposts 131 and 132.

Between Arlington and the I-82 interchange near Hermiston, the exposed basalt is part of the younger Saddle Mountains Basalt. Just east of Arlington, Lake Condon expanded over the entire Umatilla Basin, occupying a broad, low-relief area some 30 miles (48 km) across, all the way to Wallula Gap on the Washington-Oregon border. Gravel, sand, and silt were deposited in rhythmic graded beds throughout the basin as each flood backed up into a temporary lake and sediment suspended in the water settled to the bottom. The lake sediments form the fertile soils of this productive agricultural region, made possible by irrigation water derived mostly from wells that tap a regional groundwater system. Although these deposits generally don't outcrop, you can see them in a cut at milepost 145 and in the grassy ridge just west of milepost 149.

East of the I-82 interchange, I-84 gradually rises out of the Umatilla Basin and onto a plateau of Wanapum Basalt. Much of the lava is covered by windblown

View looking upstream at the Columbia River and a gravel bar deposited by the Missoula Floods near the mouth of Philippi Canyon. The bar is the smooth steep slope between the basalt cliffs in the middle ground and the paved road on the right and continues upward nearly to the skyline. The paved road leads another 0.5 mile (0.8 km) to the scablands described in the text. Parts of the gravel bar are being mined for gravel.

dust called loess, derived from glacial erosion during Pleistocene time. Beginning at about milepost 205, the road descends some 500 feet (150 m) to the Umatilla River through Pleistocene loess deposits blanketing lava flows.

US 20
Bend—Burns
131 miles (211 km)

Between Bend and Burns, US 20 follows the southeast trend of the Brothers Fault Zone, which created noticeable breaks in the Tertiary and Quaternary basalt flows. The road also passes by the volcanic vents of Pine Mountain and Glass Buttes, which erupted rhyolitic lava in Tertiary time.

For the first 15 miles (24 km) southeast of Bend, the road crosses numerous small exposures of Quaternary basalt flows from Newberry Volcano, a gigantic mountain that rises some 4,000 feet (1,200 m) in elevation about 20 miles (32 km) to the south. Pilot Butte, an amazingly symmetrical cinder cone, lies immediately north of US 20 on the outskirts of Bend. As reported by Julie Donnelly-Nolan of the US Geological Survey, Pilot Butte erupted about 188,000 years ago. The surrounding lava flows from Newberry are about 78,000 years old. Just west of milepost 1, a natural cut in the hillside reveals a cross section of the cone.

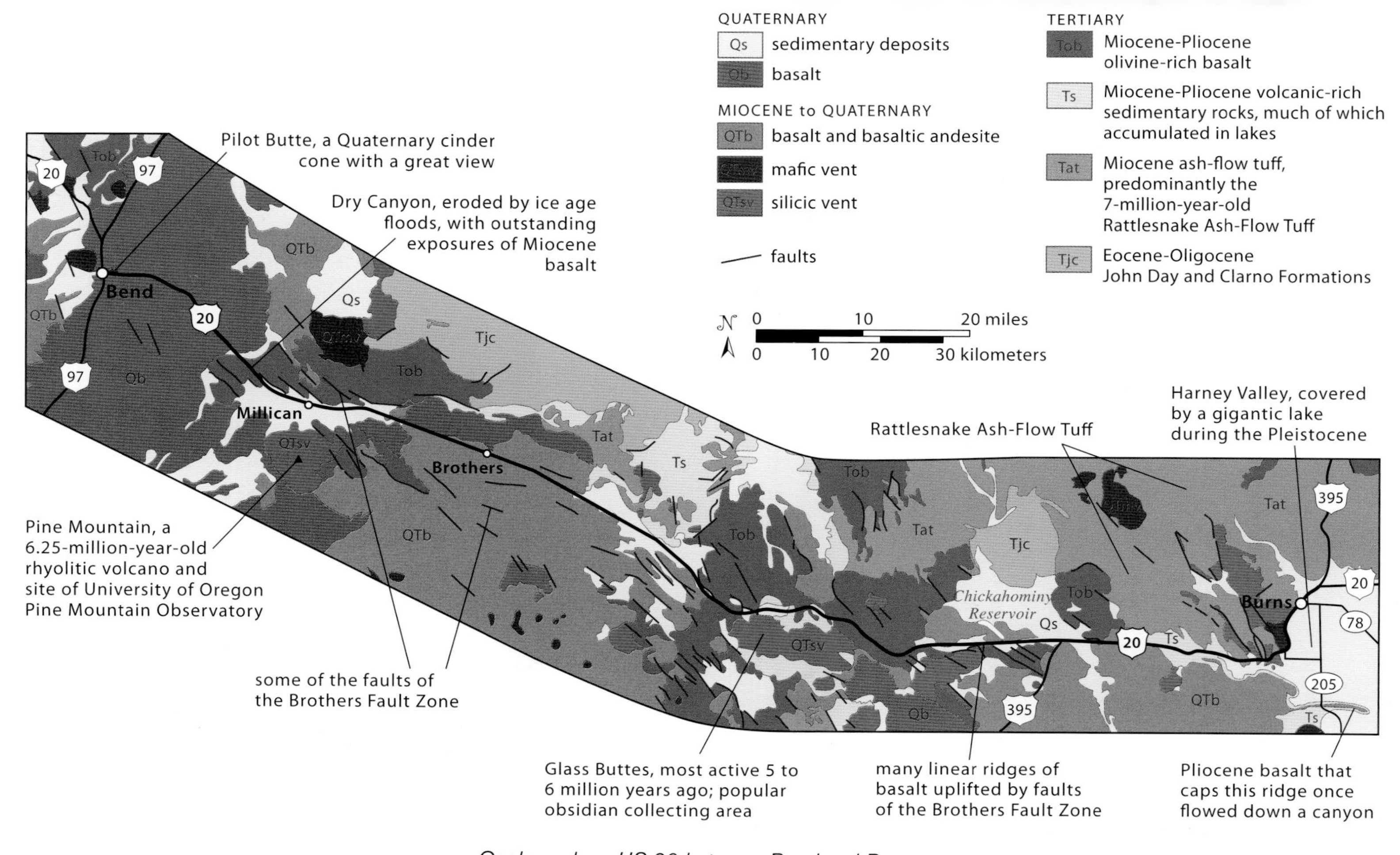

Geology along US 20 between Bend and Burns.

East of milepost 16, US 20 passes into older basalt flows, and at milepost 18 climbs along the canyon of the Dry River, which is typically dry, fed by several intermittent creeks on Newberry's eastern flank. This canyon is narrow and deep along this stretch only, which was likely formed from repeated floods during Pleistocene time, when the upper elevations of Newberry were glaciated and water backed up behind temporary ice dams. The lavas exposed in the canyon walls are about 7.5 million years old and contain interbedded flow breccias and pyroclastic materials exploded from a nearby vent. These rocks are displayed in roadcuts between mileposts 19 and 20.

From a viewpoint at milepost 19, you can see northwestward to the High Cascades, as well as the broad, nearly flat shield volcano of the Badlands, crossed high on its southern flank by US 20. The magma erupted from this volcano actually came from a source on Newberry Volcano, traveling about 10 miles (16 km) through a lava tube before erupting. Another pull-out at milepost 20, offers beautiful views of the canyon as well as an interpretive sign about the geology.

Numerous low ridges of basalt highlight the topography north and east of US 20, just east of the head of Dry Canyon. These are fault scarps of the Brothers Fault Zone, named for the small community of Brothers on US 20. The faults, where the land has been lifted on one side relative to the other side, punctuate the topography almost all the way to Burns. The road even follows a small

Dry Canyon, as seen from near the interpretive sign at milepost 20, was eroded by floods during Pleistocene time. The cliffs consist of 7.5-million-year-old basaltic flows, breccias, and near-vent pyroclastic deposits.

valley formed by these faults near Millican. Lakebed deposits crop out just west of milepost 23. Pine Mountain, a silicic vent dated at about 6.25 million years, rises to the south. The top of the mountain hosts an observatory operated by the University of Oregon physics department.

Just west of Hampton Station, near milepost 64, the road bends slightly toward the south, giving a straight-on view of Glass Buttes, a complex of volcanic vents that rise to an elevation of 6,300 feet (1,920 m), more than 1,500 feet (450 m) above the surrounding landscape. At milepost 74, the road passes close by their north side. The vents erupted mostly rhyolitic flows, tuffs, and obsidian that tended to congeal before flowing any distance, but also some basalt that seems to occupy fractures around the perimeter. Its main period of activity was between about 6 and 5 million years ago. Glass Buttes is well-known for its obsidian collecting and is traversed by numerous gravel roads; an access road intersects the highway between mileposts 74 and 75. Use extreme caution on these gravel roads, because the obsidian easily punctures car tires!

Close-up of obsidian from Glass Buttes. The reddish color, which comes from microscopic impurities gives it the name mahogany obsidian. The rock sample is just shy of 5 inches (12.5 cm) across.
–Specimen from the collection of Jim Watkins, University of Oregon

View eastward to Glass Buttes, on the right (south). The basaltic ridge on the left side of the photo is uplifted on a normal fault of the Brothers Fault Zone.

Faults of the Brothers Fault Zone are especially apparent between about milepost 84 and the turnoff for Chickahominy Reservoir, at milepost 99. US 20 follows a narrow graben, a down-dropped valley between two of these faults, at milepost 85.

West of Burns, the highway passes poorly exposed lake deposits between mileposts 114 and 120, and rhyolite and ash-flow tuff of the 7-million-year-old Rattlesnake Ash-Flow Tuff between mileposts 122 and 125. This tuff covers vast areas of southeastern Oregon and appears to have erupted from somewhere in this vicinity. At milepost 126, you can see the remains of a cinder cone, its cinders quarried to spread over snowy winter roads. Burns occupies the northwestern corner of the flat Harney Valley, which stands out on a geologic map as a huge square, floored by Quaternary lake deposits. Glacial Lake Malheur, a gigantic lake during the Pleistocene Epoch, filled the Harney Valley when the climate was cooler and wetter. Malheur Lake, which covers more than 80 square miles (207 km^2), is a remnant of this larger lake. The basin is closed, so Malheur Lake does not have an outlet. Its salinity tends to vary with climate, being more salty during drier, lower-water times, and less salty during wetter periods. During unusually wet years, it overflows into Harney Lake, normally an alkali flat at the southwestern corner of the basin.

US 20
Burns—Ontario
131 miles (211 km)

Between Burns and Buchanan, US 20 crosses the north edge of Harney Valley. You'll see farm buildings and irrigated fields occupying the flat bottom of a lake that was here during the wetter Pleistocene time. Glacial Lake Malheur, with an area of 920 square miles (2,380 km^2), was Oregon's largest glacial lake. Today's Lake Malheur is a much smaller remnant. A good roadcut of older, river-deposited sediments between mileposts 136 and 137 features crossbedding. Views to the north show flat-topped mesas, capped by the Rattlesnake Ash-Flow Tuff. Just over 7 million years old, the tuff is the youngest and largest of three gigantic ash flows that blanketed this area between about 9.7 and 7 million years ago. The Rattlesnake tuff likely covered an area larger than 10,000 square miles (25,900 km^2), centered just west of Burns. It provides an important marker bed in the rock sequence exposed in the John Day area and is well-exposed along US 395 just north of Burns, where it attains a thickness of more than 65 feet (20 m).

East of Buchanan, US 20 climbs over the Stinking Water Mountains en route to the Malheur River canyon. The headwaters of the Malheur are on the east side of the mountains; the river does not flow from Malheur Lake southeast of Burns. As US 20 climbs toward Stinkingwater Pass, it passes through a cliff of rhyolite and into a series of ash-flow and welded tuffs, all part of a Miocene-age rhyolitic dome complex to the south. Some of the welded tuff, such as the one exposed just east of milepost 157, contains bands of obsidian, formed near

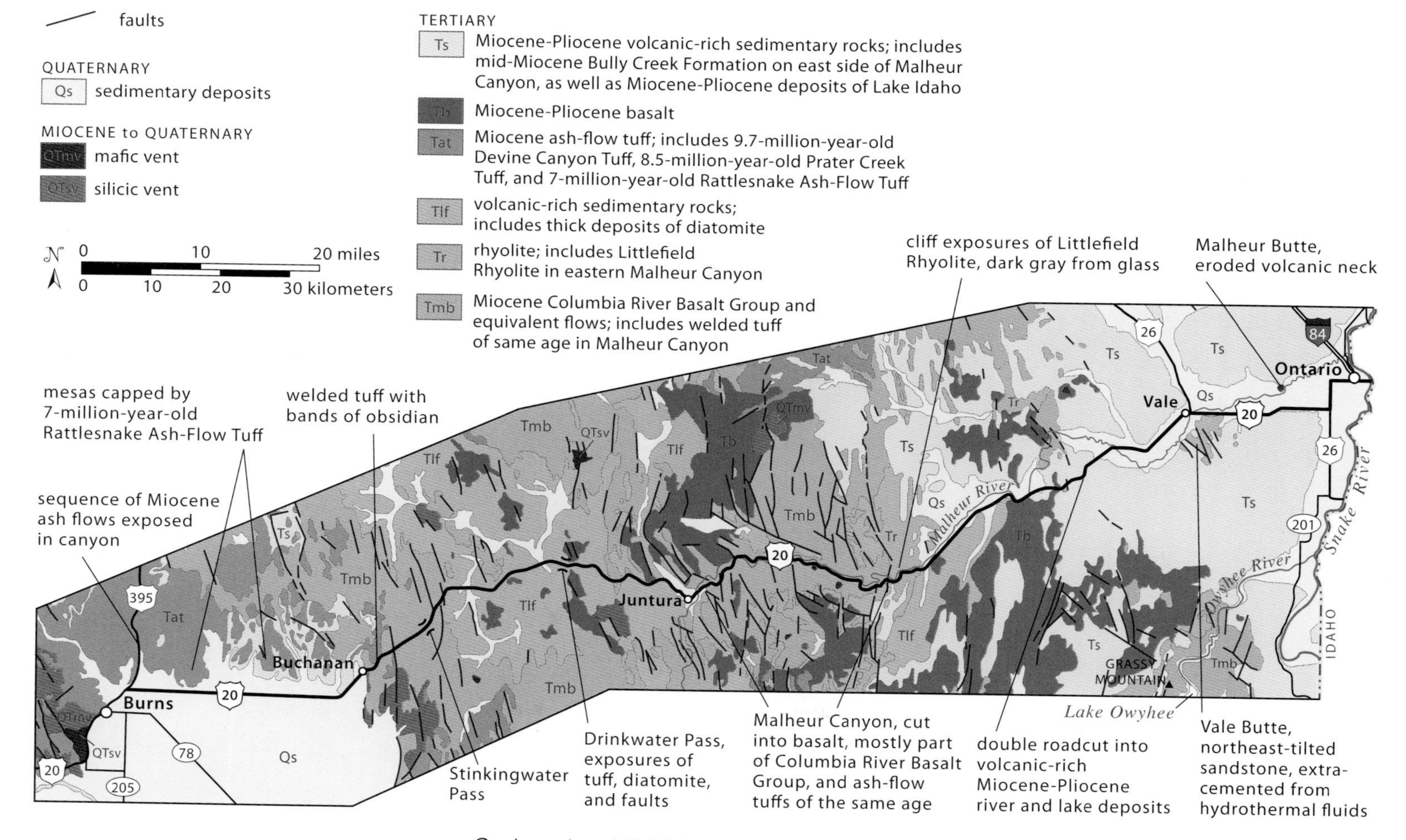

Geology along US 20 between Burns and Ontario.

the bottom of the flow where the ash, which consists of glass, was especially hot and became compressed into the dense obsidian by the weight of the overlying material. Miocene basalt flows crop out just east of the pass.

A few miles northeast of Stinkingwater Pass, the road passes into Miocene volcanic-rich sedimentary and volcanic rocks. The sedimentary rocks were mostly deposited in lakes and rivers and in many places contain beds of diatomite, a rock made of single-celled, silica-producing algae called diatoms. Some of the sedimentary rocks also contain pieces of fossilized wood, mostly from oak trees, indicators of the temperate climate in which they lived. The volcanic rocks consist mostly of fallout from rhyolitic eruptions, as well as ash-flow tuff.

The outcrop of welded tuff with bands of obsidian is about 15 feet (4.6 m) high. The most prominent band of obsidian is the red and black layer just below the middle of the photo. Notice also the scattered blebs of obsidian in the upper photo.

Numerous good exposures lie between here and Juntura. At Drinkwater Pass, between mileposts 176 and 177, you can see the tuff and underlying diatomite, as well as some minor normal faults. Except for some flows of Miocene basalt between mileposts 181 and 183, these lakebed deposits and ash flows continue all the way Juntura.

Just east of Juntura, US 20 enters Malheur Canyon, which it follows for the next 25 miles (40 km). The most prominent rocks in the canyon are two sets of Miocene basalts above and below an ash-flow tuff, called the Dinner Creek Tuff. These rocks are especially visible near milepost 190. The tuff has been dated at 15.4 and 16 million years and forms resistant ledges that show up some 200 to 400 feet (60–120 m) above the river. The basalts, the Basalt of Malheur Gorge (below the tuff) and the Hunter Creek Basalt (above the tuff), are chemically identical to basalt flows of the Columbia River Basalt Group and seem to provide a connection between the Steens Basalt to the south and the Imnaha and Grande Ronde Basalts to the north.

Malheur Canyon is an outstanding example of an incised meandering river. Winding through a series of bends and loops carved into the canyon's hard, resistant rock, the river has virtually no floodplain, just gravel bars on the inside bends of the channel where the water flows the slowest. Most meandering rivers occupy wide floodplains, moving their channels laterally as water on the outside of bends flows faster and erodes the bank. Before the Malheur River cut the canyon, it probably meandered back and forth across a wide floodplain developed on softer rock. As it eroded into the harder basalts of the canyon, however, it got stuck into a course it couldn't escape. Recent uplift of the region helped the river erode more effectively into the harder rock, cutting downward while confined to roughly the same channel. In some places, such as between mileposts 199 and 202, perched flat areas mark river terraces, former positions of the narrow floodplain. The river has cut even farther down since these terraces were deposited.

The basalt flows exposed in the canyon walls exhibit columnar jointing and entablature, as well as the occasional bright red paleosol. An especially good, and accessible, example lies on the south side of the road halfway between mileposts 214 and 215. There, the paleosol grades downward into the broken top of the underlying flow, which is unusually red from the dehydrated iron oxide. A small fault cuts these rocks adjacent to the road. Numerous other faults in the canyon noticeably offset the rocks.

Near milepost 215, the road crosses a fault, east of which is the Miocene Littlefield Rhyolite, which forms the cliffs along the road for the next several miles. These rocks look like basalt. Even their fresh surfaces are unsually dark. With the aid of a hand lens, you can see that they contain numerous scattered crystals of plagioclase, but even that suggests andesite rather than rhyolite. However, these rocks contain enough silica to be classified as rhyolite. The dark color comes from fine particles of volcanic glass dispersed through the rock. Like the basalt, these rhyolites are about 15 million years old, indicating that rhyolitic volcanism took place during Columbia River Basalt Group times. Geologists who mapped these rhyolites found that they cover hundreds of

Red paleosol, about 1 foot (30 cm) thick, between two basalt flows in a roadcut between mileposts 214 and 215. Note the rubbly, broken top of the bottom flow.

square miles and originated from north-south trending fissures along the edge of the Oregon-Idaho graben. Stretching north-south along the state border, the bounding normal faults of this graben were most active between about 15 and 10 million years ago.

Between mileposts 216 and 217, US 20 bends to the northeast and passes hills of altered ash-rich siltstone, sandstone, and diatomite, called the Bully Creek Formation. These rocks accumulated in rivers and lakes, probably also about 15 million years ago. They erode much more easily than the basalt and rhyolite, forming rounded, light-colored hills. Some especially good exposures of these rocks show up near mileposts 226 and 227. In some places, these rocks are intruded by basaltic sills, one of which lies beneath the road at milepost 219. As the road rises gradually to higher elevations, it passes through many of these rocks; the view northward shows that they form a rim along the edge of the valley. Between mileposts 227 and 229, the road passes through overlying Miocene basalt.

A double roadcut between mileposts 233 and 234 exposes river and lake sediments, likely deposited by a river system that fed Lake Idaho. A high-angle fault offsets these rocks toward the east side of the roadcut. The road descends from this roadcut onto the Malheur River's floodplain, which it follows all the way to Ontario.

Covering much of the western Snake River Plain and parts of eastern Oregon during Late Miocene and Pliocene time, Lake Idaho probably consisted of a series of large lakes and floodplains rather than a single lake for most of its

View looking westward at a meandering channel of the Malheur River.

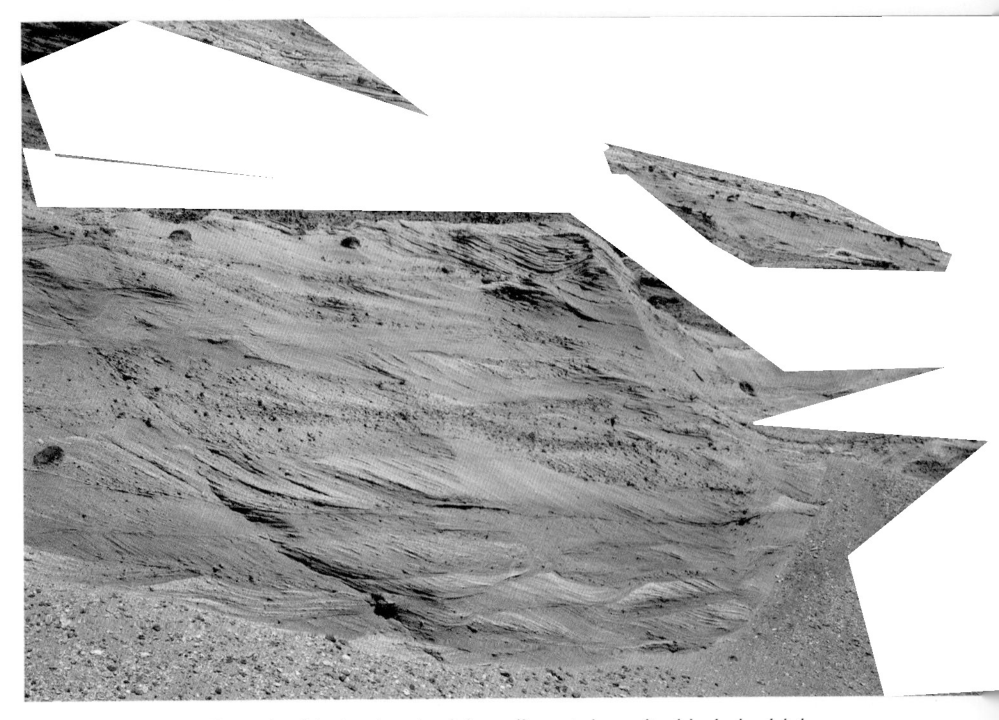

Crossbedded volcanic-rich sediment deposited in Lake Idaho in a roadcut between mileposts 233 and 234.

time. About 3 million years ago it reached a high point and overtopped a ridge that separated it from Hells Canyon. The resulting flood deepened the canyon and drained the lake. Throughout its history, Lake Idaho hosted a variety fish species, now found as fossils.

In and near Vale, unusually high-temperature water rises along fault zones to within 3,000 feet (900 m) of the surface, creating commercially viable geothermal energy. In November 2012, a geothermal plant opened at Neal Hot Springs, northwest of Vale. The plant is capable of producing 23 megawatts of electricity, enough to power seven hundred homes. The hydrothermal activity has also affected the bedrock. Vale Butte, immediately south of the highway as it heads east out of Vale, consists of faulted, northeast-tilted sandstone, some of which is unusually resistant to erosion, because it has been hardened by the precipitation of hydrothermal minerals.

Due south of Vale about 24 miles (39 km), Owyhee Dam impounds the Owyhee River to form Lake Owyhee. Grassy Mountain, some 6 miles (10 km) west of the dam, marks a large geothermal gold deposit. The mineralization, in which gold is found in quartz veins, probably occurred about 13 million years ago.

A few miles east of Vale at milepost 254, you can see directly northward to Malheur Butte, a deeply eroded Miocene volcanic neck.

US 95
Idaho Border—Nevada Border
121 miles (195 km)

US 95, a major north-south highway in the American West, snakes through the Owyhee Upland in the southeastern corner of Oregon, a remote region covered with sagebrush. The rocks are mostly basalt flows and river- and lake-deposited sediments, full of volcanic material, reflecting southeastern Oregon's nearly continuous record of volcanic activity.

A well-bedded roadcut of Miocene lake and river deposits greets the traveler only 2 miles (3.2 km) west of the Idaho border. Numerous good exposures follow over the next several miles. The river deposits are coarse-grained and display crossbedding and occasional lens-shaped channel deposits. The lakebeds tend to be fine-grained and more evenly bedded.

A minor gold rush brought people to the Jordan Valley in the 1860s. At first, they mined placer gold in the streambeds, but they eventually found its source in the Owyhee Mountains across the border in Idaho. Silver City, Idaho, sprang up there and had a population greater than 2,500 at its peak in the 1880s.

At Jordan Valley, US 95 turns westward past hills of Miocene basaltic and sedimentary rock. Between mileposts 25 and 26, the road bends southward past a rocky hill of Miocene rhyolite. Rhyolitic ash and lava flows characterize much of the Owyhee Upland part of the Lava Plateaus Province, although most lie south of the highway. Beyond here, the road follows over a low-relief surface built on Pliocene sedimentary rocks that don't crop out, except along the steep stretch west of the turnoff for Antelope Reservoir near milepost 32. As the road climbs above them, it crosses into younger basalt flows, although these too show up only sporadically over the next several miles.

A large region north of the highway is covered by Quaternary basalt associated with the Jordan Craters. The vents erupted between 9,000 and 4,000 years ago. The lava flows feature beautiful, fresh examples of pahoehoe and spatter cones.

Near milepost 52, US 95 descends steeply into the Owyhee River valley through an exposure of the Pliocene sediments. These rocks include ash-fall deposits and crossbedded sandstone and gravel deposits, as well as fine-grained lakebeds. Some of the lake deposits consist of thick beds of tuff-rich mudstone. Older, probably Miocene, lake and river deposits lie on either side of the floodplain to the north along the river's course. In some places, they are capped by a basalt flow. These rocks accumulated in rivers and lakes during Late Miocene and Pliocene time, but it is not clear how extensive or interconnected they were. Some geologists argue that they were deposited in Lake Idaho, which covered much of the western Snake River Plain during that time. Others have argued that the sediments were deposited by an early Owyhee River. Still others suggest the lake and river were more confined to the local area. Fossil discoveries in these rocks include a variety of bird and rodent bones as well as beaver teeth.

Geology along US 95 between the Idaho border and the Nevada border.

Leslie Gulch–Succor Creek Byway

Just over 2 miles (3.2 km) west of the Idaho border and just south of the Succor Creek crossing, the Leslie Gulch–Succor Creek Byway heads north about 18 miles (29 km) to Succor Creek State Natural Area, where you can collect thundereggs, Oregon's state rock. Thundereggs are semispherical rocks with minerals, typically some form of silica, on the inside. Many are the size of baseballs, but they can be much larger or smaller. They form by the precipitation of minerals in gas cavities in rhyolite lavas and tuffs. At Succor Creek, the volcanic rocks are 16 to 15 million years old and interbedded with coal-rich shale and volcanic-rich sandstone. The sedimentary rocks host fossil leaves, stem impressions, and rare wood fragments–evidence of forests that grew and were repeatedly destroyed by the volcanic eruptions.

These thundereggs, about 10 inches (25 cm) across, include two that have been cut and polished and two in their natural state.

The Leslie Gulch Road heads west off the Succor Creek Road about 10 miles (16 km) north of its intersection with US 95. It passes through a narrow canyon, up to 300 feet (90 m) deep, cut into the multicolored Leslie Gulch Ash-Flow Tuff, on the way to Lake Owyhee, a reservoir on the Owyhee River. The tuff erupted from the nearby Mahogany Mountain Caldera 15.5 million years ago. Both the Leslie Gulch and Succor Creek Roads are impassable when wet.

These lake and river deposits are beautifully exposed at the Pillars of Rome, monoliths standing some 100 feet (30 m) tall. These erosional remnants were left behind as the mesa behind them retreated. To get there, take the well-graded gravel road north from Rome for 3 miles (4.8 km) toward the Owyhee River and then west for another 1 mile (1.6 km).

US 95 crosses the Owyhee River at the lower end of Owyhee Canyon, a dramatic gorge that few people other than whitewater enthusiasts have seen. The river trip must be done in spring, when water in this desert region is high enough to float. The canyon traverses a nearly complete sampling of the area's Middle-Miocene to Pliocene rocks, including numerous ash-flow tuffs, rhyolites, basalts, and volcanic-rich sedimentary rocks. In some places, Quaternary basalt forms the canyon rim.

Near milepost 58, US 95 descends through good exposures of the lake and river sediments into Crooked Creek, which joins the Owyhee about 6 miles (10 km) to the north. On the west side, the road rises through the sediments into overlying basalt. It then follows a somewhat low-relief surface to Burns Junction, passing occasional exposures of basalt along the way. Burns Junction rests on a Quaternary-age basalt flow.

South of Burns Junction, the road mostly follows wide valleys with expansive views. Steens Mountain, uplifted by a recently active normal fault, forms the long, high ridge to the west. You can see Miocene basalt about 7 miles (11 km)

The Pillars of Rome, tall monoliths of river and lake deposits, left standing in places as the mesa erodes back.

south of Burns Junction where the road recrosses Crooked Creek and just below Blue Mountain Pass, 25 miles (40 km) farther south. Above the basalt, prominent ledges of Miocene welded tuff, erupted during the earliest phases of the Yellowstone hot spot, form many of the ridge and mountain tops. This theme of welded tuff on top of basalt continues south of the pass all the way to the Nevada border. The hills west of the highway near the Nevada border rim the McDermitt Caldera, an early source of the Columbia River Basalt Group and the earliest known caldera of the Yellowstone hot spot. The caldera also hosts some significant mercury deposits, mined between about 1920 and 1990.

US 97

Biggs Junction—Bend—La Pine

165 miles (266 km)

US 97, a major north-south route across Oregon, does not follow a major river between Biggs Junction and Bend. Highway planners avoided the steep, winding canyons of the north-draining Deschutes and John Day Rivers, choosing instead the elevated land between the two rivers. Beginning at the relatively low elevation of 220 feet (67 m) at the Columbia River, US 97 follows a small canyon and ascends steeply to the Columbia Plateau, attaining an elevation of 1,000 feet (300 m) in less than 6 miles (10 km). Beautiful cross sections of basaltic lava flows, as well as gravel and boulder deposits from the Missoula Floods, line the canyon walls. Perhaps the most dramatic Missoula Floods deposit lies only 0.75 mile (1.2 km) up the canyon behind the large pull-out on the south side.

As the road climbs and the topography flattens out, exposures of the basaltic bedrock become more widely spaced—although they continue to show up in small cuts and stream channels for the entire length of this section of US 97. About halfway between Kent and Shaniko, however, the flows change from the Wanapum Basalt of the Columbia River Basalt Group on the north to the Grande Ronde Basalt to the south. In general, the Wanapum Basalt is a slightly lighter shade of gray than the Grande Ronde Basalt, because it contains abundant visible crystals of light-colored plagioclase.

Between mileposts 74 and 75, the road drops below the Grande Ronde Basalt into light-colored ash-flow tuffs and ash-rich sedimentary rocks of the John Day Formation. Being easily weathered, the John Day Formation tends to form brush-covered slopes, although a particularly prominent red ash-flow tuff can be seen to the south near milepost 79. An outstanding exposure exists at the sharp bend in the road between mileposts 79 and 80, where the John Day Formation is faulted against the Grande Ronde Basalt. Near milepost 81, you can see these two units together in their proper order, with the basalt on top above slopes of the John Day Formation. Between milepost 81 and Madras, US 97 passes scattered exposures of the Deschutes Formation, made of volcanic-rich sedimentary rock and basalt flows that originated from the Cascade volcanoes during Late Miocene and Pliocene time.

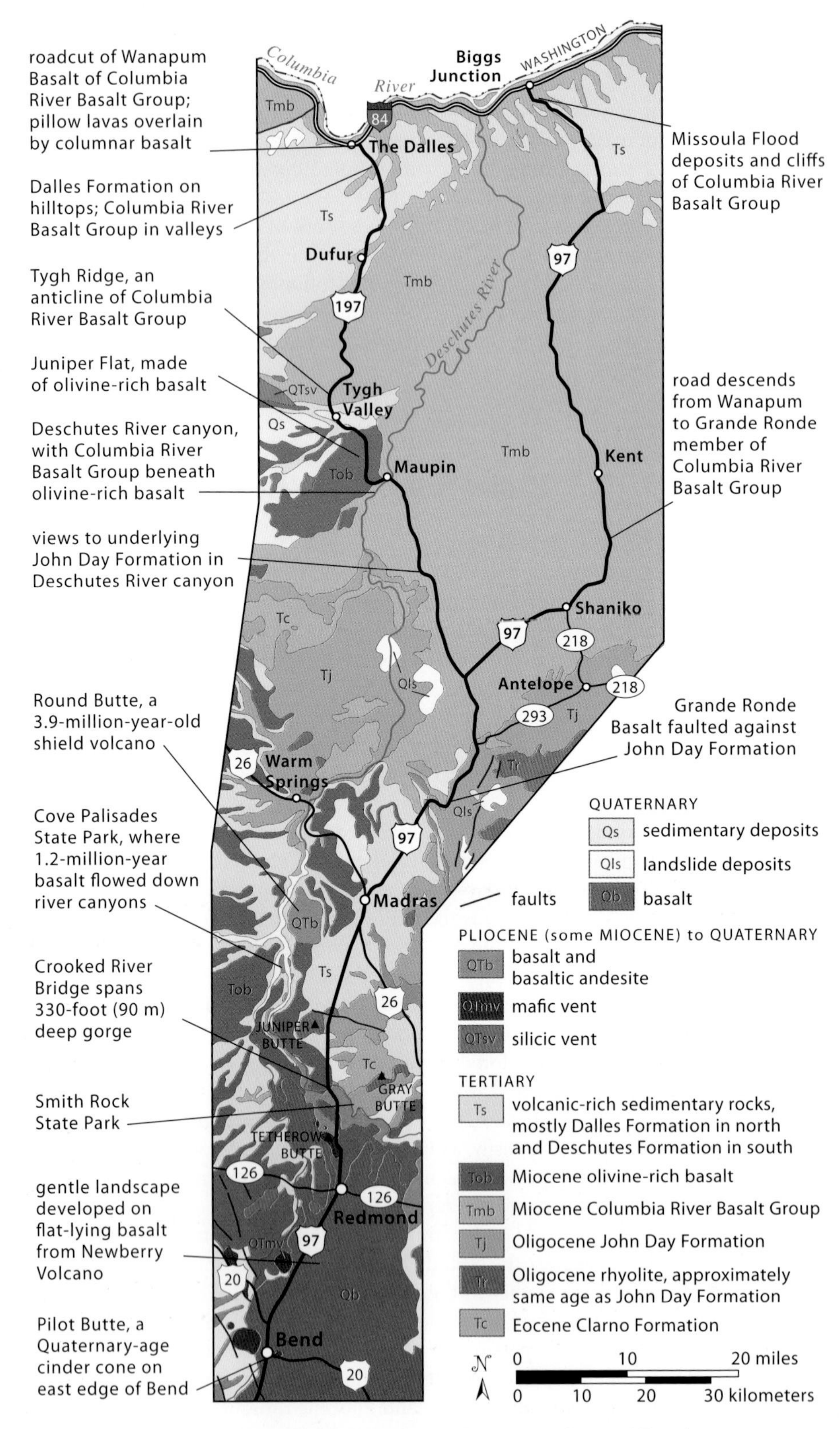

Geology along US 97 between Biggs Junction and Bend.

Deposit of gravel and boulders left by the Missoula Floods, only 0.75 mile (1.2 km) up the canyon from Biggs Junction.

The Cove Palisades State Park

At the Cove Palisades State Park, the Crooked and Metolius Rivers join the Deschutes, their confluences and canyons submerged beneath several hundred feet of Lake Billy Chinook. Round Butte Dam was constructed across the Deschutes, where basalt constricted its channel. This is not the first time the canyons have been dammed and flooded. About 1.2 million years ago, long before Newberry Volcano began to form, lava flowed down the Crooked River canyon, and then back up into the Deschutes River canyon, probably because the lava formed a temporary dam. Basalt flows that flowed down and are confined to canyons are called intracanyon basalts. The rivers have since eroded back through the basalt flows, reestablishing their courses and leaving remnants of the basalt along the canyon walls. Bishop (1990) and Smith (1998) offer more detailed descriptions of the park.

The intracanyon basalts are exposed discontinuously to the north and south of Round Butte Dam. They form cliffs, hundreds of feet high in some places, that display striking vertical columns formed by contraction of the basalt during cooling. Perhaps the most prominent exposure is the Island, a peninsula of land that separates the Crooked River Arm from the Deschutes

River Arm of the lake. Elsewhere, the basalt is exposed along the edges of the canyons, erosional remnants of when the lava practically filled the canyon. These flows originated from vents almost as far south as today's Newberry Volcano and entered the Crooked River Gorge near Smith Rock State Park.

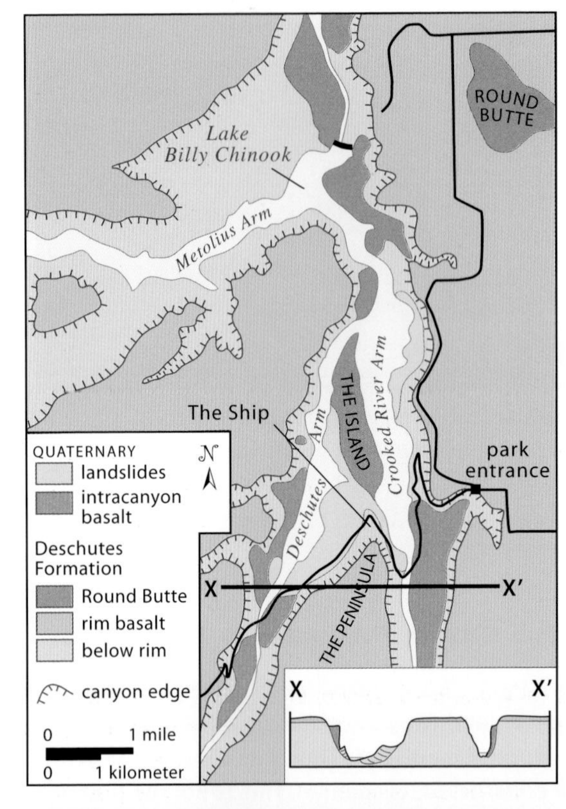

Geologic map of the Cove Pallsades State Park. –Modified from Bishop, 1990

Intracanyon flows on the Island form cliffs with columnar jointing. The Crooked River Arm occupies the foreground. Mt. Jefferson is the Cascade volcano in the background.

The Deschutes Formation forms the bedrock of the original river canyons. It consists of layers of basalt and ash-flow tuffs, as well as abundant river and flood deposits of sand and gravel, accumulated from about 8 to 4 million years ago. The volcanics were erupted from older volcanoes of the High Cascades, and the sediments were eroded from the growing mountains. The Deschutes Formation, which reaches 2,300 feet (700 m) in thickness, makes up much of the older bedrock between Redmond and Warm Springs.

A long roadcut along the steep grade just below the park entrance displays the spectrum of rock types in the Deschutes Formation, from basalt flows to ash-flow tuffs to crossbedded sandstone and conglomerate. A feature called the Ship, which consists of a thick white ash-flow tuff and darker-colored sandstone and conglomerate, forms an interesting promontory just south of the Island. Round Butte, only a few miles north of the park entrance on the east side, is a low shield volcano that produced the youngest basalt flows of the Deschutes Formation 3.9 million years ago. A dirt road climbs to the top of the cinder cone that caps Round Butte.

As a testament to the ongoing erosion in the river canyons, numerous talus cones and landslide deposits occupy the narrow areas between the reservoir waters and the cliffs. Many of these deposits extend well into the canyons and underwater. The most recent landslide, visible near the bridge at the south end of the Crooked River Arm, occurred during the winter of 1988.

Between Madras and Redmond, the road passes the turnoff for the Cove Palisades State Park. Round Butte, which overlooks much of Cove Palisades, forms a small, 3.9-million-year-old shield volcano north of the park.

South of Madras, US 97 crosses a low-relief surface of Deschutes Formation until the road crosses a small fault at the base of a hill between mileposts 105 and 106. From there, the road rises southward to a pass between Juniper and Haystack Buttes. Juniper Butte, to the west, is a rhyolite dome that formed just outside the margin of the Crooked River Caldera, which erupted 29.5 million years ago. Haystack Butte consists mostly of tuff that was produced during the caldera-forming eruption.

The road crosses the 330-foot-deep (100 m) Crooked River Gorge between mileposts 112 and 113. Peter Skene Ogden State Scenic Viewpoint on the south side of the gorge provides a rest area and a trail along the canyon edge. The canyon walls consist of basalt flows that originated from Newberry Volcano some 400,000 years ago. A short distance downstream, the river cuts into older, 5.3-million-year-old basalt flows from Tetherow Butte, near Redmond. It seems that the Crooked River has cut this canyon at least twice. About 1.2 million years ago, it cut the canyon through the Tetherow Butte basalt, providing the pathway for the intracanyon basalt flows of the Cove Palisades. Then, 400,000 years ago, the canyon filled with the Newberry basalt, and the Crooked River had to cut the canyon to its present configuration.

View looking upriver at the Crooked River Gorge and US 97 bridge from Peter Skene Ogden State Scenic Viewpoint. Basalt cliffs originated from now-buried vents on the flanks of Newberry Volcano, south of Bend. At about 400,000 years old, they are some of Newberry's oldest flows.

Smith Rock, a series of spires eroded from tuff of the Crooked River Caldera, lies east of milepost 114; the turnoff to Smith Rock State Park is at Terrebonne at milepost 116. Between mileposts 116 and 119, you can see a few red cinder cones west of the highway, the red color deriving from oxidation of the iron-rich, basaltic cinders. Cinder cones are mined by the Oregon Department of Transportation for their cinders, which are used on icy winter roads for traction.

Between Redmond and Bend, US 97 crosses a gently sloping landscape developed on flat-lying basalt flows from Newberry Volcano. Almost all of this basalt is younger than 75,000 years old. At Bend, these flows are broken by a series of normal faults that are not apparent from the highway but seem to connect with the High Cascades graben to the north. Pilot Butte, a cinder cone in Bend, erupted 188,000 years ago and is now surrounded by the younger basalt from Newberry Volcano.

Between Bend and La Pine, US 97 continues over Newberry basalt flows and provides good views of the massive volcano to the west. Newberry Volcano shows both basaltic and rhyolitic volcanism and last erupted only 1,300 years ago. Its most accessible feature is Lava Butte, a cinder cone that rises more than

400 feet (120 m) above the road between mileposts 149 and 150. The cone erupted 7,000 years ago from a rift zone that cuts northwestward across the volcano's western flank. The area hosts interpretive displays and a bookstore at Lava Lands Visitor Center. It also gives opportunities to explore Lava River Cave, a large lava tube that actually runs under the highway about 2 miles (3.2 km) to the south.

Smith Rock State Park

Smith Rock State Park, a mecca for hikers and rock climbers, hosts spires and cliffs, many of which have weathered into fantastic shapes. The rock is volcanic tuff of the John Day Formation, erupted during a caldera-forming explosive event just over 29 million years ago. This eruption, as well as others from the same volcanic field, likely also produced the tuff for many of the colorful beds elsewhere in the John Day Formation, such as at Painted Hills.

The Crooked River separates the tuff of Smith Rock from a flat bench of basalt immediately to the south. This lava flow originated about 400,000 years ago from Newberry Volcano, some 40 miles (64 km) south of Smith Rock, and is one of the volcano's most far-traveled flows. The basalt abuts the high towers of Smith Rock, indicating the tuff formed a barrier to its flow.

Researchers used to think that the tuff formed as part of an isolated volcano, but they now recognize it as part of the much larger Crooked River Caldera, which measures some 15 miles (24 km) across at its narrowest

The Crooked River flows past 29-million-year-old tuff at Smith Rock State Park. A bench in the left foreground is Newberry basalt. The tower on the right side of the photo is a rhyolite dike.

and about 25 miles (40 km) across at its widest. The caldera dwarfs Crater Lake, which measures only about 6.2 miles (10 km) at its widest. Smith Rock State Park occupies the northwestern edge of the caldera, and its tuff can be followed along the edge of the caldera to the southeast more than 10 miles (16 km) beyond Prineville. Altogether, the tuff's volume measures about 200 cubic miles (830 km^3). Researchers argue that the tuff formed during collapse of the caldera because they observe that the thickest portions of it lie within the caldera.

Tuff contains pumice, which is frothy glass expelled violently during an eruption. Much of the Smith Rock tuff contains unusually large fragments of pumice, a consequence of forming within the actual caldera and not having traveled far. In some places, the tuff also contains lens-shaped layers with abundant rock fragments, a few of which consist of Permian-age limestone. The basement rock here may be Permian limestone, with these limestone fragments being brought to the surface by the erupting magma.

The towers and unusual rock shapes at Smith Rock form by differential erosion. Parts of the tuff are more resistant than others and form promontories while the surrounding material erodes. In some places the rock is more resistant because it was hotter and welded together more strongly. In other places, it's more resistant because it has fewer fractures for water to infiltrate. Elsewhere, a weathering process called case hardening made the rock more resistant. This process involves the precipitation of a protective material—zeolite minerals at Smith Rock—over the exposed bedrock. In still other places, more-resistant rhyolite dikes intrude the tuff and form spines and towers. These dikes show that volcanism continued after caldera collapse. Numerous rhyolite domes intruded the caldera around its margin. These domes include Gray Butte, which forms the skyline behind Smith Rock, and Powell Buttes, visible from OR 126 between Redmond and Prineville.

Climbers ascend the tuff at Smith Rock.

Newberry Volcano

Newberry Volcano poses a series of enigmas. First, it's Oregon's largest recently active volcano, but it's so large that it's hard to recognize as a volcano. Second, its broad domal shape, covered by basaltic lava flows, resembles a large shield volcano, but its central area consists of a rhyolitic caldera. Finally, the volcano contains elements of both the High Cascades and the High Lava Plains. The volcano was named a national volcanic monument in 1990, administered by the Forest Service.

The volcano stretches some 40 miles (64 km) in a north-south direction and covers nearly 1,200 square miles (3,100 km^2). Many of its lava flows extend well beyond the volcano. To the north, for example, a 400,000-year-old flow forms the basaltic bench at Smith Rock as well as the towering cliffs of the Crooked River Gorge, nearly 10 miles (16 km) north of Redmond. The volcano contains a caldera that exceeds 15 square miles (39 km^2) in area and is partially filled by two lakes. The surrounding caldera rim reaches an elevation of 7,984 feet (2,433 m) at Paulina Peak, just slightly lower than the highest point on the rim of Crater Lake, Oregon's more famous caldera.

Its immense size notwithstanding, Newberry is fairly young, with some features erupted 1,300 years ago. Its oldest known feature is a rhyolite dome with an age of only about 400,000 years. Its oldest basalt flows are also about 400,000 years old, and its caldera-forming eruption occurred only 75,000 years ago. The bulk of the mountain consists of basaltic lava flows that emanated from vents well to the north or south of the present caldera. The numerous rhyolitic flows, which originated at or near the caldera, can be further divided into those that predate or postdate the catastrophic eruption of Mt. Mazama that formed Crater Lake about 7,700 years ago. Those that predate the eruption lie beneath Mazama Ash; those that postdate it lie above it. On Newberry's southern flank, these ash deposits can reach 6 feet (1.8 m) thick.

The youngest lavas, for the most part, are concentrated within the caldera or along a northwest-trending rift zone within the boundaries of the national monument. Of these, rhyolitic lavas occupy the caldera, while basaltic lavas erupted from the rift zone. Caldera lavas include the Big Obsidian Flow, the 1,300-year-old rhyolite that is Newberry's most recent eruption. This flow and many of its features are accessible by a trail within the caldera. The northeast-trending rift zone of young volcanic activity includes some twelve separate flows of both aa and pahoehoe lava, erupted from cones and fissure vents about 7,000 years ago. Lava Butte, site of the monument's visitor center and museum along US 97, is the largest of these vents and produced a flow that forced the Deschutes River to move its channel farther west. All told, some four hundred cinder cones and fissure vents of pre- and post-Mazama ages dot Newberry's surface.

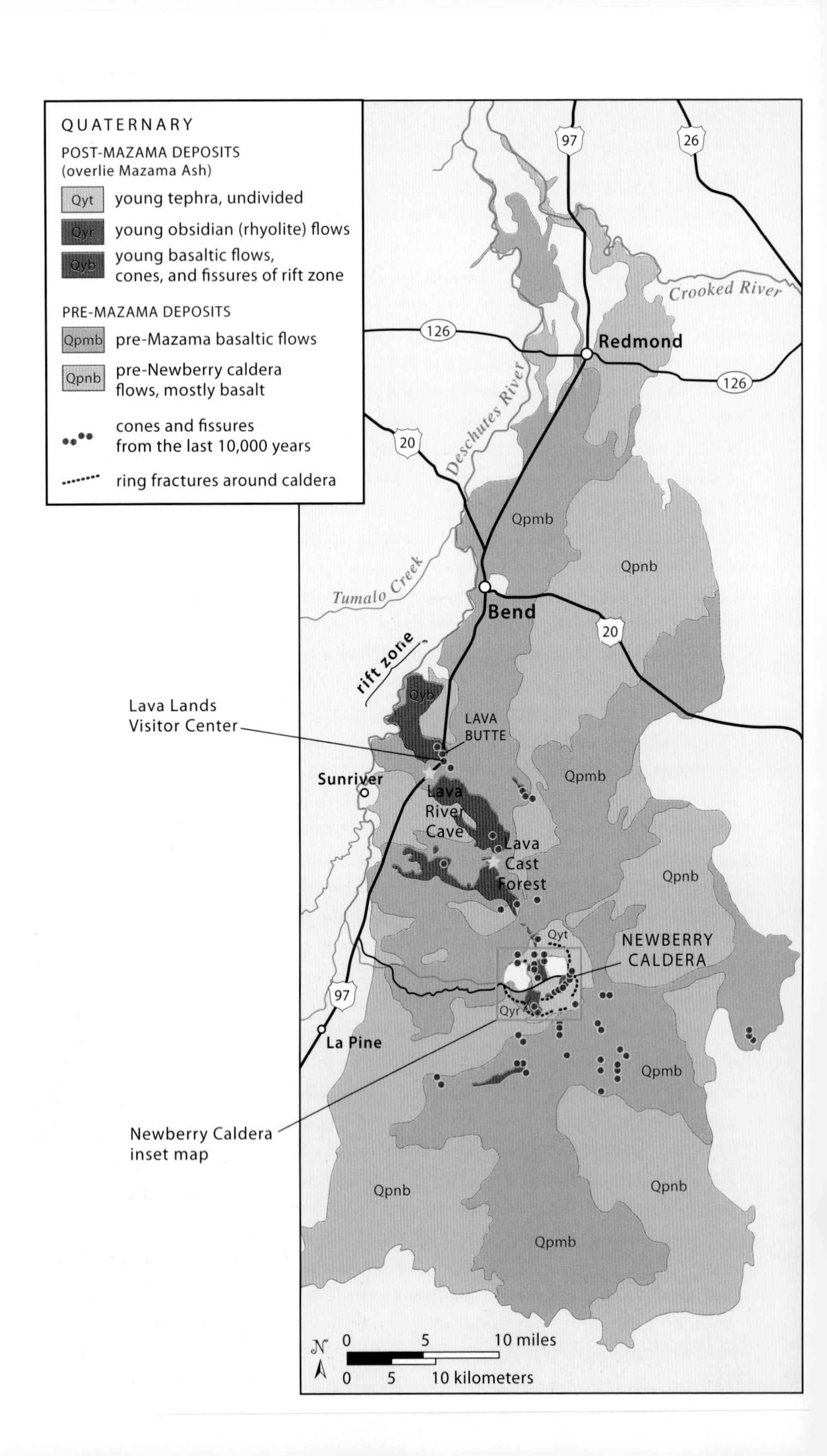
QUATERNARY
POST-MAZAMA DEPOSITS
(overlie Mazama Ash)
Qyt
young tephra, undivided
Qyr
young obsidian (rhyolite) flows
Qyb
young basaltic flows, cones, and fissures of rift zone
PRE-MAZAMA DEPOSITS
Qpmb
pre-Mazama basaltic flows
Qpnb
pre-Newberry caldera flows, mostly basalt
cones and fissures from the last 10,000 years
ring fractures around caldera
97
26
Crooked River
126
Redmond
126
Deschutes River
20
Qpmb
Qpnb
Tumalo Creek
Bend
20
rift zone
Qyb
Lava Lands Visitor Center
LAVA BUTTE
Sunriver
Lava River Cave
Lava Cast Forest
Qpmb
Qpnb
Qyt
NEWBERRY CALDERA
97
Qyr
La Pine
Qpmb
Newberry Caldera inset map
Qpnb
Qpnb
Qpmb
0
5
10 miles
0
5
10 kilometers

Because so many of Newberry's flows are young, they preserve examples of tree molds and lava tubes, features that often collapse, erode, and otherwise degrade with time. Tree molds form when lava inundates a forested area and engulfs its trees. The trees burn, but sometimes the lava that initially chilled around the tree trunk preserves the trunk's mold as a cylindrical hole in the basalt. If the hole were later filled with sediment, the sediment would form a cast. The Lava Cast Forest at Newberry offers a variety of tree molds left by ponderosa pines and few, if any casts. The molds are visible from a 1-mile-long (1.6 km) paved trail through one of the smaller post-Mazama lava flows of the northwest rift zone. The trail can be reached at the end of the 9-mile (14 km) gravel road that intersects US 97 about 2 miles (3.2 km) south of the turnoff for Sunriver.

Newberry Volcano. –Modified from MacLeod and others, 1995; Donnelly-Nolan and others, 2011

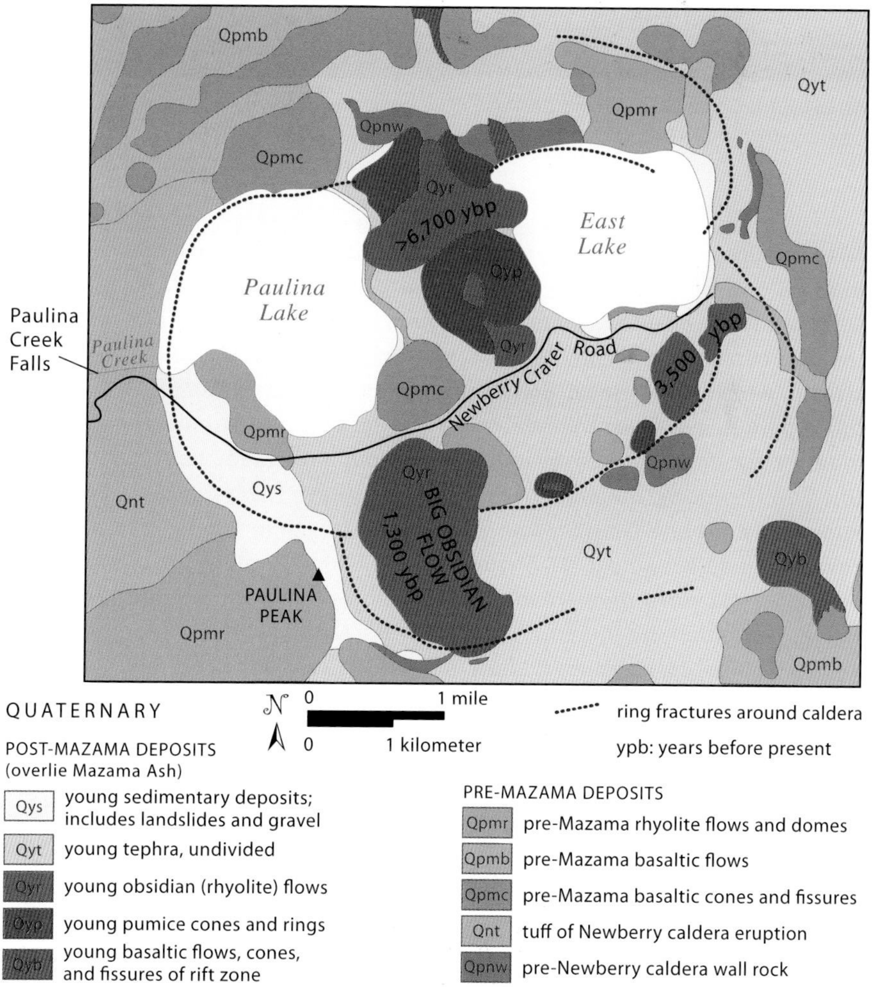

Qpmb
Qpmr
Qyt
Qpnw
Qpmc
Qyr
>6,700 ybp
East Lake
Paulina Lake
Qyp
Qpmc
Paulina Creek Falls
Paulina Creek
Newberry Crater Road
3,500 ybp
Qpmr
Qpnw
Qys
Qnt
BIG OBSIDIAN FLOW 1,300 ybp
Qyt
Qyb
PAULINA PEAK
Qpmr
Qpmb
QUATERNARY
0 1 mile
0 1 kilometer
ring fractures around caldera
ypb: years before present
POST-MAZAMA DEPOSITS (overlie Mazama Ash)
Qys young sedimentary deposits; includes landslides and gravel
Qyt young tephra, undivided
Qyr young obsidian (rhyolite) flows
Qyp young pumice cones and rings
Qyb young basaltic flows, cones, and fissures of rift zone
PRE-MAZAMA DEPOSITS
Qpmr pre-Mazama rhyolite flows and domes
Qpmb pre-Mazama basaltic flows
Qpmc pre-Mazama basaltic cones and fissures
Qnt tuff of Newberry caldera eruption
Qpnw pre-Newberry caldera wall rock

Aerial view southward of the Big Obsidian Flow within Newberry Volcano. The high point on the right (west) side is Paulina Peak. Snow accentuates the flow ridges.

Lava tubes are the tube-shaped caves that sometimes form during basaltic eruptions. Because flowing lava cools most quickly on its outer surfaces, it can develop crusts that insulate the flowing lava underneath. As the lava cools inward from its sides, the flow may concentrate into a single stream, which flows down and eventually out of its self-made cave. Lava tubes permit lava flows to travel much greater distances from their sources than if they simply flowed overland. Numerous lava tubes have been identified on Newberry, the most accessible being Lava River Cave, the longest known lava tube in Oregon, with a length greater than 1 mile (1.6 km). The cave crosses beneath US 97 about 1 mile (1.6 km) south of the Lava Lands Visitor Center. Lava River Cave formed in flows that postdate formation of Newberry Caldera but predate eruption of the Mazama Ash.

The floor of Newberry's caldera is anything but flat, filled with a range of rhyolitic features, most of which erupted in the 7,700 years since the Mazama eruption. Several obsidian flows and pumice or tuff cones and rings are covered with a blanket of loose material, collectively called tephra, consisting mostly of rock fragments, pumice, and ash. Glass from the obsidian flows was a valuable resource for prehistoric people who ventured into the caldera during warm summer months to quarry it for tool making. The caldera also hosts several small, mostly pre-Mazama basaltic flows and cones.

Two lakes, East Lake and Paulina Lake, fill part of the caldera, separated by a prominent pumice cone that erupted about 6,700 years ago. Some researchers suggest that prior to formation of the pumice cone, the lakes were joined as one. Both lakes lack inlets, obtaining water through snowmelt and subsurface flow. Paulina Lake empties westward into Paulina Creek, which spills some 100 feet (30 m) over Paulina Creek Falls, a short distance below the lake and easily accessed from the road. The falls are made of welded tuff erupted during the caldera-forming eruption 75,000 years ago. Paulina Lake is one of Oregon's deepest, with a maximum depth of about 250 feet (76 m).

Several hot springs, which reach temperatures of up to 149°F (65°C), exist near the lakes, and a steam vent was reported along the northeastern margin of the Big Obsidian Flow. In the late 1970s and early 1980s, the US Geological Survey drilled three exploratory wells between the Big Obsidian Flow and East Lake to evaluate Newberry's geothermal energy potential. The deepest well reached a temperature of 509°F (265°C) at a depth of just over 3,000 feet (900 m). These features imply the presence of a shallow heat source, perhaps fueled by shallow magma. In 2012, researchers from the University of Oregon and Michigan Tech published geophysical findings that confirmed the presence of a magma body beneath Newberry. They argued that it is between 0.38 and 1.92 cubic miles (1.6–8 km^3) in volume and resides between 1.9 and 3.7 miles (3–6 km) beneath the caldera floor.

Crater at the top of the Lava Butte cinder cone, showing red cinders in the foreground and Newberry Volcano in the left background toward the southeast. Note the many cinder cones on the flanks of Newberry.

US 197
THE DALLES—US 97

62 miles (100 km)

See map on page 238

US 197 climbs from the Columbia River to the Columbia Plateau on the east side of the Cascades, passing through lavas of the Columbia River Basalt Group, topped by sediments of the Dalles Formation and even younger, olivine-bearing basalt flows. US 197 has more interesting exposures than the parallel US 97, and it can be blissfully solitary, with magnificent views of the High Cascades and the Deschutes River canyon.

Near the beginning of the trip, near milepost 1 at the intersection of US 197 and US 30, an amazing roadcut of Wanapum Basalt of the Columbia River Basalt Group lies on the east side of the road across from a large pull-out. This exposure shows pillow basalt overlain by columnar basalt. The pillows are surrounded by palagonite, a nondescript yellow-brown mineral. Both the pillows and the palagonite formed because the lava flowed into a lake. The outer skin of basaltic lava chills rapidly in the water, while the inside remains hot and plastic, bulging into pillows. Some explosions occur when the lava enters the water, generating fragmented volcanic glass or ash, which is almost instantly altered to palagonite. Note that the pillows are somewhat ellipsoidal in shape, and their long axes point down to the southwest, the direction the lava poured into the lake.

Note how the pillows seem to grade into the overlying columns. They are both part of the same flow. This large lava flow flowed toward the southwest and inundated a lake, probably on an early floodplain of the Columbia River. Early parts of the flow poured into the lake, forming the pillows, and once the lake was filled, subsequent parts of the flow cooled slowly on the land surface, forming the columns.

US 197 heads steeply out of the Columbia Gorge through more outcrops of Columbia River Basalt Group. It encounters overlying sedimentary rock of the Dalles Formation between mileposts 2 and 3. The Dalles Formation consists mostly of volcanic material eroded from the High Cascades between about 7.5 and 5 million years ago and deposited as debris flows or stream deposits. For about the next 10 miles (16 km), the highway rises over hills topped with the Dalles Formation and dips into valleys that expose the Columbia River Basalt Group. From just south of Dufur, the highway gradually climbs over gently north-dipping Columbia River Basalt Group toward the summit of Tygh Ridge, just south of milepost 22. Tygh Ridge marks an anticline that trends approximately east-west, with a gentle north limb and a steeper south limb. On the south side of the ridge, the road descends steeply through a canyon cut into the basalt, which dips southward near the canyon mouth. Consistent with the folding, Tygh Valley is floored by the overlying Dalles Formation. The Dalles Formation is well exposed on the slopes of the mesa that marks the south edge of the valley.

Tygh Creek and the White River join in Tygh Valley, en route to the White River's confluence with the Deschutes River. At White River Falls State Park, the

White River drops a total of 137 feet (42 m) over two shelves of Columbia River Basalt Group on its descent into the Deschutes River canyon.

Juniper Flat is developed on basalt flows that overlie the Dalles Formation. They are lighter in color than the Columbia River Basalt Group and contain

Wanapum Basalt at the intersection of US 197 and US 30. The lower part of the flow consists of pillow basalt and palagonite, which formed in water, while the upper part exhibits a colonnade, typical of forming on land.

View south of the Deschutes River canyon and Maupin, with Tygh Ridge on the horizon. Nearly all the rock in the photograph belongs to the Columbia River Basalt Group except for the prominent cliff of olivine-rich basalt in the middle background. Note the angular unconformity between it and the gently dipping Columbia River Basalt Group beneath.

scattered crystals of white plagioclase and green olivine. Near milepost 43, the road descends steeply into Maupin and the Deschutes River canyon. The road passes several exposures of the Dalles Formation, as well as basalt flows of the Grande Ronde unit of the Columbia River Basalt Group. Where the road crosses the Deschutes River, look upstream to see cliffs of the young olivine-bearing basalt at the top of the canyon and Grande Ronde Basalt at the bottom.

South of Maupin, US 197 climbs through gently north-dipping Grande Ronde and Wanapum Basalt, both of the Columbia River Basalt Group. As it climbs, look back at Maupin to see a subtle angular discordance between the tilted Grand Ronde Basalt and the nearly flat olivine-bearing basalt. Between mileposts 58 and 59, the road offers some intriguing views westward into the Deschutes River canyon. There, directly below the Grande Ronde Basalt are exposures of ash-flow tuffs and ash-rich sedimentary rock of the John Day Formation. Criterion Summit, just south of milepost 62 at an elevation of 3,360 feet (1,020 m), offers more views, with a semicircular map that locates many of the peaks of the High Cascades. A quarry on the east side of the highway mines the Grande Ronde Basalt.

US 395
John Day—Burns
70 miles (113 km)

This section of US 395 climbs over the Aldrich Mountains through the Malheur National Forest. The route begins in the Blue Mountains Province and finishes in the Lava Plateaus. I included the road in this chapter because it passes through Devine Canyon, which exposes three major ash-flow tuffs of the Lava Plateaus, including the Devine Canyon Tuff.

For about the first 2 miles (3.2 km) south of John Day, the road passes recently deposited river gravels. Behind the gravels are exposures of basaltic rocks of the Strawberry Volcanics, a collection of mostly andesitic lava flows that erupted about 15 million years ago and are probably related to the Columbia River Basalt Group.

Just south of Canyon City, the road crosses into the Canyon Mountain Complex, oceanic lithosphere of the Baker terrane, likely the roots of a volcanic arc. The main rocks visible in the canyon are amphibolites, gabbros, and serpentinites. The amphibolites and gabbros are metamorphosed and unmetamorphosed oceanic lithosphere, respectively. The amphibolites exhibit a metamorphic layering called foliation, whereas the gabbros do not. The serpentinites are from a deeper, ultramafic section of the oceanic lithosphere and are also metamorphosed. The serpentinite tends to be green or dark brown with a scaly, somewhat shiny foliation. Good examples of serpentinite crop out near the canyon entrance, and gabbro is found just north of milepost 4. In addition, some of the outcrops show a great deal of fracturing and veining, such as the fine gabbro directly across the gravel quarry near milepost 4.

The road passes into Malheur National Forest near milepost 10, with exposures of Triassic sandstone. Good exposures of similar sandstone, as well as shale, show up along the road for the next 5 miles (8 km) as the road climbs into the Aldrich Mountains. These rocks are part of the Izee terrane, deposited over the top of the Baker terrane but likely before the two were accreted to North America. The interbedding of sandstone and shale through here suggests that many of these rocks originated as turbidites. They are also highly deformed, with variable orientations of bedding from roadcut to roadcut, and zones of intense fracturing and contorted beds. Some especially good exposures exist between mileposts 12 and 14.

Between mileposts 15 and 16 the road crests the Aldrich Mountains at an elevation of 5,152 feet (1,570 m) and descends through sparse outcrops to an open valley floored by Late Tertiary ash-rich volcanic and sedimentary rocks as well as recently deposited alluvium. From the valley, you can see the Strawberry Mountains to the northeast, formed of resistant rock of the Baker terrane.

South of Seneca, US 395 closely follows Silvies Creek, which eventually empties into the closed basin of Malheur Lake, south of Burns. The road passes occasional outcrops of Jurassic sandstone, also part of the Izee terrane. Late

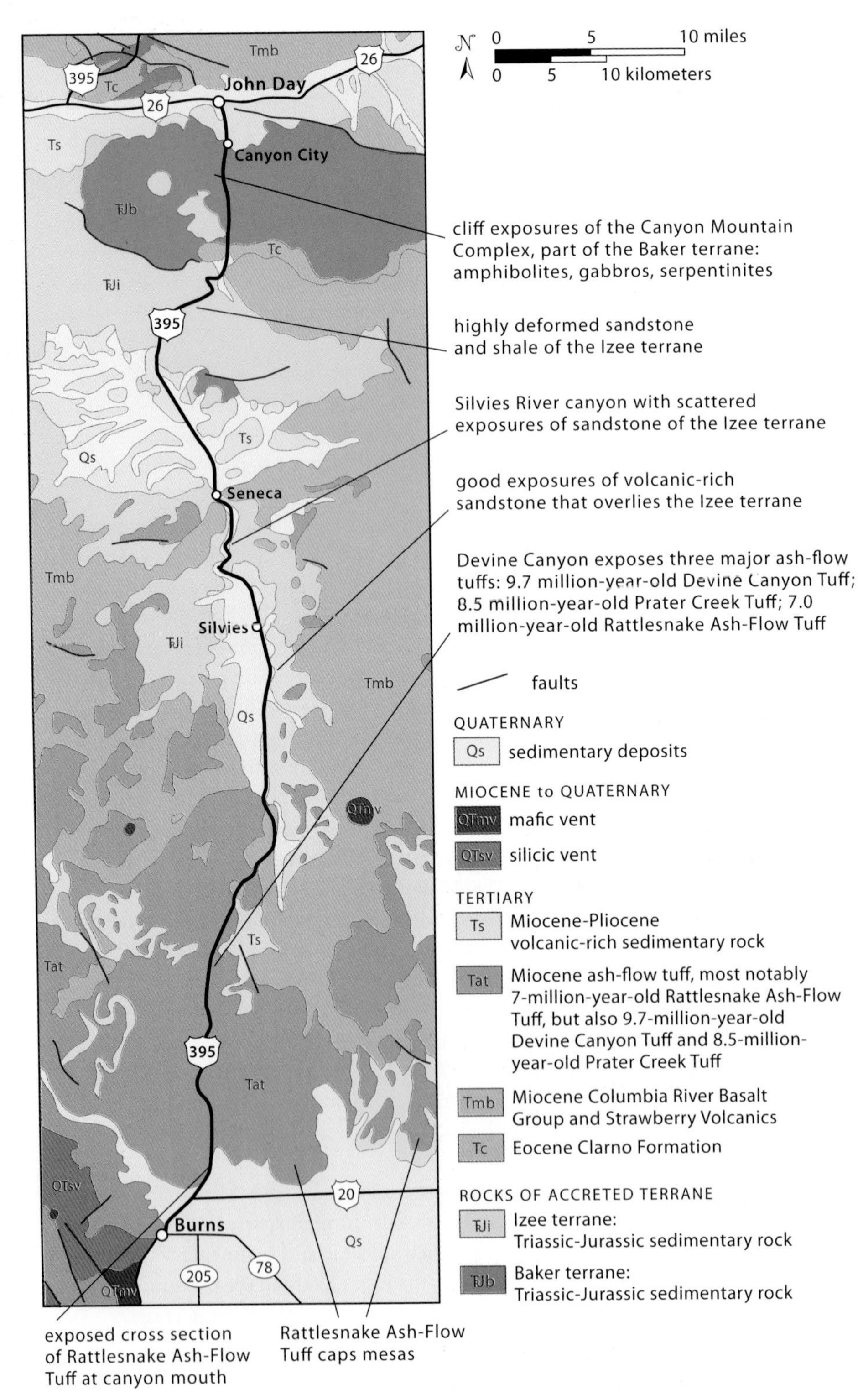

Geology along US 395 between John Day and Burns.

The lighter beds of sandstone are interbedded with gray shale in turbidites of the Izee terrane.

Tertiary ash-rich sedimentary rocks overlie the sandstone, some of which show up particularly well 2 miles (3.2 km) south of Silvies. The presence of alkali in local drain ditches reflects the arid environment, in which rainwater leaches salts from the bedrock and then evaporates, leaving a salty coating on the surface.

Near milepost 44, the road climbs into a pine-covered upland, with occasional ledges of ash-flow tuffs. At about milepost 51, the road begins its descent down Devine Canyon, a geologically divine place where you can see the entire three-part stratigraphy of Late Tertiary welded tuffs. Near road level, the 9.7-million-year-old Devine Canyon Tuff crops out, above that is a cliff made of 8.5-million-year-old Prater Creek Tuff, and above that is the 7-million-year-old Rattlesnake Ash-Flow Tuff, which also forms a cliff. Because the rocks tilt gently southward, the road gradually crosses into younger flows as it continues southward. Exposures of the ash flows and intervening ash-rich sedimentary rocks line the road. At the end of the canyon near milepost 65, you can see a nearly complete cross-sectional view of the Rattlesnake Ash-Flow Tuff on the west side of the road.

Each ash-flow tuff covers vast areas of central Oregon and is approximately centered on Burns. When it erupted, the Rattlesnake tuff covered more than a tenth of the state! Considering the thickness of the ash flows—the Rattlesnake

View of the three welded tuff units in Devine Canyon. The lowest cliff consists of the Devine Canyon Tuff. Above that, close to the top, is the Prater Creek Tuff. At the top is the 7-million-year-old Rattlesnake Ash-Flow Tuff.

Ash-Flow Tuff in this area reaches 75 feet (23 m) thick—you can sense the incredible size of the volcanic eruptions that produced them. Almost instantaneously, a huge section of Oregon was incinerated beneath a superheated flow of ash!

OR 78
Burns—Burns Junction
92 miles (148 km)

To reach Oregon's lonely southeast corner, you must take OR 78 southeast from Burns. You cross the vast Pleistocene lakebed of the Harney Valley and then pass over youthful lava flows and the northern tip of the Steens Mountain fault block. Glacial Lake Malheur, which was Oregon's largest Pleistocene lake, filled the Harney Valley, covering an area of 920 square miles (2,380 km^2). Today, Malheur Lake is its largest remnant and provides important wetlands for a wide variety of migratory birds. Near milepost 22, the road passes the north side of Warm Springs Butte, which consist of a surprising variety of rock. Most obvious are the red deposits of Quaternary-age cinder cones, but the butte also hosts Pliocene basalt and volcanic-rich sedimentary rock. During Pleistocene time, the butte would have been an island in Glacial Lake Malheur, rising well above the 35-foot-deep (10 m) lake.

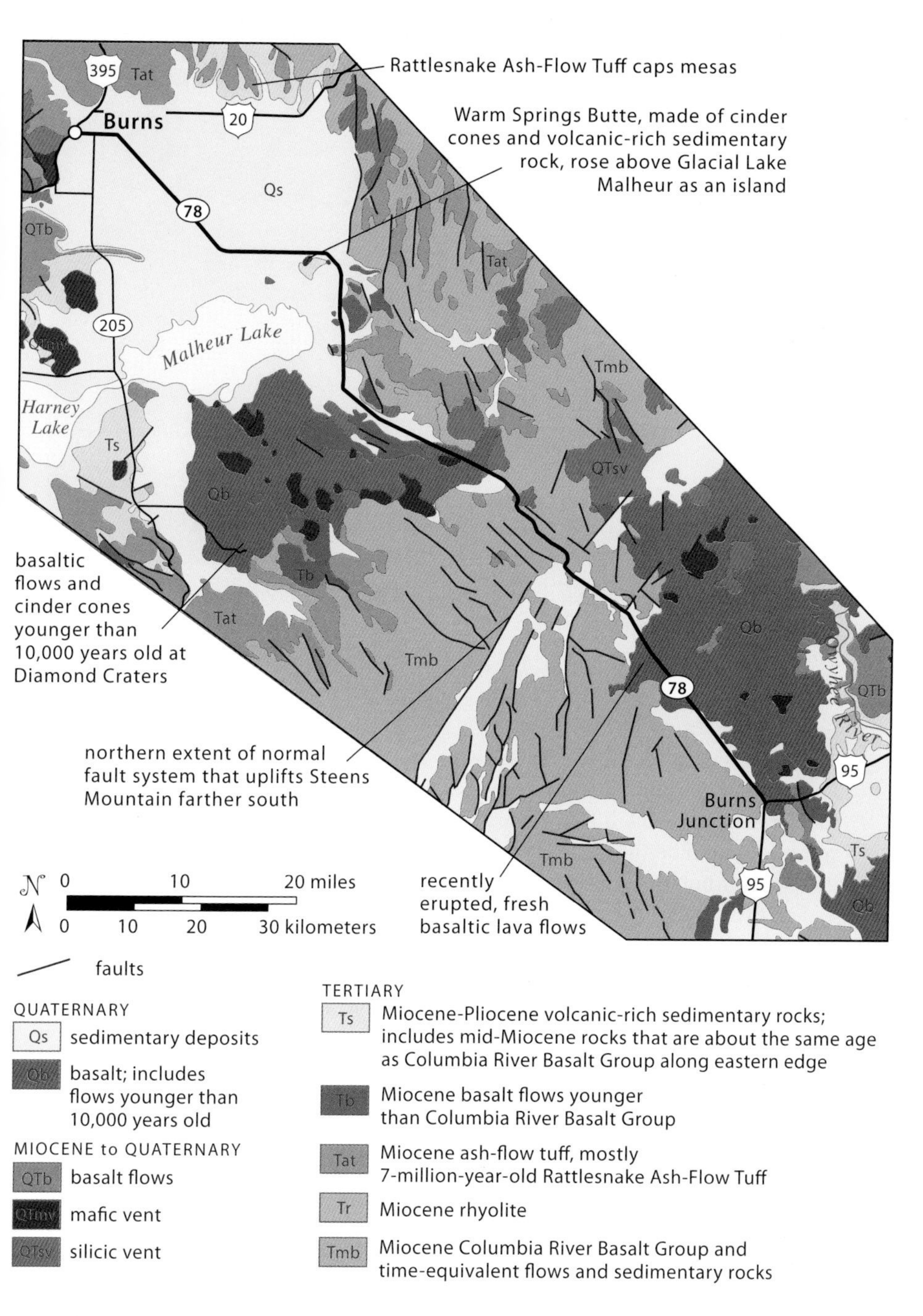

Geology along OR 78 between Burns and Burns Junction.

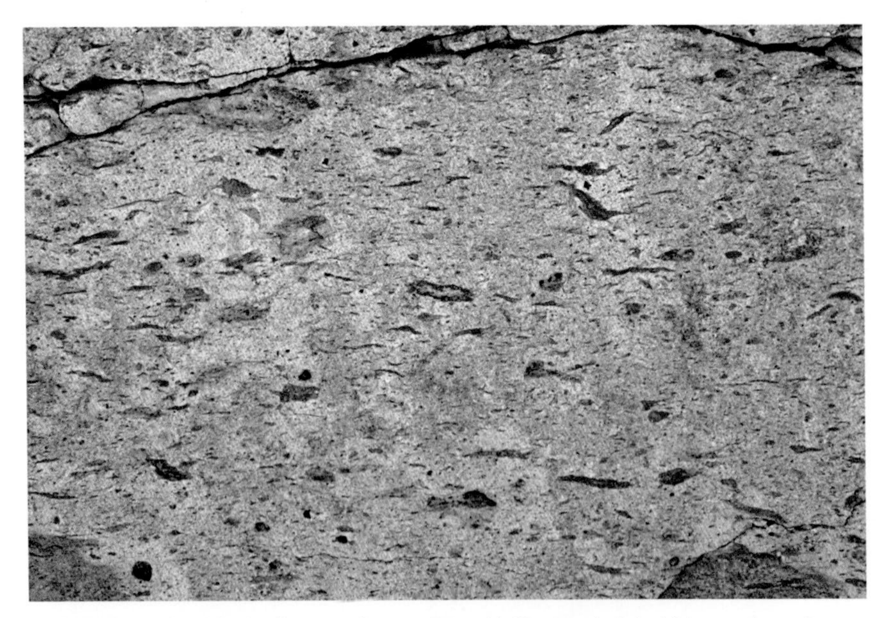

Close-up of welded tuff near milepost 35, with flattened air bubbles and pumice, as well as rock fragments. The photo is about 12 inches (30 cm) across.

Near milepost 35, the road rises gently to pass between two flat-topped hills. Pliocene basalt caps the hill to the west, and the 9.7-million-year-old Devine Canyon Tuff caps the hill to the east. The long roadcut between the two hills consists of an older welded tuff. A close look at the rock shows that it is full of flattened air bubbles and fragments of pumice and rock.

For the next 26 miles (42 km) to a small pass near milepost 61, the road passes Miocene- through Quaternary-age basalt flows, punctuated by occasional dacite vents. A quarry exposes one of these vents at milepost 51. East of the quarry, Malheur Cave Road heads north to Malheur Cave, a 3,000-foot-long (910 m) lava tube.

The road crosses deeply weathered dacite between mileposts 59 and 60. For the first few miles south of the pass, the road follows a narrow canyon in Miocene basalt formed along a northwest-trending fault zone. It turns abruptly southward between mileposts 63 and 62 and leads across the fault and a ridge of the basalt, entering a wide, square-shaped valley that formed at the north end of the Steens Fault Zone.

The view southward across this valley is unusually instructive, as it shows several normal faults related to Basin and Range extension. The prominent ridge on the west side of the valley continues southward at progressively higher elevations until it becomes Steens Mountain, which has been lifted up by a normal fault along its base. The southeastern corner of the valley is framed by another faulted ridge that rises to the east. The valley is a graben, a block dropped down between the Steens Fault Zone on the west and this other fault on the east. In between, the down-dropped block itself is faulted. The low hills at the southern edge of the valley contain a prominent ledge, the same rock unit that lies at different elevations along the valley's walls because of faulting. All of the bedrock is part of the Steens Basalt, the earliest flows of the Columbia River Basalt Group, which erupted about 16.6 million years ago.

The road rises gently out of the valley to the southeast and then drops into a smaller valley that is completely surrounded by higher land. Because it has no outlet, this smaller valley occasionally floods during wet periods. Notice the dry lakebed that covers its floor. Many similar dry lakebeds dot southeastern Oregon. As the road climbs out of this valley, it passes hills made of basalt and then crosses another normal fault just before reaching another pass. Immediately north, you can see faulted and tilted basalt flows.

Several miles farther south, OR 78 crosses the Late Pleistocene Saddle Butte lava field, its youthful surface relatively uninhabited by vegetation and consisting of highly angular, black, fresh-looking basalt. The lava field provides a nice contrast to the slightly older basalt farther on. From there to Burns Junction, the road skirts the edge of this older flow.

View southwestward along the northern extent of Steens Fault Zone, which runs along the base of the cliffs to the right. The ridge on the left side of the photo is uplifted to the east by another fault. The low area in between is a graben, which is faulted in several places, giving rise to the hills in the middle.

Blue Mountains

The Blue Mountains Province displays some of Oregon's most diverse scenery and geology, traversed by uncrowded, sometimes lonely highways. In addition to the Blue Mountains, the region also encompasses the Ochoco, Strawberry, and Wallowa Mountains, some of which sustain snowfields throughout the year. Many peaks within these ranges exceed 8,000 feet (2,400 m) in elevation, with Sacajawea Peak in the Wallowas reaching 9,838 feet (2,999 m). The Wallowa and Strawberry Mountains were glaciated during the Pleistocene Epoch. The ice eroded U-shaped valleys and glacial cirques and deposited glacial moraines.

The region hosts several of Oregon's accreted terranes and one of the most prolific mammal fossil collections in the world in the John Day Basin. Lava flows of the Columbia River Basalt Group, described in detail in the chapter

Wallowa Lake is dammed by a moraine (left) from a Pleistocene glacier that flowed out of the Wallowa Mountains. Granite boulders, similar to those in the foreground, make up most of the moraine and were carried out of the mountains by the glacier.

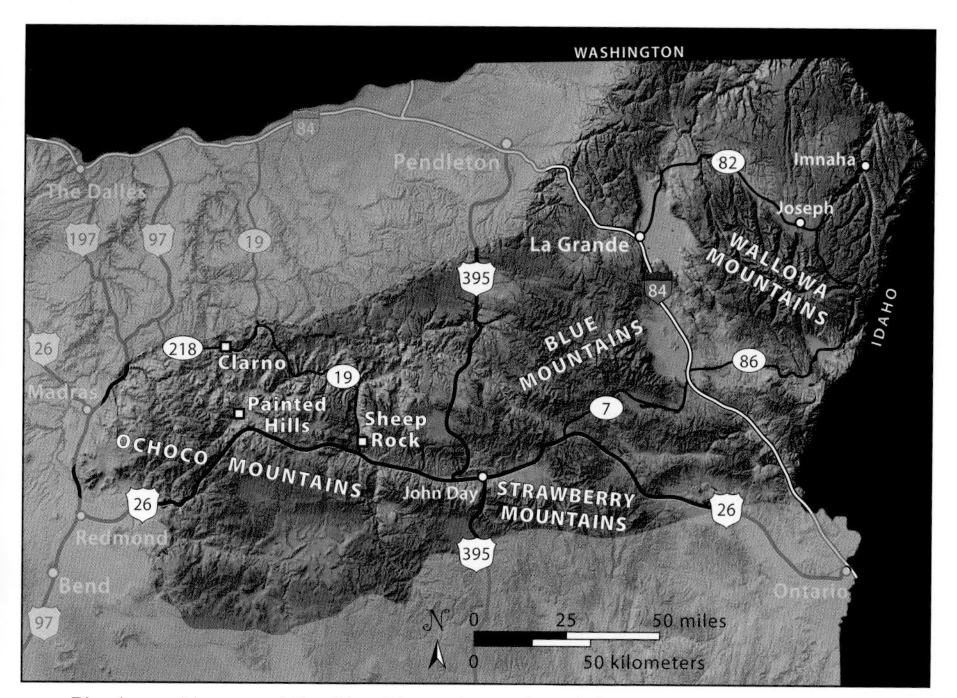

Physiographic map of the Blue Mountains region of Oregon. –Base image from US Geological Survey, National Elevation Data Set Shaded Relief of Oregon

on the Lava Plateaus, cover much of the Blue Mountains Province, with the underlying rocks of the John Day Basin and accreted terranes only visible where the basalt has been eroded away. Scattered remnants of the basalt flows cap numerous ridges and peaks. In other places, rivers have cut valleys into the basalt to reveal the older rock below, perhaps the most dramatic example being Hells Canyon.

Accreted Terranes

Along with the Klamath Mountains in southwestern Oregon, the Blue Mountains reveal Oregon's history of accreted terranes, a history that mostly preceded the Cenozoic Era. Rock of the accreted terranes originated in oceanic volcanoes, subduction zones, and related sedimentary basins of the ancient Pacific Ocean, in some cases many hundreds of miles away from their present locations. From north to south, the accreted terranes of the Blue Mountains are called the Wallowa, Baker, Izee, and Olds Ferry terranes. Granitic intrusions, which may or may not have North American origins, intruded across terrane boundaries. They are called stitching plutons because they appear to stitch the terranes together. Younger sedimentary rock was deposited over the terranes in environments that ranged from streams and rivers to shallow and deep oceans.

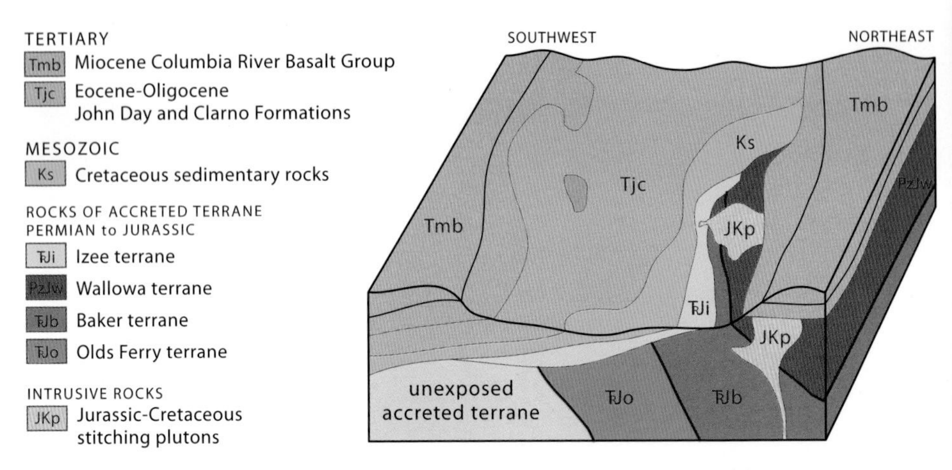

Simplified block diagram from southwest to northeast of part of the Blue Mountains, illustrating general age relations. Accreted terranes underlie everything. The Izee terrane was deposited over the top of them after they came together but possibly before accretion. In turn, the Izee is intruded by the stitching plutons. The plutons are overlain by Cretaceous sedimentary and younger rock, all of which formed on North America.

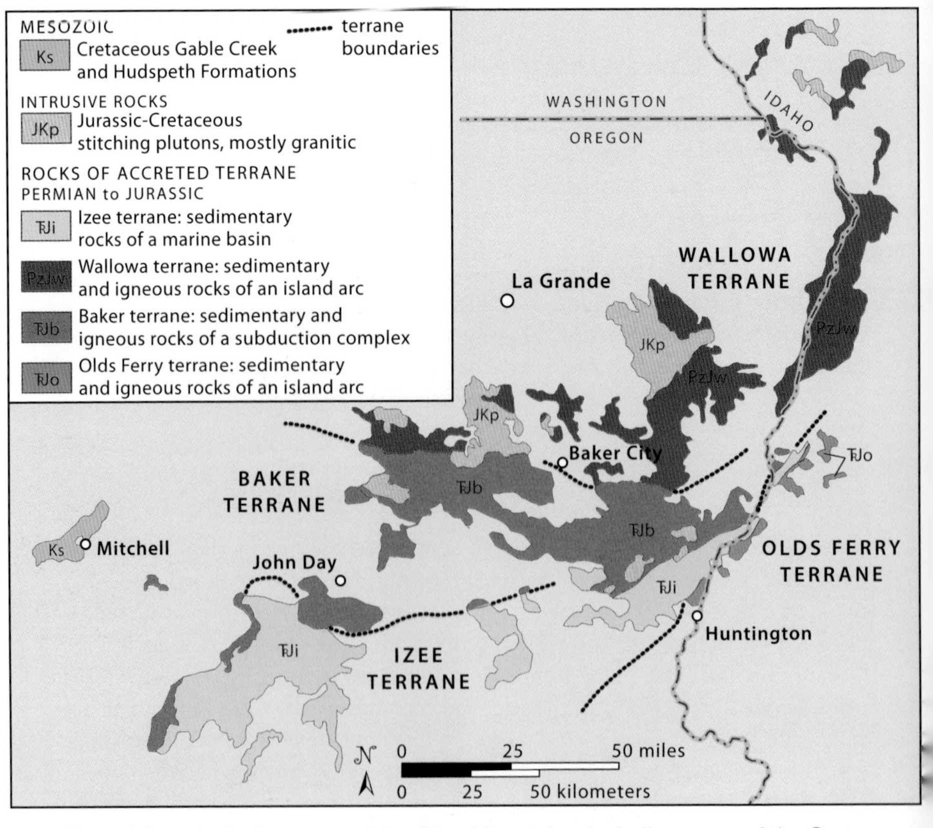

Map of the principal terranes of the Blue Mountains, including some of the Cretaceous sedimentary rocks that were deposited after the accretion. –Modified from LaMaskin and others, 2011

Wallowa Terrane

The Wallowa terrane occupies most of northeastern Oregon, including much of the lower elevations of Hells Canyon, the Wallowa Mountains, and the scattered hills visible from I-84 just north of Baker City. The Wallowa terrane formed in an island arc setting, far from North America during Permian to Early Jurassic time. For the most part, the Wallowa terrane rocks consist of marine shale, chert, sandstone, and minor limestone, as well as basaltic and andesitic volcanic rocks and minor intrusive rocks. The granite of the Wallowa Mountains, however, is not a part of this terrane. It intruded the terrane as a stitching pluton.

Baker Terrane

The Baker terrane marks the remainder of one or more subduction zones. The terrane contains an enormous variety of rock types mixed together as mélange. Much of the mélange appears scaly or sheared, as if it had been caught in a vice whose grips moved from side to side or up and down. Besides the nearly ubiquitous marine sedimentary and volcanic rock, the Baker terrane includes a sequence of rock that likely formed at great depths beneath a volcanic arc, as well as serpentinite and blueschist. These two distinctive rock types give further clues to its subduction zone origin because they form by metamorphism of the iron- and magnesium-rich rock of the oceanic lithosphere. Serpentinite forms when oceanic lithosphere interacts at moderate temperatures with seawater, typically near spreading centers. Blueschist forms by metamorphism of the rock in the high-pressure and low-temperature environments that characterize subduction zones. Rock of the Baker terrane is beautifully exposed along I-84 from Baker City some 30 miles (48 km) southeastward to the town of Weatherby. It also forms a band that stretches westward nearly as far as Madras.

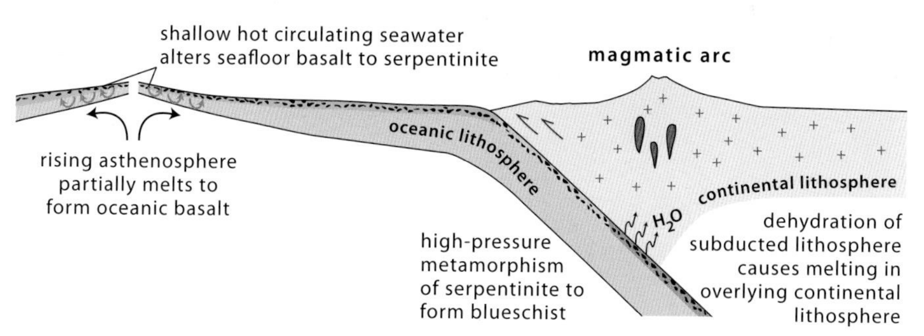

Plate tectonic settings for the formation of serpentinite and blueschist.

Olds Ferry Terrane

The Olds Ferry terrane, which formed in an island arc setting during Triassic time, consists mostly of volcanic rock. It is found only as scattered exposures from Huntington northeastward into Idaho.

Izee Terrane

The Izee terrane, a thick sequence of sedimentary and minor volcanic rock, was deposited in a marine basin during Triassic and Jurassic time. Some of the Izee deposits contain material that was eroded from and deposited directly on rocks of the adjacent Baker terrane. This relationship has led some researchers to argue that the Izee sequence of strata does not represent an independent accreted terrane but instead formed by deposition on older crust of the Baker terrane.

Stitching Plutons and Sedimentary Rocks

Some of the most dramatic Mesozoic rocks of the Blue Mountains are the stitching plutons, which intruded the accreted terranes during Late Jurassic and Early Cretaceous time. Generally granitic in composition, these rocks form upland areas with abundant outcrops, because the rock is resistant to erosion. The Eagle Cap Wilderness of the Wallowa Mountains, which is almost entirely within the Wallowa pluton, is perhaps the most spectacular example, but several smaller plutons grace the roadsides of some of the highways of the Blue Mountains. As these plutons crystallized, they caused circulation of hot fluids, which precipitated a variety of metals, most notably gold and silver. The Blue Mountains are riddled with mine workings in and adjacent to the plutons.

The granitic outcrops tend to be noticeably rounded, typical of most granite landscapes. Granite is a homogeneous rock, so it weathers uniformly.

The Goose Rock Conglomerate in the Sheep Rock Unit of John Day Fossil Beds National Monument was deposited in river systems that flowed westward toward the ocean during Cretaceous time.

Because any protruding edges of an outcrop expose more surface area to attack by weathering processes, the irregularities are destroyed, and the outcrop maintains a rounded shape. Fracture surfaces enhance weathering by giving water access to the outcrop's interior, so fractured granite outcrops are not as rounded.

The youngest Mesozoic rocks of the Blue Mountains are the Cretaceous sedimentary rocks exposed mostly around Mitchell. These rocks contain particles that came from the Idaho Batholith, a huge granitic intrusion that covers much of west-central Idaho. As no one doubts the Idaho Batholith intruded North America, we can infer that these Cretaceous rocks mark some of the first deposits in Oregon that are truly North American—sediments eroded from rock in North America and deposited in North America.

John Day Fossil Beds

The colorful Cenozoic rocks of the Blue Mountains have yielded tens of thousands of fossil plant and animal specimens of hundreds of different species. The rocks also record important changes in Oregon's plate tectonic setting and climate. Much of this rock, including volcanic lavas, ashes, and tuffs, accumulated in a series of basins in central Oregon from Early Eocene to Late Miocene time, a period of some 40 million years. Many of the finer-grained deposits are preserved as ancient soils called paleosols. These rocks, collectively called the John Day Fossil Beds, include the Clarno, John Day, Mascall, and Rattlesnake Formations. Sandwiched between some layers are lava flows of the Columbia River Basalt Group. The Mascall and Rattlesnake Formations each contain a distinctive tuff layer, the Mascall Tuff and the Rattlesnake Ash-Flow Tuff, respectively.

The Clarno Formation was deposited directly by eruptions from the Clarno volcanoes, an early volcanic arc. The arc originated from a subduction zone probably east of but about parallel to the modern one off Oregon's coast. The formation consists of up to 6,000 feet (1,800 m) of basaltic and andesitic lava flows, coarse-grained sedimentary rocks, and lahars. The transition from the Clarno Formation to the John Day Formation indicates an important change in the position of the arc. It shifted to the west by the beginnings of John Day time, some 37 million years ago. We know this because rocks of the John Day Formation are much finer-grained, derived from volcanoes much farther away. The rocks of the John Day Formation consist mostly of rhyolitic tuffs, weathered to form distinctive paleosols, as well as some lava flows and minor amounts of conglomerate. The volcanic deposits likely originated from the Western Cascades and other calderas in central Oregon, including the Crooked River Caldera. Occasional lakebeds and gravel-filled channel deposits speak to floodplain and localized lake environments for the John Day, Mascall, and Rattlesnake Formations.

Established in 1974, John Day Fossil Beds National Monument protects nearly 14,000 acres (57 km^2) of the outcrop area of the John Day Basin. The monument consists of three different park units, each of which is comparatively small but beautifully illustrative of a different part of the Cenozoic rock sequence. In addition, through a federal interagency agreement, the monument oversees more than 750 fossil localities scattered throughout the John Day Basin, an area of more than 10,000 square miles (26,000 km²). The fossil beds have yielded a phenomenally detailed view of animal and plant life in central Oregon from Eocene through most of Miocene time. Since the discovery of the first fossil bones and teeth in 1861, tens of thousands of specimens have been recovered and described, and hundreds of new species have been recognized.

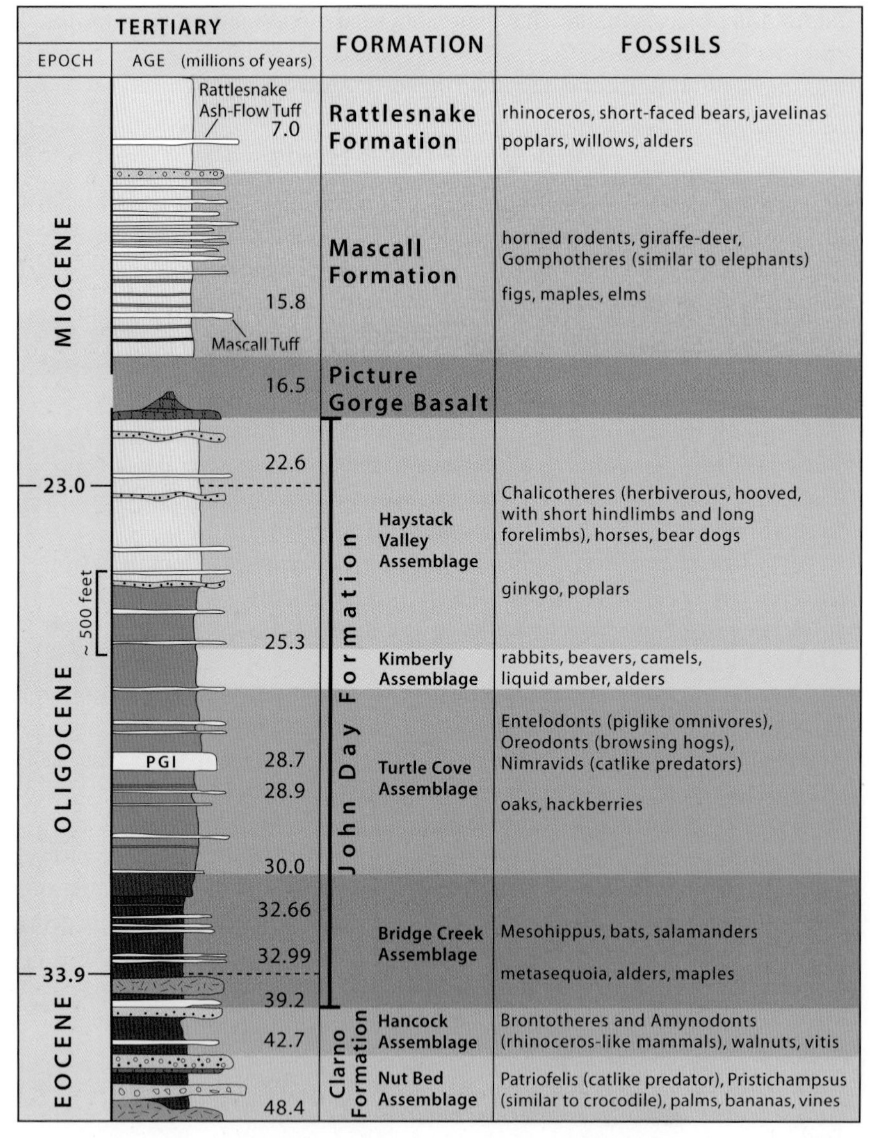

Generalized column of Eocene through Miocene rocks of the John Day Basin, including some representative fossils of each assemblage. The thin white layers, some with dates in millions of years, are prominent volcanic ash beds. PGI is the Picture Gorge Ignimbrite, a volcanic tuff. –Modified from Dillhoff and others, 2009

In addition to providing well-documented evolutionary lines, the change of species through time tracks changes in climate.

Over the course of 40 million years of deposition of the John Day fossil beds, the climate changed several times. The Clarno Formation formed during Eocene time in a tropical to subtropical environment, similar to southern Mexico today, whereas the John Day Formation formed during the Oligocene in temperate woodland settings that evolved to wooded grasslands. The overlying Mascall Formation records a brief return to warm and wet conditions in the Middle Miocene, followed by the Rattlesnake Formation, which records increasingly cooler and arid conditions at the end of Miocene time.

The transition from tropical to temperate climates is expressed in the rock record in several ways. Most directly, the types of plant fossils changed over time. The Clarno fossils consist of cycads (palms), banana tree seeds, and tropical varieties of sycamore. Those of the John Day consist largely of oak and alder near the base and increasing amounts of sagebrush, live oaks, and grass toward the top of the formation. Mammal fossils, too, suggest this transition. The Clarno contains fossils of mostly short-legged animals that could thrive in dense, tropical forests, whereas the John Day Formation contains fossils of longer-legged animals that were better suited for running over open land.

After deposition of each layer of fine sediment in the John Day Basin, chemical processes set to work, converting the sediment to soil. For the most part, the original material consisted of volcanic ash, which either settled out of the atmosphere directly onto the landscape or was transported by streams and deposited on floodplains or in lakes. Enough time elapsed between the deposition of individual layers that complete soils developed, one after the other. These ancient soils, called paleosols, indicate a change from tropical to temperate environments. Many of those in the Clarno Formation are thick and red, evidence of the deep, oxygenated weathering typical of tropical settings. Other paleosols of the Clarno apparently formed in swamp environments, because they exhibit gray colors from iron particles in a reduced (nonoxidized) state. As the climate dried, however, soils tended to show less extreme weathering and contain less clay and red color. Thus, paleosols in the John Day Formation are red near the base and grade upward to yellow and cream. The green colors that are so prominent higher in the John Day Formation originated from an originally brown soil. The green mineral celadonite, which forms through alteration of a specific type of volcanic soil, is disseminated throughout those soils.

Because each park unit lies some 30 to 40 miles (48–64 km) apart, they also represent different parts of the John Day Basin. Typical of rock deposited on land, the Clarno Formation has highly variable layers, which accumulated amidst the Clarno volcanoes. In the Painted Hills Unit, for example, the Clarno Formation consists mostly of andesite and rhyolite lava flows, but in the Clarno Unit, it contains lahars, conglomerate, tuff, and basalt, as well as andesite. The Clarno Formation is largely absent in the Sheep Rock Unit, perhaps eroded away. The younger John Day Formation was deposited directly on the Cretaceous-age Goose Rock Conglomerate.

The paleosols in all the units weather and erode into badlands topography, marked by a near-absence of vegetation and numerous steep-walled gullies cut into the easily eroded bedrock. The high clay content of many of the paleosols inhibits plant growth and promotes the runoff of surface water, leading to increased erosion. The type of clay also affects its style of erosion. Smectite, which characterizes the younger, less deeply weathered paleosols, expands when it is wet and contracts as it dries, leading to the formation of mounds with a crackled appearance on their bare surfaces. Kaolinite, common to the deeply weathered soils of the Eocene part of the John Day and Clarno Formations, does not expand and contract nearly as much, leading to the development of small cliffs.

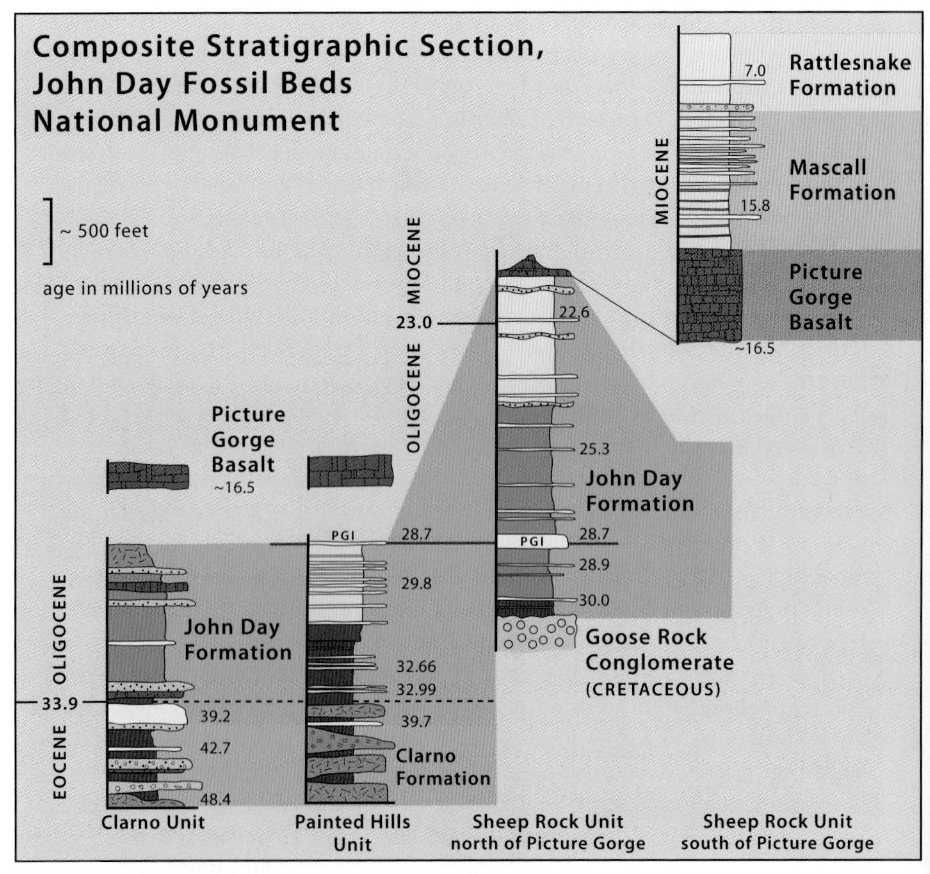

Detailed composite stratigraphic section for individual units of John Day Fossil Beds National Monument. Ages of individual ash layers are shown in millions of years. Different colors indicate colors of paleosols. –Modified from Dillhoff and others, 2009

The Palisades, eroded lahar deposits of the Clarno Formation, exposed in the Clarno Unit of John Day Fossil Beds National Monument along OR 218.

GUIDES TO THE BLUE MOUNTAINS

INTERSTATE 84
PENDLETON—ONTARIO
167 miles (269 km)

Interstate 84 follows much of the route of the Oregon Trail across the Blue Mountains. After fording the Snake River upstream (south) of Hells Canyon, the emigrants followed the valley that now hosts Baker City and La Grande, between two large stitching plutons of Late Jurassic and Cretaceous age. The wagon trail then crossed a relatively flat part of the Blue Mountains, capped by flat-lying lava flows of the Columbia River Basalt Group. This road guide approximately follows that route from north to south as far as Baker City.

South of Pendleton, I-84 follows a level surface on top of Pliocene river deposits and Pleistocene loess before crossing a series of faults, at which point the road climbs some 2,500 feet (760 m) up into the Blue Mountains. Good exposures of multiple basalt flows and intervening paleosols show up on both sides of the road. These flows belong to the Grande Ronde and overlying Wanapum Basalt units of the Columbia River Basalt Group. The road levels

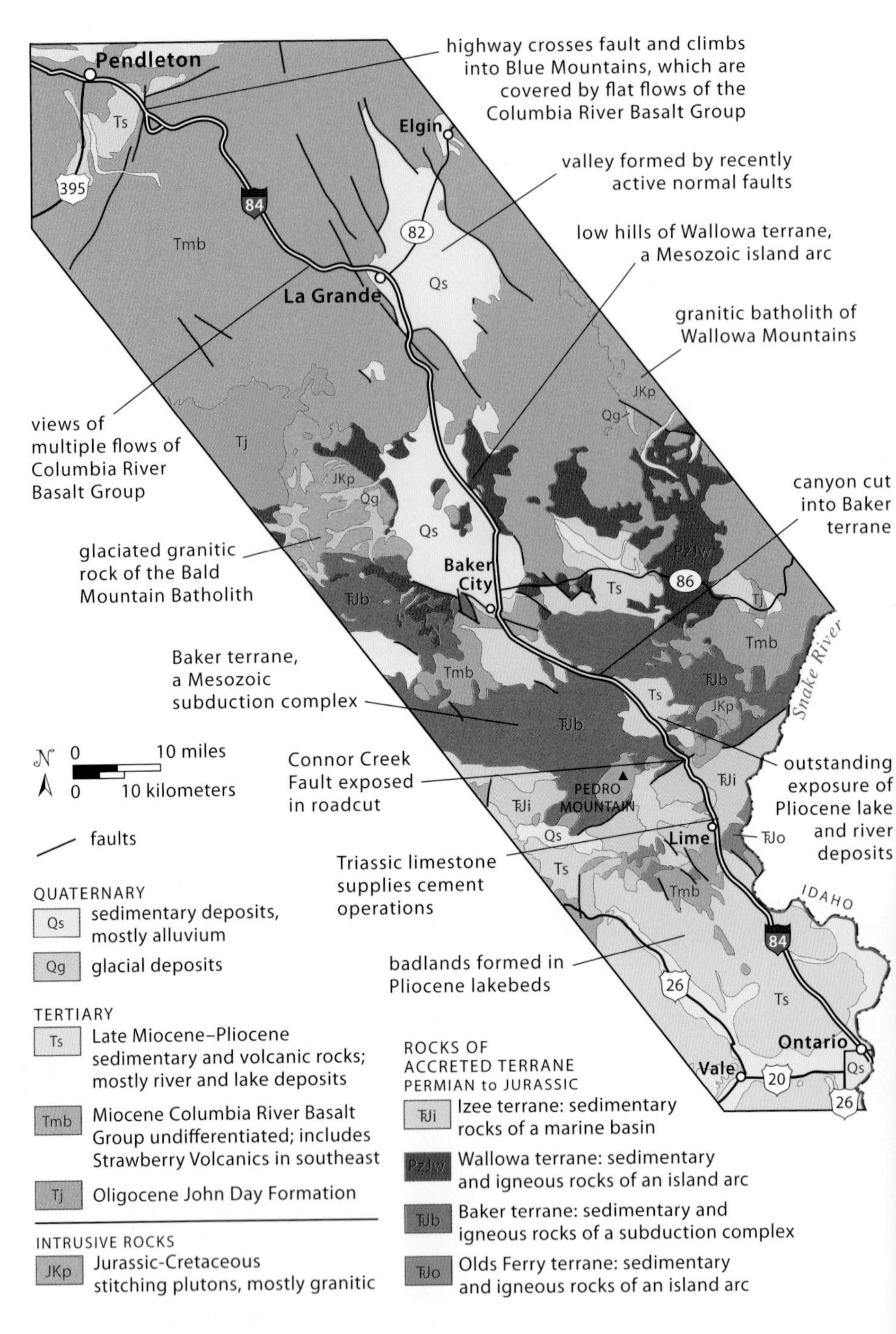

Geology along I-84 between Pendleton and Ontario.

out by about milepost 225, and you can get a sense of the overall flatness of this part of the Blue Mountains. The basalt flows form numerous roadcuts between mileposts 230 and 235 and are distinctly rubbly. Netting covers the cuts to mitigate rockfall danger to motorists. About 0.5 mile (0.8 km) east of milepost 247, a cut on the north side of the road exposes Grande Ronde Basalt for about 0.5 mile (0.8 km). Look for a red paleosol between the flows. Another good exposure of a paleosol exists near milepost 251 on the south side of the road. While driving down the grade into La Grande, you can see outstanding views of the multiple flows of the Columbia River Basalt Group.

La Grande lies on the western side of a valley between the Wallowa Mountains to the east and rest of the Blue Mountains to the west. While driving into La Grande, note how the valley sides end abruptly at the range front, a sign of recently active faulting. The valley is a graben, a block of land dropped down between normal faults that dip toward each other. The highway exits the valley at its southwest edge, passing into exposures of Saddle Mountains Basalt of the Columbia River Basalt Group.

Near milepost 276, I-84 begins a short descent to Clover Creek and then into Baker Valley. There, you get the first hint of the Late Paleozoic and Mesozoic rocks that make up the accreted terranes of the Blue Mountains. Just north of milepost 289, low hills to the east consist of Permian and Triassic volcanic and sedimentary rock of the Wallowa terrane, formed in an island arc. The Elkhorn Mountains rise to the west of the valley, held up by the large Bald Mountain

The upper surface of the Blue Mountains is relatively flat, because it consists mostly of flat-lying lava flows of the Columbia River Basalt Group.

Batholith, a granitic stitching pluton that caused a great deal of gold mineralization in the surrounding area. During Pleistocene time, mountain glaciers flowed off the Elkhorn Mountains and left behind numerous glacial moraines. Baker City, near milepost 304, sits on the northwestern edge of the Baker terrane, the remnants of a subduction zone. These rocks are mostly covered by Pliocene lakebed and river deposits and the Columbia River Basalt Group but are exposed in a canyon between mileposts 317 and 322. Between mileposts 326 and 327, an especially good exposure of the Pliocene rocks, containing an ash layer and some minor faults, exists on the east side of the highway.

Between mileposts 330 and 345, I-84 passes through a canyon with some exposures the Baker terrane in the north and the Izee terrane in the south. The contact between them is the Connor Creek Fault, exposed in a roadcut on the east side of the highway near milepost 336. The Triassic-age Burnt River Schist of the Baker terrane is pushed southward over Jurassic-age phyllites of the Weatherby Formation of the Izee terrane. Roadcuts on both sides of the road at milepost 337 expose folded Weatherby Formation intruded by basaltic dikes. For southbound travelers, this roadcut is especially conspicuous, because it hosts a pair of railroad tunnels, marked by a brown basaltic dike at the north entrances.

Marble forms prominent exposures in roadcuts and quarries on both sides of the Connor Creek Fault. The marble is quarried for lime. North of the fault, a good exposure of Triassic marble exists between mileposts 331 and 332. South of the fault, Jurassic marble figures prominently between mileposts 342 and 344, near the exit for Lime. Just north of milepost 344, mafic dikes cut the limestone.

The Connor Creek Fault, exposed along the interstate, cuts diagonally through the photo. The Baker terrane (top left) moved southward along the fault over the Izee terrane. Note the dike that cuts nearly vertically through the Izee terrane on the right.

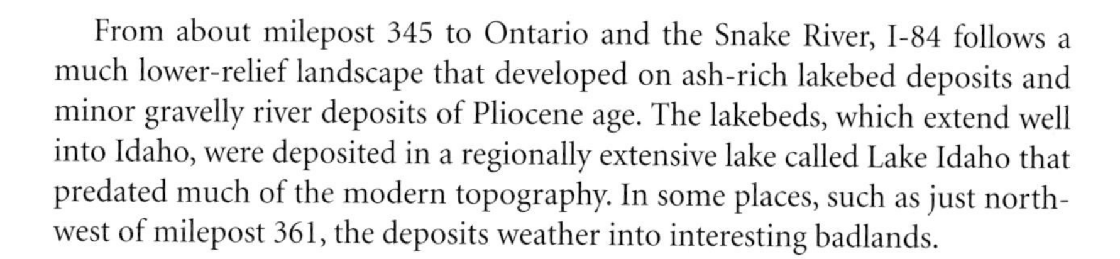

From about milepost 345 to Ontario and the Snake River, I-84 follows a much lower-relief landscape that developed on ash-rich lakebed deposits and minor gravelly river deposits of Pliocene age. The lakebeds, which extend well into Idaho, were deposited in a regionally extensive lake called Lake Idaho that predated much of the modern topography. In some places, such as just northwest of milepost 361, the deposits weather into interesting badlands.

US 26

Madras—John Day

146 miles (235 km)

Probably more than any other highway in Oregon, US 26 samples rocks that formed after terrane accretion but before eruption of the Columbia River Basalt Group. This scenic route climbs over the Ochoco Mountains, then drops into the John Day River drainage basin, which it follows to the town of John Day. For almost the entire route, the road passes through Cretaceous sedimentary rocks deposited directly on the accreted terranes and Eocene through Early Miocene rocks of the John Day Basin, including the Clarno and John Day Formations.

Between Madras and Prineville, US 26 crosses a flat surface developed on Late Tertiary tuff-rich sedimentary rocks and, near Prineville, Pleistocene river and lake deposits. Near milepost 12, however, the road crosses into the Crooked River Caldera, a roughly oval-shaped feature that continues nearly 40 miles (64 km) from here to the southeast. The caldera erupted 29.5 million years ago and produced the tuff of Smith Rock State Park near Redmond and many of the ash deposits in the John Day Formation. Gray Butte and Grizzly Mountain, the two prominent peaks west and east of the road, respectively, originated as rhyolite domes that intruded along the margin of the caldera after it formed. Gray Butte can be identified by its radio towers. Several other rhyolite domes, including Powell Buttes to the south, intruded elsewhere along the caldera margin. The hill just southwest of the highway near milepost 15 consists of tuff of the John Day Formation, and the long mesa approaching Prineville near milepost 22 consists of Late Tertiary basalt.

East of Prineville, the road follows Ochoco Creek through scattered outcrops of the John Day Formation. Pleistocene-age basalt rims the valley to the south. Ochoco Reservoir fills the valley behind Ochoco Dam for just over 3 miles (5 km) along the road. The dam was built privately just after World War I for irrigation and flood control but had to be replaced by the Bureau of Reclamation in 1949, because persistent leaking posed a potential failure hazard. Mill Creek Road, on the east side of the reservoir at milepost 28, leads north into the 40-million-year-old Wildcat Mountain Caldera and to Steins Pillar, a 300-foot-tall (90 m) monolith eroded in welded tuff from that eruption.

For the rest of the way to Ochoco Summit, andesite flows and some lahars and ash-rich sediments of the Clarno Formation make up the rock exposures. In general, the andesite flows consist of fairly uniform, solid rock, whereas the lahars consist of jumbled rock debris. A double roadcut halfway between mileposts 40 and 41 provides a good example of andesite flows. A good example of a lahar lies just east of Ochoco Summit between mileposts 50 and 51.

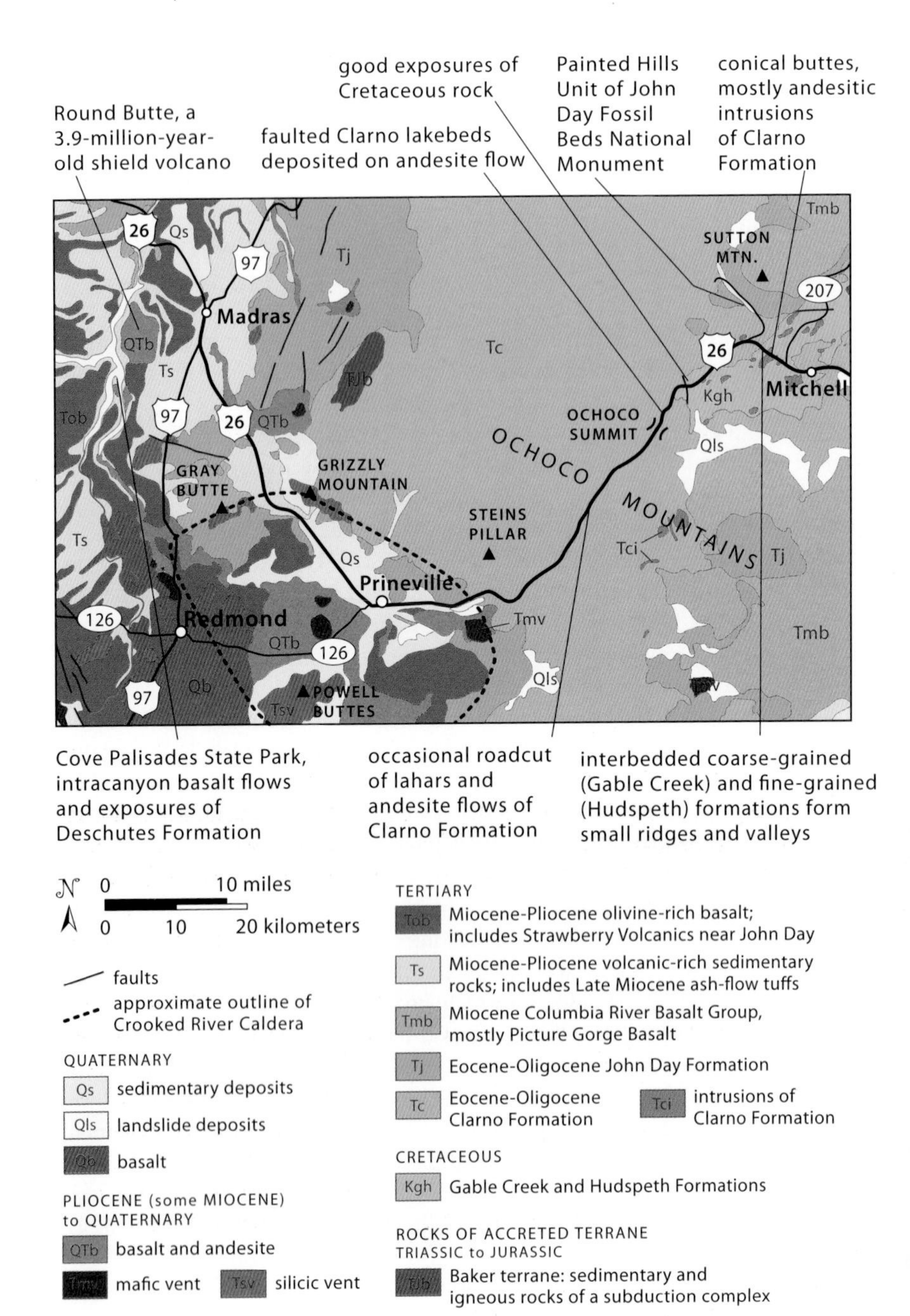

Geology along US 26 between Madras and Mitchell.

Lakebeds of the Clarno Formation (left) were deposited against andesite (right), also of the Clarno Formation, just west of milepost 53. Although faults displace these rocks elsewhere in the roadcut, the diagonal feature (center) is the depositional contact and not a fault.

Lakebed deposits of the Clarno Formation overlie a dark-colored andesite in a double roadcut next to a large pull-out just west of milepost 53. Note how the lakebed sediments were also deposited against the side of the andesite, rather than only on top of it. From a distance, this buttress unconformity looks like a fault, because the younger lakebeds terminate against the contact. The orientation of this unconformity differs from one side of the road to the next, indicating some variation in the lake bottom topography. Several well-exposed faults cut these rocks. Just under 1 mile (1.6 km) farther east, the road crosses into Cretaceous sedimentary rock.

Between here and Mitchell, outcrops and roadcuts consist of Cretaceous sedimentary rocks, with some lahars, lavas, or shallow intrusions of the Clarno Formation. The Cretaceous rocks consist mostly of highly deformed shale and mudstone of the Hudspeth Formation and conglomerate of the Gable Creek Formation. In some places, these rocks interlayer with each other, indicating they formed at the same time, most likely on different parts of a submarine fan complex. The coarse-grained Gable Creek Formation accumulated in active channels, and the fine-grained Hudspeth accumulated between the channels and in the more distant reaches.

The area around Mitchell is punctuated by numerous conical buttes, the erosional remnants of 46- to 40-million-year-old shallow intrusions of the Clarno Formation. They range in composition from basalt to rhyolite, although most are andesite, similar in composition to most of the lava flows in the Clarno. Black Butte and White Butte, the prominent buttes south of the highway and west of Mitchell, are two of the larger andesitic intrusions.

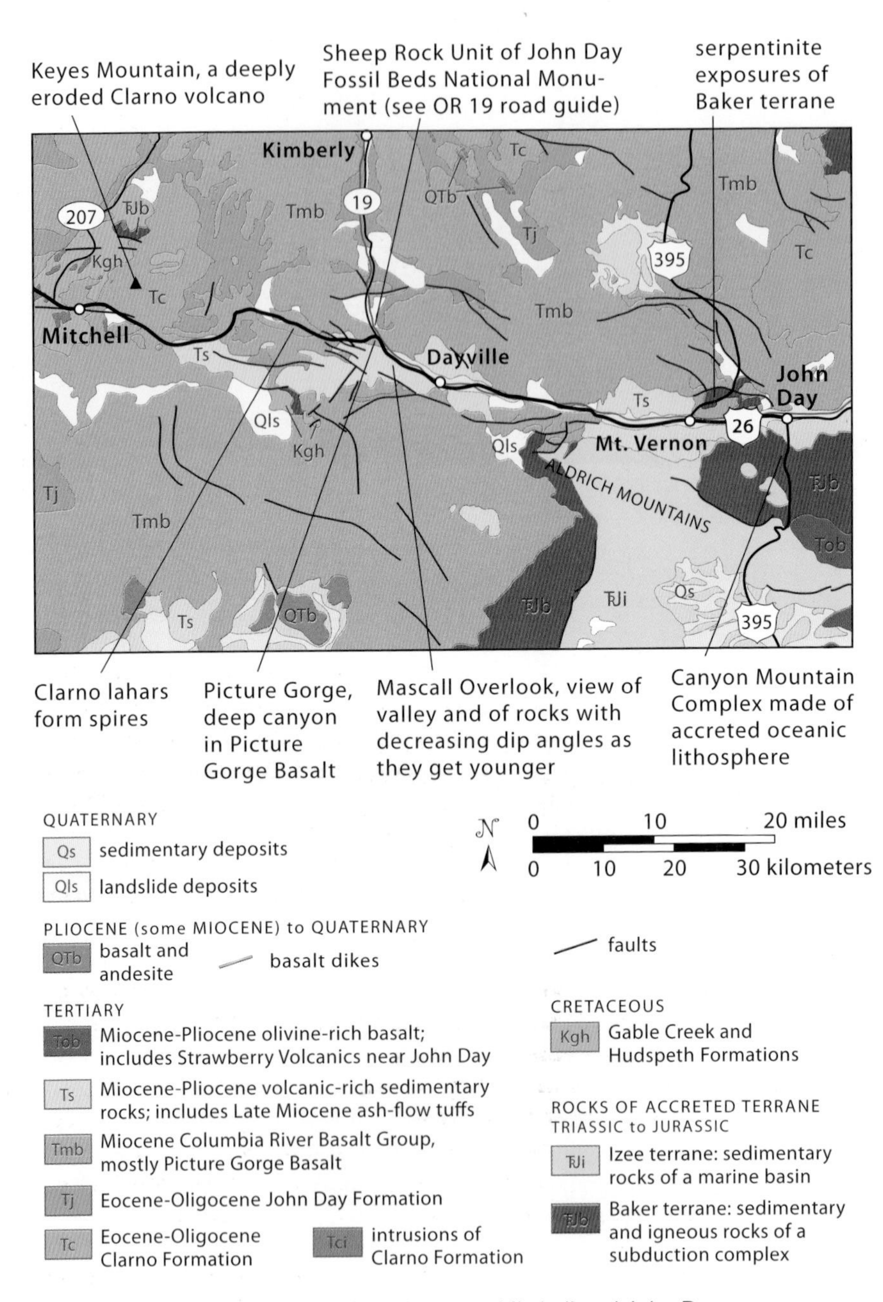

Geology along US 26 between Mitchell and John Day.

Sargent Butte, a rhyolitic intrusion, and the other mostly andesitic intrusions of this part of the Clarno Formation form conical buttes, because they are more resistant to erosion than the surrounding rock.

Sargent Butte, a rhyolitic intrusion, looms almost 1,000 feet (300 m) above the north side of the highway between mileposts 61 and 62. On its east side, the Bridge Creek Road intersects the highway, providing a route to the Painted Hills Unit of John Day Fossil Beds National Monument. At the turnoff, US 26 passes through good exposures of Clarno andesite flows. Immediately to the east of the turnoff, they lie on top of west-dipping Cretaceous rocks. There, the Gable Creek and Hudspeth Formations are clearly interbedded, with the conglomeratic and more resistant Gable Creek Formation holding up small ridges, and the easily eroded Hudspeth occupying valleys. Just west of Mitchell, at milepost 65, the road passes directly north of Bailey Butte, another andesitic intrusion. Part of the eroded intrusion forms a smaller pinnacle on the north side of the road.

Mitchell lies astride a geologic contact, with the west side of town resting on Cretaceous rocks and the east side on basalt flows of the Clarno Formation. Just south of the highway lies the Mitchell Fault, a large east-trending structure that shows evidence for both dip-slip and strike-slip movement. Keyes Mountain, a deeply eroded Clarno volcano, rises to an elevation of 5,704 feet (1,738 m), about 6 miles (10 km) to the east-northeast of Mitchell. For the next several miles east, the highway passes through numerous well-exposed lahars and basalt flows of the Clarno, likely derived from Keyes Mountain.

East of Keyes Summit, US 26 passes flows of the Picture Gorge Basalt of the Columbia River Basalt Group. From there east to Picture Gorge, the road passes mostly through Clarno and John Day Formations with overlying Picture Gorge

Painted Hills Unit of John Day Fossil Beds National Monument

The Painted Hills Unit of the John Day Fossil Beds National Monument is likely the most photographed part of central Oregon because of its colorful badlands. The badlands are eroded into paleosols of the lower third of the John Day Formation and the underlying Clarno Formation. The upper John Day and overlying flows of the Columbia River Basalt Group are visible just east of the park boundaries on the slopes and ridgeline of Sutton Mountain. The paleosols are red, brown, or yellow, suggestive of the climatic conditions during their formation. In general, the red colors indicate deep weathering under warm, wet conditions, typical of tropical to subtropical environments, while yellow and tan colors result from moderate weathering under increasingly dry, cooler conditions, typical of temperate forests.

Some of the most dramatic exposures of the paleosols lie along the park road between the park entrance and the Painted Hills Overlook. These paleosols exhibit mostly tan colors, although two prominent red beds exist near the middle, suggesting a temporary return to deep weathering conditions. Several beds near the bottom are dark gray, suggesting soil formation in nonoxygenated conditions, such as occurs locally in wetlands or swamps. You can see that these paleosols are folded into a broad anticline and broken by small normal faults, indicating both crustal compression and extension in the time after their formation.

Red and tan paleosols of the Painted Hills Unit, with overlying Picture Gorge Basalt in the distance. View from the Painted Hills Overlook Trail.

We know the approximate age of the paleosols, because they contain several relatively unaltered tuff layers. Tuff is a volcanic rock made from consolidated ash and pumice, which can be dated using radiometric methods. The prominent hill immediately north of the park entrance road, consisting mostly of tan paleosols, shows a thin but prominent tuff at its top, an ash-flow deposit called the Picture Gorge Ignimbrite. It is the youngest rock in the Painted Hills, with an age of 28.7 million years. The Picture Gorge Ignimbrite forms an even thicker layer at the Sheep Rock Unit.

In the Painted Hills, the rocks get older, and so redder, toward the west. A short trail at Painted Cove allows a close-up view of some of the John Day Formation's deep red paleosols. The Leaf Hill Trail approaches the park's richest fossil leaf locality, where tens of thousands of leaf, seed, and cone specimens have been recovered from approximately 33-million-year-old lakebeds. Called the Bridge Creek Flora, these plants grew in the cooler and somewhat drier conditions of Early Oligocene time.

Basalt. In general, the John Day Formation consists of poorly exposed, deeply weathered ash-fall deposits, whereas the Clarno consists of much more resistant andesitic lava and lahars. Between mileposts 90 and 91, lahars of the Clarno erode into some towers and spires along the highway. Just west of Picture Gorge, between mileposts 95 and 96, the basalt dips beneath well-exposed younger sedimentary and volcanic rocks of the Mascall and Rattlesnake Formations. The 7-million-year-old Rattlesnake Ash-Flow Tuff shows up as a thin, nearly horizontal cap above the dipping basalt.

Picture Gorge, a short but deep canyon carved by the John Day River, displays beautiful south-dipping lava flows of the Picture Gorge Basalt. Many of these flows exhibit a wide range of features typical of large lava flows, including a paleosol at the base and a well-developed colonnade and entablature. Some of the best exposures are at the confluence of Rock Creek and the John Day River, at the OR 19 intersection. At this point, the milepost numbering jumps to milepost 125. Between the junction and the canyon's east entrance 1.3 miles (2.1 km) upstream, US 26 appears to descend as it heads south, even though it maintains a steady uphill gradient. This optical illusion results from the southward inclination of the basalt.

The canyon presents an enigma: why would the river cut through the hard basalt rather than simply flow around it? Most likely, the river's course became established in softer, overlying material that has since mostly eroded. Then, during regional uplift, the river became trapped in its channel and had to cut through the much harder basalt.

Mascall Overlook, reached by a marked road on the south side of the valley between mileposts 127 and 128, provides visible evidence of long-term uplift of the region. From here, you can see the Mascall and Rattlesnake Formations, the John Day River entering Picture Gorge, and a variety of dip angles in bedding

Columnar jointing in flows of the Picture Gorge Basalt at junction of OR 19 and US 26 in Picture Gorge on the John Day River.

Westward view of angular bedding relations from US 26 just below Mascall Overlook. The Picture Gorge Basalt dips about 20 degrees to the south at the east entrance to Picture Gorge (right). The overlying cream-colored Mascall Formation (straight down the highway) dips about 10 degrees southward. The Rattlesnake Ash-Flow Tuff, which caps the ridges, dips only about 5 degrees southward.

that speak to the uplift and folding. The Picture Gorge Basalt and the Mascall Formation beneath the overlook dip moderately southward. Looking upward through the Mascall Formation, however, you can see that the bedding gradually dips less steeply, suggesting the area was being folded as the Mascall was being deposited. The base of the Rattlesnake Formation, marked by a thin conglomerate unit, dips just slightly southward, as does the overlying Rattlesnake Ash-Flow Tuff, which caps the ridges. These low dip angles suggest that some folding continued even after those rocks were deposited.

US 26 follows the John Day River for the 35 miles (56 km) between Picture Gorge and John Day. The first 6 miles (10 km) or so offer good views of the Picture Gorge Basalt to the north, dipping southward beneath the valley. Most of the valley is filled with recent river sediments deposited on the Mascall and Rattlesnake Formations. Between mileposts 145 and 146, the road narrows where it passes through Picture Gorge Basalt, which in some places shows well-developed paleosols. Near milepost 146, the valley opens back up into a lovely floodplain. To the south, you can see old gravelly river terraces appearing well above the present road level, with the Aldrich Mountains rising steeply behind to elevations of 7,000 feet (2,100 m), more than 4,000 feet (1,220 m) above the valley floor. These mountains consist of Paleozoic and Triassic accreted rock of the Baker terrane, overlain by mostly Jurassic rock of the Izee terrane. Similar rocks make up the hillsides north of the highway between Mt. Vernon and John Day.

Kam Wah Chung State Heritage Site in John Day offers the opportunity to reflect briefly on the area's human past. During the late 1800s, John Day was home for over one thousand Chinese immigrants, most of whom were gold miners. The building that houses the museum functioned as a general store, medical clinic, and social center.

US 26
John Day—Vale
116 miles (187 km)

Beginning at an elevation of 3,000 feet (900 m) at John Day, US 26 stays at relatively high elevations for almost the whole trip to Vale, crossing four passes in the Blue Mountains, three of which exceed 4,500 feet (1,370 m) in elevation. Between the passes lie high valleys, each with expansive views to near and distant peaks.

Between John Day and Prairie City, the road follows the floodplain of the John Day River. To the south lie the Strawberry Mountains, which rise abruptly from the valley floor along a recently active fault. On their west side near John Day, the Strawberry Mountains consist of Permian and Triassic accreted rock of the Baker terrane. To the east by Prairie City, they consist mostly of andesite and basalt of the Miocene-age Strawberry Volcanics. North of the highway,

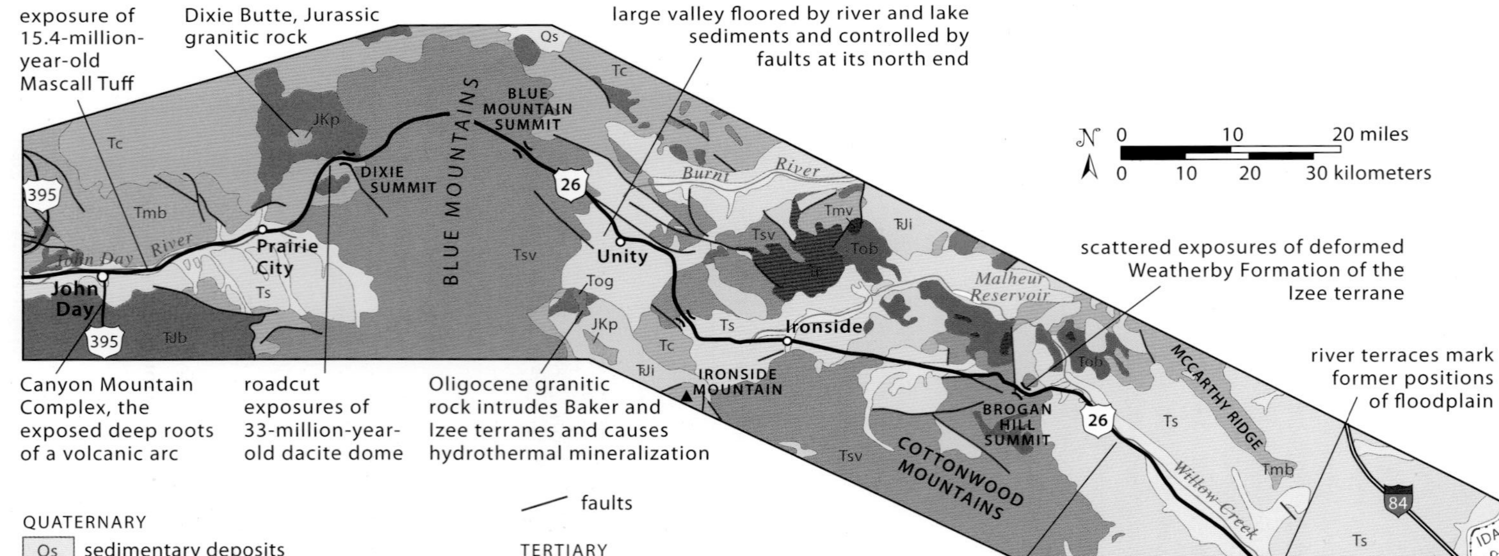

Geology along US 26 between John Day and Vale.

View southward to the Strawberry Mountains from across the valley north of Prairie City. The abruptness of the northern range front is caused by a recently active fault zone along its base.

prominent olivine-bearing basalt flows form the tops of ridges, and below them lie scattered light-colored ledges of weathered volcanic ash of the Mascall Formation. The tuff near the bottom of the hill about 7 miles (11 km) east of John Day is called the Mascall Tuff and, at another locality, yields an age of 15.8 million years. The Mascall Tuff, part of the Mascall Formation, is one of several widespread ash flows that erupted during the early stages of the Columbia River Basalt Group eruptions, a reminder that rhyolite was erupting at the same time as the basalt. In some places, this tuff is about 30 feet (9 m) thick!

East of Prairie City, US 26 begins its steep ascent to Dixie Summit, climbing through scattered outcrops of Mascall Formation and overlying basalt. Between mileposts 182 and 183, part of a 33-million-year-old dacite dome is exposed in three consecutive roadcuts on the west side of the highway. These rocks consist of light-colored lava flows and breccias and are cut by several basaltic dikes. Still older andesitic rocks, about 42 million years in age, display columnar jointing just east of Dixie Summit. A few miles to the northeast, Dixie Butte rises to 7,592 feet (2,314 m), more than 2,000 feet (610 m) in elevation above the summit. It consists of a Jurassic granitic stitching pluton that intruded Paleozoic rock of the Baker terrane.

East of Dixie Summit the highway passes into the Strawberry Volcanics, which erupted between about 17 and 12 million years ago from numerous vents between Unity, some 15 miles (24 km) to the southeast, and the area south of Prairie City. Although mostly andesite, the Strawberry Volcanics range

Roadcut exposure of 33-million-year-old brecciated dacite and a basaltic dike between mileposts 182 and 183. The dacite is part of a larger dacite dome that formed either during the end of Clarno times or immediately afterward.

from basalt to rhyolite. Some researchers argue that they are related to the Columbia River Basalt Group, because most of them erupted about 15 million years ago, at the same time as the extensive basalt flows. Moreover, many of the basalt flows in the Strawberry Volcanics contain similar chemistries to the Steens Basalt, an early phase of the Columbia River Basalt Group that makes up most of Steens Mountain to the south. Good outcrops exist near milepost 189, some of which contain andesites with unusually large plagioclase crystals. In other exposures, such as many between Blue Mountain Summit and milepost 202, the andesites display closely spaced fractures, giving the rock a distinctly platy appearance.

Near milepost 208, the road enters the Unity basin, a valley controlled by normal faults on its northern side. Three forks of the Burnt River join here and flow east to the Snake River. Just north of the road, a ridge of andesitic breccia, uplifted by one of these faults, runs parallel to the road. River and lake deposits cover the valley floor, some of which were deposited at the same time as the younger parts of the Strawberry Volcanics.

Between Unity and Ironside, the road passes through a transition zone between the Baker terrane to the north and the Izee terrane to the south. The mountains to the south consist of oceanic lithosphere of the Baker terrane and highly deformed sandstone and siltstone of the Izee terrane. Both are intruded by Oligocene granitic rock, which caused a great deal of hydrothermal alteration and mineralization. Much of the Izee terrane consists of the Weatherby Formation, likely deposited in the region between a subduction zone and its volcanic arc. The best exposures along the road of the Weatherby, however, don't show up until farther east, near milepost 249.

From Ironside, you can look southwestward about 10 miles (16 km) toward the imposing Ironside Mountain. It consists of Clarno Formation, whereas its northwest and southeast sides consist of folded Izee terrane. About 20 miles (32 km) to the northeast, Pedro Mountain rises slightly above the surrounding ridges. It is a granitic stitching pluton of Late Jurassic or Cretaceous age, part of a series of intrusions into the boundary between the Baker and Izee terranes.

Between mileposts 243 and 249, you can see southward to the Cottonwood Mountains, uplifted along a normal fault and capped by basaltic lava flows, likely part of the Grande Ronde unit of the Columbia River Basalt Group. Similar basalts are exposed near the road near Brogan Hill Summit. Just east of the summit, near milepost 249, is an outcrop of the Weatherby Formation,

Outcrop of Weatherby Formation near milepost 249. The bedding in these rocks angles down to the left, whereas the slaty cleavage dips steeply down to the right. The cleavage indicates these rocks are strongly deformed by compression.

which tends to break along slatelike planes at a high angle to the bedding. This feature, called slaty cleavage, is caused by partial dissolution of the rock because of compressive stresses oriented perpendicular to the planes.

Between Brogan Hill Summit and Vale, US 26 gradually descends through scattered exposures of eroded Pliocene-age deposits of Lake Idaho, which extended eastward from here over much of the Snake River Plain. The basalt-capped Cottonwood Mountains form a long ridge to the southeast, whereas Miocene river and lake deposits, overlain by Columbia River Basalt Group, form layers in McCarthy Ridge to the northeast. The road passes down through river terraces, visible on the northeast side between mileposts 272 and 273, and onto the floodplain of Willow Creek, which it follows the rest of the way to Vale.

US 395
Pendleton—Mt. Vernon
119 miles (192 km)

Although US 395 crosses most of the Blue Mountains, the route lies almost entirely on the Columbia River Basalt Group. The accreted terranes and their complex geology are exposed only where the overlying Columbia River Basalt Group and Clarno and John Day Formations have been eroded away.

Heading south on US 395 from I-84, the road crosses some low hills eroded into the Wanapum Basalt of the Columbia River Basalt Group; look for roadcuts between mileposts 4 and 5. It then follows river terraces in the valley of McKay Creek. These terraces mark former positions of the river's floodplain but have been left stranded by downcutting of the channel. The valley lies within Pliocene lake and river deposits, exposed in some places between mileposts 9 and 11. McKay Creek Reservoir, impounded for irrigation in the late 1920s, doubles as a national wildlife refuge, providing habitat for migratory birds.

The town of Pilot Rock gets its name from the cliff of Grande Ronde Basalt west of town, which was visible to emigrants on the Oregon Trail. US 395 begins a long steady climb up a canyon cut into the Grande Ronde Basalt, with numerous roadcuts and basaltic cliffs adorned with beautiful columns. Near milepost 25, the road leaves the canyon and continues climbing a gently sloping surface developed on top of the lava flows. At milepost 37, the road crests the hill and turns abruptly eastward. Across from a large pull-out, the Columbia River Basalt Group rests on top of granite, with a red paleosol between them. Although not visible from the road, this same granite intrudes Triassic volcanic rocks of the Wallowa terrane. The intrusion likely occurred about 140 million years ago near the end of Jurassic time, soon after the Wallowa and Baker terranes came together.

Scattered roadcuts over the next few miles to Battle Mountain Summit expose more of the granite. South of the pass, the road passes back into Grande Ronde Basalt, which it follows, along with some overlying river and lake deposits, south as far as Ukiah, near milepost 50.

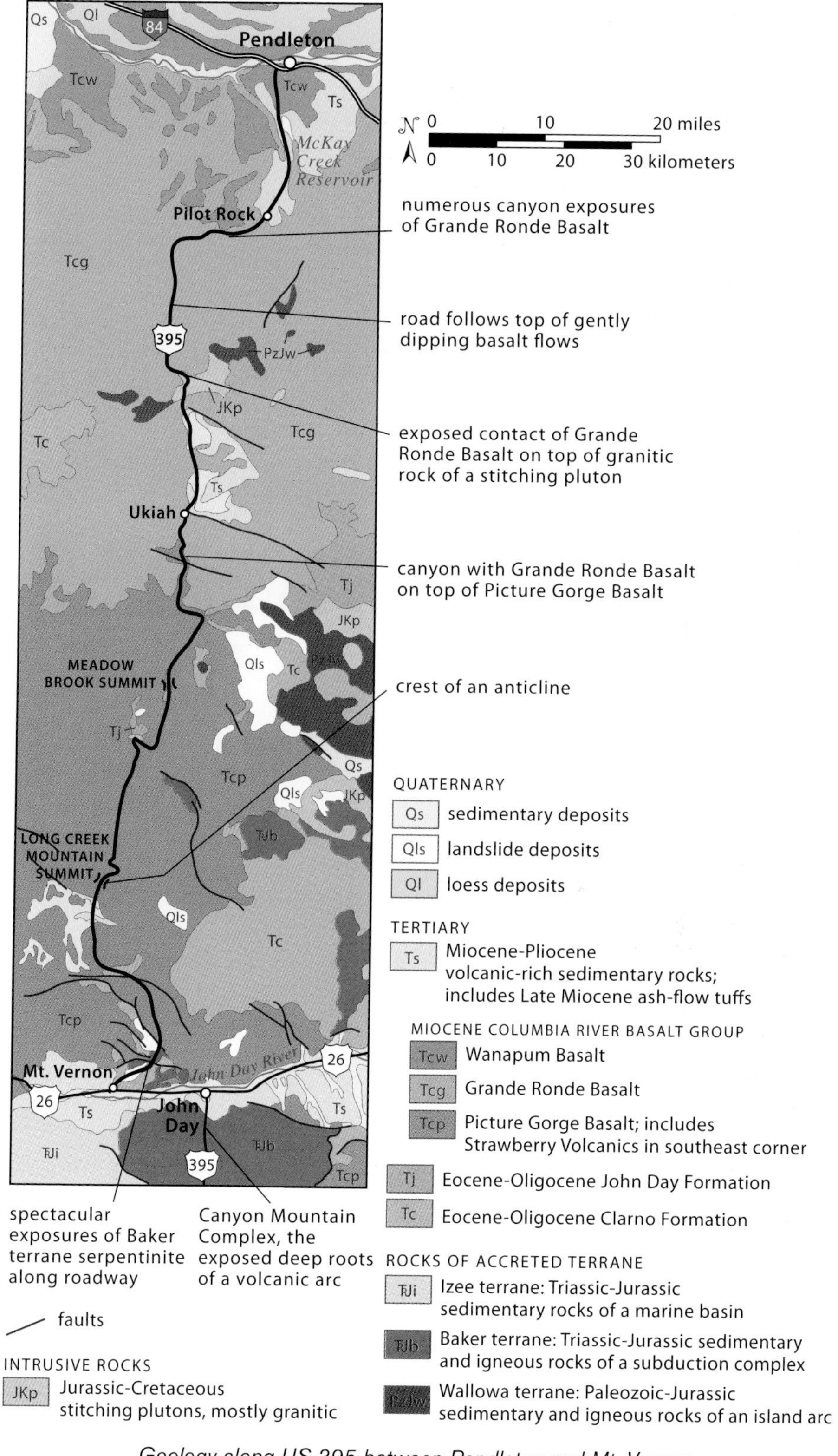

Geology along US 395 between Pendleton and Mt. Vernon.

Contact of the Columbia River Basalt Group (left) with Jurassic-age granitic rock (right) at milepost 37. The reddish zone is a paleosol developed on top the granite and baked when it was covered by the hot basalt flow.

South of Ukiah, US 395 enters a beautiful canyon cut into two members of the Columbia River Basalt Group: the Picture Gorge Basalt and the overlying Grande Ronde Basalt. From here to Mt. Vernon, the road lies within the Picture Gorge Basalt, except for two stretches of road where erosion cut into older rocks. The first of these older exposures, on either side of Meadow Brook Summit, consist of poorly exposed John Day Formation. One good roadcut, with overlying Picture Gorge Basalt, exists 0.4 mile (640 m) south of the pass.

The Picture Gorge Basalt has been gently folded into some east-trending anticlines and synclines that are not very apparent from the roadway, but the summit of Long Creek Mountain, near milepost 96, marks the crest of an anticline. The basalt flows dip away from the crest in either direction.

Near milepost 105, the road enters the southernmost stretch of older rocks, uplifted here relative to the basalt by a series of north- and east-trending fault zones. The first rocks are lahars of the Clarno Formation. More Clarno andesite and lahars, as well as Picture Gorge Basalt, lie farther down the road, and just west of milepost 115, the highway passes an exposure of serpentinite of the Baker terrane. More serpentinite exposures, with their characteristic green color, show up over the next 2 miles (3.2 km). The route finishes by passing back through the overlying rocks: first, lahars of the Clarno Formation, and then columnar-jointed Picture Gorge Basalt.

Serpentinite of the Baker terrane exposed along US 395 about 5 miles (8 km) north of Mt. Vernon.

OR 19

Arlington—US 26 at Picture Gorge

122 miles (196 km)

OR 19 heads south onto the Columbia Plateau toward the Blue Mountains, roughly following the trend of the John Day River. Lying deep in a canyon to the west of the road, the river is not in sight until well into the mountains. Beginning at Arlington, as OR 19 climbs Alkali Canyon, it passes through gravel deposits left by the Missoula Floods until just south of milepost 10, with the best exposures between mileposts 1 and 4. The highest gravels along the road are just below 900 feet (270 m) elevation, but they reach elevations above 1,000 feet (300 m), about 800 feet (240 m) above the Columbia River. Gravel deposits can be found to similar elevations up many of the Columbia's other tributaries in this area, because the Missoula Floods backed up behind constrictions, called hydraulic dams, in the Columbia River channel.

South of milepost 11, the road reaches the top of the plateau, here developed on the Wanapum and Grande Ronde units of the Columbia River Basalt Group. Between the river and Condon (milepost 37), the basalt mostly belongs to the Wanapum Basalt. South of Condon, it drops into the older Grande Ronde Basalt. Fine windblown sediment called loess blankets much of the basalt on the plateau. The loess originated as fine glacial sediment during Pleistocene

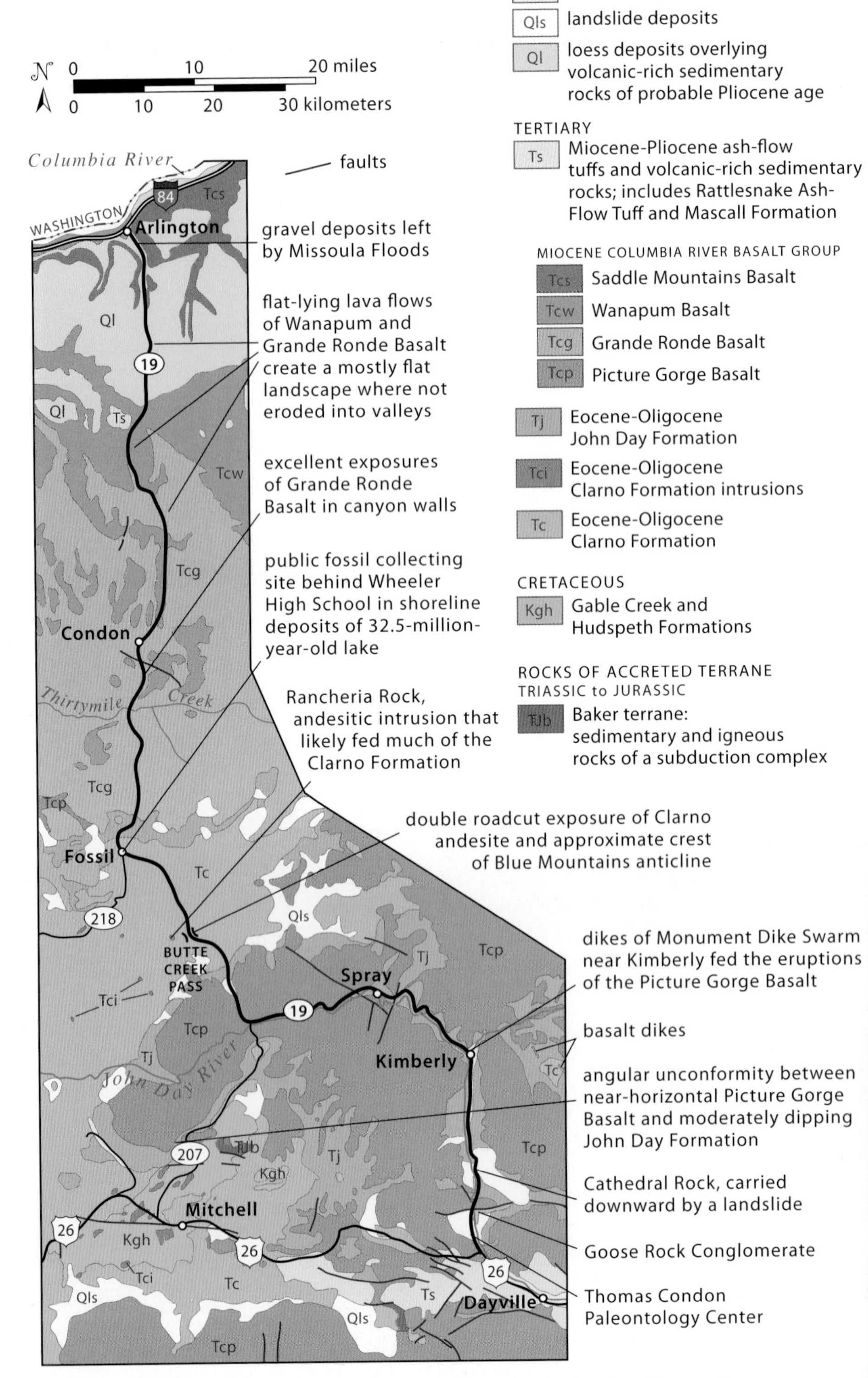

Geology along OR 19 between Arlington and US 26 at Picture Gorge.

time and now covers much of southeastern Washington and northeastern Oregon. Where the road dips into valleys, such as near milepost 16, the basalt is exposed beneath the loess. Between mileposts 17 and 18, the hexagonal tops of some basaltic columns are well exposed in a stream channel on the west side of the road. Just north of milepost 21, some well-developed soils in the loess are exposed. Were it not for the loess, this would be a tough place to farm.

Just south of Condon, the road descends into a tributary of Thirtymile Creek, itself a tributary of the John Day River, some 12 to 15 miles (19–24 km) to the west. The canyon exposes beautiful cross sections of Grande Ronde flows, many exhibiting well-developed colonnade and entablature. After climbing back to the plateau, the road passes southward into the underlying John Day and Clarno Formations just south of milepost 53. The John Day, a fairly nondescript, tan, ashy slope former, outcrops first. Within about 1 mile (1.6 km), it gives way to andesite of the underlying Clarno Formation, which forms a much more variable topography.

The town of Fossil hosts the Oregon Paleo Lands Center, which provides information about the area's geology, including the John Day Basin. The center administers a public fossil collecting site behind Wheeler High School. Abundant plant debris, including metasequoia, as well as fossilized fish bones, can be found near the bottom of the John Day Formation, deposited at the edge of a lake some 32.5 million years ago.

Fossilized and modern leaves of metasequoia, also known as dawn redwood. Photo is about 4 inches (10 cm) across.

South of Fossil, OR 19 passes through more Clarno andesite as it climbs steadily through ponderosa pine forest to Butte Creek Pass, near milepost 67. An outstanding double roadcut at the pass exposes the andesite. About 2 miles (3.2 km) to the west and most visible from the north side of the pass, the conical Rancheria Rock, with radio towers, rises more than 1,000 feet (300 m) in elevation above the pass. It is the erosional remnant of an andesitic intrusion that

OR 207 to Mitchell

At Service Creek, the road meets the John Day River, where you have the option to take OR 207 south to Mitchell, a geologically rewarding side trip. OR 207 passes through the Picture Gorge Basalt and back into the John Day Formation. The contact between these units is marked by an angular unconformity in some places, particularly near milepost 12. The road also passes through Cretaceous conglomerate of the Gable Creek Formation and shale and sandstone of the Hudspeth Formation. Both of these units are well exposed on the downgrade into Mitchell, south of milepost 13. The last few miles north of Mitchell allow wonderful views southward, including several of conical buttes that are shallow, 46- to 40-million-year-old andesite intrusions.

Angular unconformity near milepost 12 on OR 207 between Service Creek and Mitchell. Note how the green-colored John Day Formation is tilted, whereas the overlying Picture Gorge Basalt is relatively flat, indicating the John Day Formation was deformed before the eruption of the basalt.

likely fed many of the Clarno flows, similar to many of the buttes that distinguish the topography around Mitchell. Butte Creek Pass marks the crest of the large-scale anticline that trends northeastward across much of the Blue Mountains. The rocks north of the pass dip northwestward, while those south of the pass dip southeastward. Travelers go into younger rocks in either direction away from the pass, even though they are going downhill. For example, about 4.6 miles (7.4 km) south of the pass, near milepost 73, the road crosses back into the John Day Formation, and the overlying cliffs consist of basalt of the Picture Gorge unit of the Columbia River Basalt Group. By the time the road reaches Service Creek, it lies entirely within the Picture Gorge Basalt—higher in the rock sequence but more than 1,500 feet (450 m) lower in elevation.

The John Day Highway follows the John Day River upstream to its intersection with US 26 in Picture Gorge about 45 miles (72 km) away. Cliffs of Picture Gorge Basalt rim light-colored, ashy slopes of John Day Formation nearly the whole route. The John Day Formation, which takes on green and even red colors, consists of ash-fall deposits, many of which accumulated on floodplains but were then converted to soils as weathering processes took over. These ancient soils, called paleosols, take on a variety of colors depending on the intensity of weathering. In general, lighter colors result from less weathering, yellow colors result from deeper weathering, and red colors result from the deepest weathering. The green colors are the result of a later period of alteration of what was originally a brown soil.

Between Service Creek and Spray, numerous gravel bars mark the inside bends of river meanders, where water flow is slower and sediment gets deposited. Spray sits on the John Day Formation beneath cliffs of the Picture Gorge

At Spray, on the John Day River, the contact between the light-colored John Day Formation and the overlying Picture Gorge Basalt lies just behind the white buildings.

Basalt. Just east of the intersection with the road to Heppner Junction, variable bedding orientations in the John Day show that it is deformed. Because the overlying basalt stays pretty flat, this deformation must have occurred after the John Day was deposited between 39 and 22 million years ago but before eruption of the basalt 16.5 million years ago.

One mile (1.6 km) northwest of Kimberly, you can see a normal fault juxtaposing the basalt and John Day Formation across the river to the south, as well as a basaltic dike. More dikes are visible on both sides of the river 1 mile (1.6 km) south of Kimberly. These dikes fed the massive eruptions of the Picture Gorge Basalt. They trend roughly north-south and form narrow ridges, because they are more resistant to erosion. They belong to the Monument Dike Swarm,

Sheep Rock Unit of John Day Fossil Beds National Monument

Between mileposts 113 and 114, the road enters the Sheep Rock Unit of John Day Fossil Beds National Monument. In addition to its stark beauty, the Sheep Rock Unit reveals evidence for the presence of ancient topography in the John Day Basin. For example, at Goose Rock near milepost 120, the Goose Rock Conglomerate lies directly beneath the John Day Formation, with no intervening Clarno Formation. The missing rocks were either never deposited or were removed by erosion after being deposited. In a similar fashion, an ancient topography was eroded into the John Day Formation before the eruption of Picture Gorge Basalt, so the basalt overlies different parts of the John Day Formation. In addition, the Mascall and older rocks are tilted some 20 degrees toward the south, whereas the Rattlesnake Formation is tilted only about 5 degrees. This change in dip indicates long-lived deformation, because it occurred before and after the Rattlesnake deposition.

The Sheep Rock Unit consists of three subareas: Foree, Cathedral Rock, and Blue Basin. The Foree area offers a couple of short hiking trails through beautiful outcrops of the upper and middle parts of the John Day Formation, exposed beneath cliffs of Picture Gorge Basalt. At milepost 116, the road passes by Cathedral Rock, the smallest of the subareas. Cathedral Rock, a landslide block, dropped down from the cliffs immediately to the west. It consists of colorful green and tan paleosols of the John Day Formation capped by the more resistant, reddish Picture Gorge Ignimbrite. Above the ignimbrite, you can see yellow paleosols. In the distance, you can see Picture Gorge Basalt at road level, some of which is tilted, confirming the landslide origin. The Picture Gorge Ignimbrite formed from a giant ash flow 28.7 million years ago and is an important marker bed throughout the region.

Blue Basin is the southernmost and largest subarea of the Sheep Rock Unit. It offers hiking trails into the badlands of the middle John Day

Formation, the rock green from its alteration. Much of the area west of the river consists of landslide deposits of the Picture Gorge Basalt jumbled with the softer John Day Formation. Just south of milepost 120, the road crosses a fault that uplifts the Cretaceous-age Goose Rock Conglomerate on the south side. These rocks are equivalent to the Gable Creek Formation exposed near Mitchell, part of the assemblage of Cretaceous rocks deposited over the accreted terranes of the Blue Mountains. Much of the area to the west and south of Goose Rock consists of more landslide deposits.

South of Goose Rock, OR 19 passes reddish beds of the lower John Day Formation, their red color originating from the deep tropical weathering that formed the soils. At milepost 122, the road passes the Thomas Condon Paleontology Center, opened in 2005. Thomas Condon, Oregon's first state geologist, was also an Irish immigrant, an Irish Congregational minister, and the University of Oregon's first geology professor. The paleontology center holds a museum, sample preparation lab, and more than fifty thousand fossil specimens from the John Day Basin.

Sheep Rock forms the distinctive pyramid-topped peak 0.5 mile (0.8 km) southeast of the Paleontology Center. Although it mostly consists of the uppermost John Day Formation, it is capped by an erosional remnant of Picture Gorge Basalt. Near the middle of the rock lies the prominent Picture Gorge Ignimbrite, offset by a normal fault.

Just south of the entrance to Picture Gorge, the basalt exhibits beautiful columnar jointing up the slope to the west and a well-defined transition from columnar jointing to entablature next to the river on the east.

Sheep Rock (high peak at right) of the Sheep Rock Unit. Note that Sheep Rock and the adjacent hills are capped by Picture Gorge Basalt, with the upper John Day Formation underneath. The 28.7-million-year-old Picture Gorge Ignimbrite forms the prominent brown ledge between the green and tan beds of the John Day Formation. A normal fault offsets the ignimbrite in the cliffs below and slightly left of Sheep Rock.

named for the town of Monument, which is about 10 miles (16 km) northeast of Kimberly on OR 402. Many of these dikes cut the John Day Formation between Kimberly and Monument.

For several miles south of Kimberly, the road lies on top of well-developed river terraces, which mark former positions of the river channel. Notice the exposures of river gravel at road level.

OR 218
Shaniko—Antelope—Fossil
42 miles (68 km)

Only 42 miles (68 km) long, OR 218 gives an instructive sampling of the John Day and Clarno Formations, as well as the overlying Columbia River Basalt Group. Besides beautiful landscape, the road passes through an enormous landslide complex and the Clarno Unit of John Day Fossil Beds National Monument.

From Shaniko, the road leads southward over a plateau of Columbia River Basalt Group. Little outcrop exists, except where the road crosses small valleys.

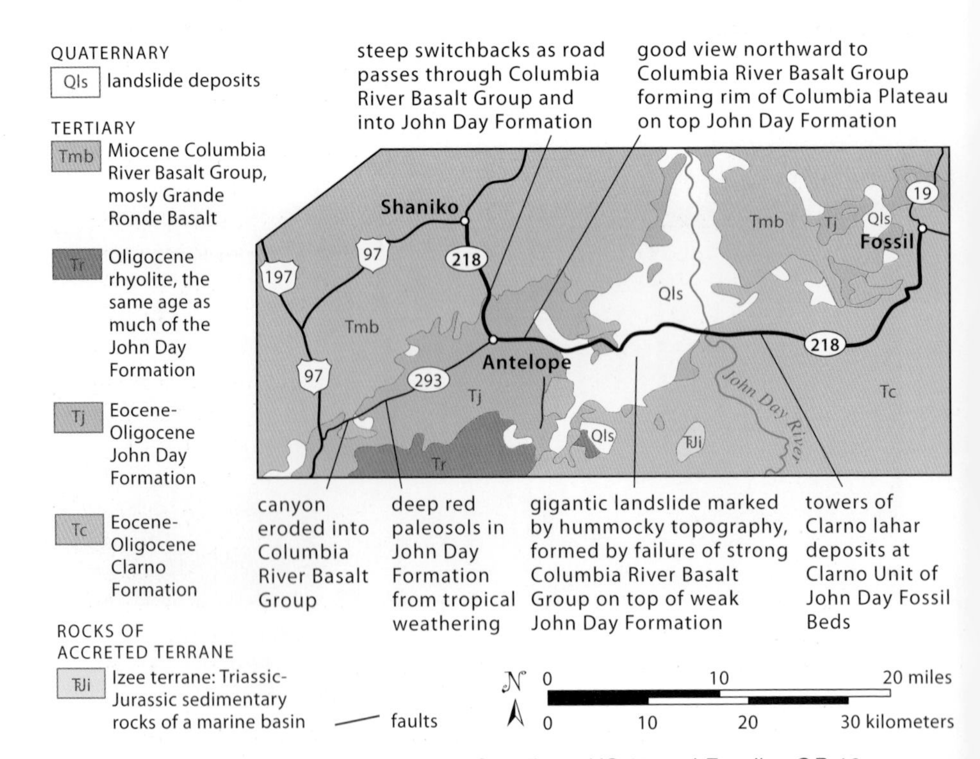

Geology along OR 218 between Shaniko at US 97 and Fossil at OR 19.

Antelope Highway

The alternate route to Antelope, which follows Antelope Creek eastward from US 97, passes almost entirely over deposits and flows of the John Day Formation, exposed below the Columbia River Basalt Group. An exception exists near the west end of the road where it passes through a canyon eroded into the Columbia River Basalt Group. East of milepost 4, however, basalt flows belong to the John Day Formation. Numerous roadcuts reveal ash deposits, flows, and even flow breccias, which form because lava flows tend to fragment as they cool. Between mileposts 7 and 8, some deep red paleosols from an older part of the John Day Formation crop out south of the road. Deep weathering in a tropical climate formed these soils.

Near milepost 4, the road begins its steep, switchbacking descent into the valley of Antelope Creek and crosses downward through the basalt into the upper part of the John Day Formation. These rocks are mostly bedded tuffs and tuff-rich sandstone but also include some andesite and basalt flows. Much of the John Day Formation consists of ancient soil deposits, which reflect the environmental conditions at their time of formation. These upper beds tend to be cream colored to white, indicating arid conditions. Older beds of the John Day tend to be yellow or even red colors, indicating deeper weathering during wetter and warmer conditions.

East of Antelope, OR 218 climbs through lava flows and tuffs of the John Day Formation to a pass between mileposts 14 and 15. Two to three miles (3.2–5 km) east of the pass, you gain a sweeping view eastward across a gigantic landslide complex that extends from here to beyond the John Day River, some 8 miles (13 km) to the east. Landslides are especially common where weak, ash-rich beds of the John Day Formation are exposed beneath the Columbia River Basalt Group. To the northeast, the Columbia River Basalt Group forms high cliffs, and the John Day Formation forms white, weathered slopes beneath them. The road winds its way through the landslide complex to the river, passing numerous exposures of fragmented John Day Formation and basalt.

At Clarno, cliffs of Clarno Formation line the John Day River to the south. They consist mostly of andesite and basalt flows, as well as lahars. Above the cliffs, rounded hills consist of John Day Formation. From here to Fossil, nearly all easily accessible rock exposures along the road consist of the Clarno Formation. Near milepost 25, you can see a good example of Clarno andesite flows on the north side of the road.

Between mileposts 37 and 38 the road crosses a pass and drops into the drainage of Cottonwood Creek. Although not visible from the highway, Columbia River Basalt Group directly overlies Clarno Formation just to the north. The lack of intervening John Day Formation indicates that a period of

Clarno Unit of John Day Fossil Beds National Monument

At milepost 26, OR 218 passes some unique cliffs eroded into Clarno lahars. Called the Palisades, these cliffs are part of the Clarno Unit of the John Day Fossil Beds National Monument and are easily accessed by maintained trails. Individual towers become isolated from the main cliff through erosion, which is concentrated along the many vertical fractures that cut through the exposure. What distinguishes a lahar from an ordinary conglomerate is the general lack of arrangement of its particles. Lahars transport all sizes and shapes of material at once, in a fast-moving medium of mud, sand, and water, similar to flowing cement. Their deposits consist of randomly oriented boulders and cobbles surrounded by a much finer-grained matrix. The Palisades contain dozens of these flows, each one marked by a single layer.

The lahar deposits contain abundant fossil leaves, most notably sycamore, picked up by the lahars as they overran a tropical forest. It is definitely worth stopping at one of the two parking areas along the highway to inspect

This boulder of Clarno Formation below the Palisades is covered in fossilized plant fragments. The fossils appear as innumerable thin lines that reach lengths of 12 inches (30 cm) or more.

fallen blocks of these rocks close up. Visitors can walk any or all of the several short trails that meander through the boulders or ascend to the cliffs themselves. In the cliffs lie some fossilized tree trunks. Please remember that, as a national parkland, the rocks and fossils belong to everybody, so collecting is not allowed without a permit.

The Palisades represent only one part of the Clarno Formation. To the north, the rocks become younger and consist of a whole range of sedimentary and volcanic rock types, as well as many intervening deep red paleosols. These rocks have yielded many important fossils, including the many specimens from the Hancock Mammal Quarry and the Clarno Nut Beds. Animal fossils include Brontotheres and Amynodonts, which bore some resemblance to modern rhino, and Patriofelis, a catlike predator. Many of the exposures can be accessed from the Hancock Field Station, a science camp operated by the Oregon Museum of Science and Industry. Travelers who wish to visit the field station need to first contact the museum.

---

erosion, likely brought on by uplift, occurred before the eruption of the basalt. The John Day and Clarno beds tend to be tilted, whereas the Columbia River Basalt Group in this area remains fairly flat-lying, further evidence of uplift and erosion prior to the eruptions of the Columbia River Basalt Group.

OR 82, WALLOWA MOUNTAIN ROAD, AND OR 86
La Grande—Joseph—Baker City
187 miles (301 km)

This road guide circles the Wallowa Mountains and visits Hells Canyon on the Idaho border. The Wallowa Mountains host Oregon's largest wilderness area, its greatest concentration of alpine lakes, and the greatest number of peaks over 8,000 feet (2,400 m). Its highest point, Mt. Sacajawea, rises to 9,838 feet (2,999 m), an elevation higher than anywhere else in the state outside of the Cascade volcanoes. The Wallowa Mountains attained these elevations through a long period of crustal uplift, which continues today because the range is bound by a number of active fault zones.

The Wallowa Mountains consist mostly of granitic rocks of the latest Jurassic-Cretaceous Wallowa Batholith, one of Oregon's largest intrusions. It intruded rocks of the Wallowa and Izee accreted terranes after they were joined together. The Wallowa Batholith consists of five different plutons, which intruded between 140 and 120 million years ago. The geochemistry of the older plutons suggests they formed in a magmatic arc environment. The younger

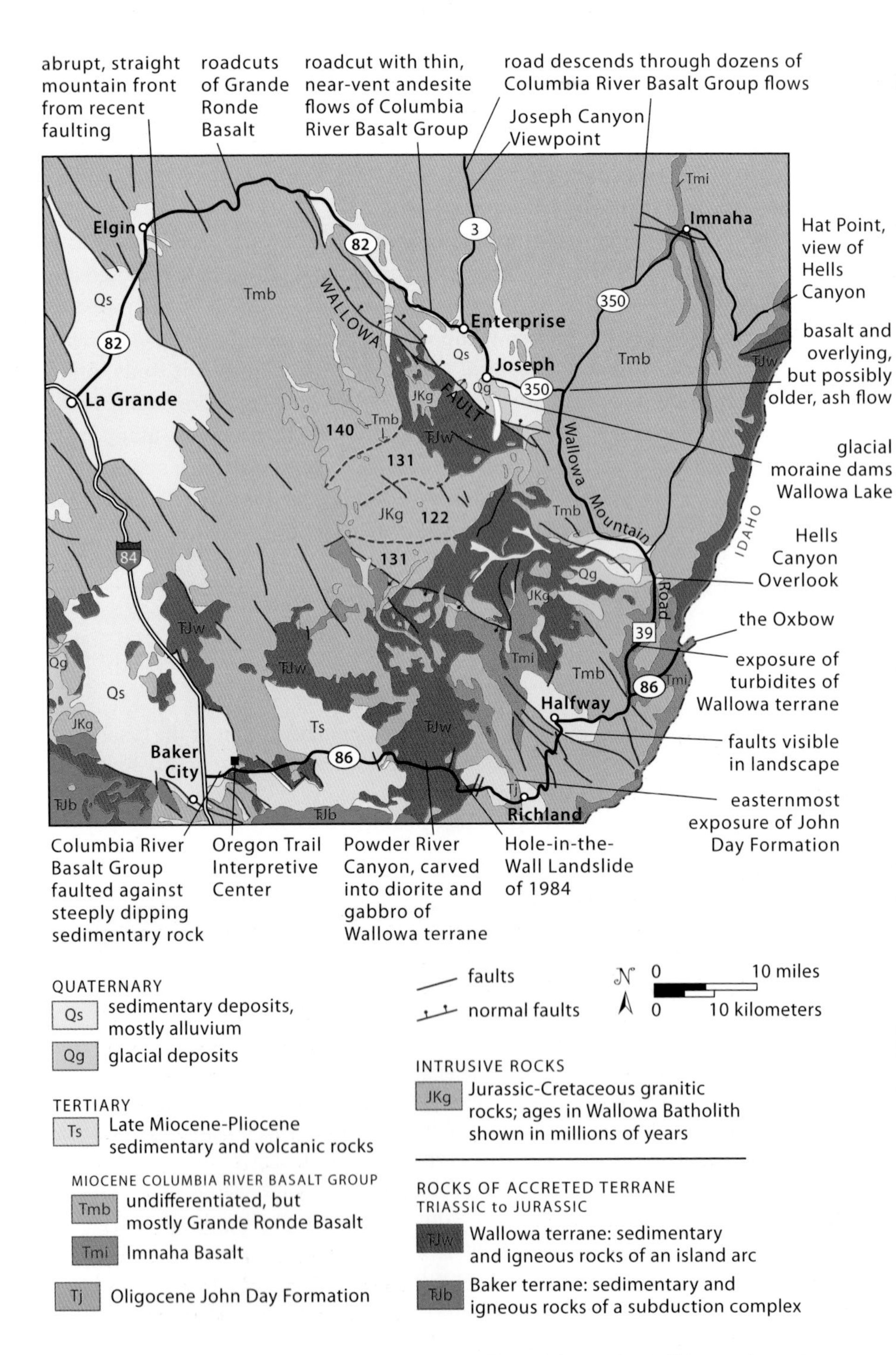

Geology of the Wallowa Mountains. The Wallowa Batholith consists of five plutons, ranging in age from 140 to 122 million years.

Some of the Wallowa's high peaks illustrate the bedrock geology, with granitic rock of the Jurassic-Cretaceous Wallowa Batholith intruding the Triassic-age Martin Bridge Limestone of the Wallowa terrane. Overlying all these rocks is the Columbia River Basalt Group, erupted in Miocene time.

ones, however, formed from melting brought on by extreme crustal thickening, which likely occurred during the actual terrane accretion.

In some places, lava flows of the Columbia River Basalt Group lie on top the granite and accreted terrane rock, and some 1,800 basaltic dikes cut through the older rock. These dikes are part of the Chief Joseph Dike Swarm, which fed eruptions of the Columbia River Basalt Group. Lava flows of the Columbia River Basalt Group surround the range at low elevations, but those within the range lie at much higher elevations, indicating uplift occurred after about 16 million years ago, when the basalt erupted. The uplift was accomplished through normal faulting and gentle folding. Two fault systems lie on either side of the range, the largest of which is the Wallowa Fault, on the northeast. It runs the length of the range and has a maximum slip of 6,600 to 7,900 feet (2,000–2,400 m). The fault creates an imposing mountain front and cuts Quaternary-age deposits, suggesting it is still active. In addition, the basalt is gently arched over the range, indicating the entire range is folded.

At least some of the uplift of the Wallowa Mountains probably took place during eruption of the Columbia River Basalt Group. Lava flows of the Grande Ronde Basalt are thinner over the range than they are at low elevations, suggesting the area was somewhat elevated at the time. Geophysical studies of the upper mantle some 50 to 100 miles (80–160 km) beneath the Wallowas show that it consists of a region of especially high seismic velocities, likely caused by a depletion of magma during the massive eruptions of the Columbia River Basalt Group. Mathematical modeling of this scenario finds that the eruptions, by depleting their mantle source area of magma, caused the source area to be more buoyant, uplifting the range during and just after the eruptions.

In Pleistocene time, the high elevations of the Wallowas supported glaciers, which carved numerous cirques and U-shaped valleys. At the mouths of the valleys are glacial moraines, ridge-like landforms of rock debris carried and deposited by the ice. The largest, most accessible, and visually striking of these moraines is the one that dams Wallowa Lake next to Joseph, but other sizable moraines exist at the mouths of other valleys.

Road Guide to the Wallowas

OR 82 crosses the diamond-shaped Grande Ronde Valley for the first 10 miles (16 km) north of La Grande. Views to the west, north, and east reveal that the valley is bound by faults on those three sides. The abruptness and linear nature of the range fronts suggest the faults have been active recently. To the east, the range front displays triangle-shaped facets, another indication of recent faulting. The somewhat flat top of the range reflects its flat-lying basalt bedrock.

At about milepost 14, the road crosses over the trace of the eastern fault and follows the Grande Ronde River, which flows northeast to the Snake River. Good outcrops of Grande Ronde Basalt line the road near milepost 17, where the road follows a lovely canyon cut by the river. Near milepost 21, the road begins to climb out of Elgin toward a pass just over 3,600 feet (1,100 m). A roadcut just west of milepost 26 shows ash deposits beneath a flow of Grande Ronde Basalt. The ash attests to the presence of explosive volcanism at the same time as the Columbia River flood basalts.

Near milepost 29, the road begins a downhill grade with exposures of Grande Ronde Basalt on the west side of the road. Ample pull-outs on the road allow parking for closer inspection. These exposures display several individual flows with intervening reddish paleosols. Looking northeastward into the canyon of the Wallowa River, you can see more flows of the Grande Ronde Basalt. Notice that the rock is uniformly fine-grained and nearly black, typical of the Grande Ronde. You can also see basal breccia, colonnade, and entablature. Most of the exposures near the top of the grade are covered by fencing to mitigate rockfall.

The road follows the Wallowa River through a canyon cut into successive flows of the Grande Ronde Basalt, which show up as ledges in the canyon wall. Just south of milepost 42, the road enters an open valley, the western edge of which coincides with a normal fault. From here you gain the first good view to the southeast along the northwest side of the Wallowa Mountains. The range rises steeply behind a system of large, active normal faults. The flat area along the river to the south is a river terrace, part of the floodplain that was abandoned as the river deepened its channel.

The front of the range changes abruptly south of Lostine. To the northwest, the range crest is fairly even because it is made of gently dipping basalt flows. To the southeast, the range crest is much more rugged, because it consists of a variety of other rocks, including accreted rocks of the Wallowa terrane and granitic rocks of the Wallowa Batholith. While viewing the range en route to Joseph, look for bedding in the accreted sedimentary rocks, glacial cirques, and deep glacial valleys.

A long roadcut in basalt lies on the northeast side of the road near milepost 61. Look for a small fault near its southeast end, separating a thick basalt flow from a series of thin flows that likely formed near a vent. Each thin flow consists of a dense, black interior, separated by irregular reddish layers that are full of

Roadcut of Grande Ronde Basalt, showing a distinct reddish paleosol between two of the flows.

Joseph Canyon Viewpoint

From Enterprise, you can take OR 3 north to Washington. The highway rises to a plateau of Grande Ronde Basalt. Joseph Canyon Viewpoint provides an amazing view into Joseph Canyon, exposing nearly 2,000 feet (610 m) of flows of the Grande Ronde Basalt. Just north of there, OR 3 rises onto the younger Saddle Mountains unit of the Columbia River Basalt Group and then descends steeply toward the Washington state line and the Grande Ronde River.

air bubbles. Despite their dark color, they are andesitic in composition, even though they are still part of the Columbia River Basalt Group.

From Enterprise, you can take Hurricane Creek Road to a trailhead, which in addition to the longer hikes, includes a short walk to a waterfall on Falls Creek over Triassic-age Martin Bridge Limestone of the Wallowa terrane.

OR 82 ends in Joseph, right at the base of the fault-bounded Wallowa Mountains. Less than 1 mile (1.6 km) south of town, the road to Wallowa Lake passes a state heritage site called Iwetemlaykin, which means "at the edge of the lake" in Nez Perce. The site offers views and trails near the bottom of the terminal moraine that rises some 300 feet (90 m) immediately to the south. The site also hosts the gravesite of Old Chief Joseph, the father of the better known Chief Joseph. Only 1 mile (1.6 km) beyond, the road climbs over the terminal moraine and allows a wonderful view of Wallowa Lake, spilling out of the mountains and hemmed in by moraines. The lateral moraine, accessed by a hiking trail near the north end of the lake, gives an idea of the thickness of the ice that deposited it. From its position nearly 300 feet (90 m) over the lake where it joins the terminal moraine, it steadily climbs to more than 800 feet (240 m) above the lake near the mountain front. Considering that Lake Wallowa's maximum depth is just over 300 feet (90 m), the ice was probably about 1,100 feet thick (340 m) where it exited the mountains!

Falls Creek spills over Triassic-age Martin Bridge Limestone of the Wallowa terrane. Note the two basaltic dikes on the right side of the photo. These dikes are part of the dike swarm that fed the Columbia River Basalt Group eruptions.

To continue the circumnavigation of the Wallowa Mountains, take OR 350 eastward from Joseph. Views of the Wallowas rise behind the large moraine of Wallowa Lake to the south, and views of the Seven Devils Mountains of Idaho rise to the east. The Seven Devils make up a large portion of the Wallowa terrane. Good exposures of basalt show up along the road beginning at about milepost 7 and continue all the way to the intersection with Wallowa Mountain Road at about milepost 8.

A roadcut of basalt mixed with ash flow exists across from a large pull-out about 0.25 mile (400 m) up from the intersection. This exposure demonstrates that silicic eruptions occurred during the same time interval as the Columbia River Basalt Group. The irregular, somewhat intrusive nature of the contact between the two rock units has been interpreted in two completely different ways. Either the ash flow came over the basalt and, in the process, ripped up large fragments of the basalt and included them into the flow. Or, the basalt flow came after the ash flow but somehow managed to burrow into it and then, in some places, intrude and break into fragments as it moved upward into the ash flow.

Heading southward from the intersection, Wallowa Mountain Road follows Little Sheep Creek past scattered exposures of Grande Ronde Basalt toward Salt Creek Summit, 10 miles (16 km) away. You can see better views from the pass

Roadcut near the intersection of OR 350 with the Wallowa Mountain Road, shows complex contact relations between an ash flow and a flow of the Columbia River Basalt Group.

Hat Point

At its intersection with Wallowa Mountain Road, a side trip follows OR 350 north and east to the small community of Imnaha, some 21 miles (34 km) away on the Imnaha River. The road descends more than 2,000 feet (610 m) through multiple lava flows of the Columbia River Basalt Group, which form striking cliffs on both sides of the road. For about the first 15 miles (24 km) these cliffs consist of the Grande Ronde Basalt of the Columbia River Basalt Group, but for the last 6 miles (10 km) they are the underlying Imnaha Basalt. From Imnaha, you can take the well-graded Hat Point Road to views of Hells Canyon and beyond. See pages 309–11 for a more detailed description of Hells Canyon.

by driving a short distance down one of the side roads. More scattered roadcuts of Grande Ronde Basalt show up on both sides of the road south of the pass.

An interesting exposure of basalt and an ash-flow tuff exists at a hairpin turn 11.5 miles (18.5 km) south of the pass. The basalt here is full of air bubbles called vesicles, which form in lavas that are high in dissolved gases. The ash-flow tuff lies between different flows of basalt. The spot also affords a beautiful view to the Seven Devils Mountains in Idaho.

The 3-mile (4.8 km) road to Hells Canyon Overlook lies 5.8 miles (9.3 km) south of the intersection with the Imnaha Road. You can see dozens of basaltic lava flows belonging to the Grande Ronde (above) and Imnaha (below) members of the Columbia River Basalt Group. Below them lies the Triassic-age Martin Bridge Limestone, folded into an anticline and syncline. Note that the overlying

View of Hells Canyon and Seven Devils Mountains from Hells Canyon Overlook. Note how the folded Martin Bridge Limestone lies beneath the Columbia River Basalt Group, indicating the strong deformation that occurred long before the eruption of the basalt flows.

basalt is not folded, an indication that the fold formed before the basalt erupted. The folds probably formed sometime during accretion of the rock in Cretaceous time. On the other side of the river, the Seven Devils Mountains rise to elevations above 9,000 feet (2,700 m).

About 5 miles (8 km) south of the overlook turnoff, numerous exposures of Imnaha Basalt, which lies beneath the Grande Ronde Basalt, line the road. They exhibit rubbly bases, developed as the flow cooled. Only 1 mile (1.6 km) or so beyond, accreted Mesozoic greenstones and turbidites appear on both sides of the road for about 0.25 mile (400 m). The distinctive rock weathers differently from the basalt and in one place exhibits steeply dipping sedimentary layering. A close inspection of the sedimentary rocks shows that some individual beds are graded, with the coarsest particles near the base and becoming finer upward. In another few miles, the canyon begins to open up, and you can see even more basalt in the distance.

Where County Road 39 tees into OR 86, a left turn leads northeastward about 8 miles (13 km) to Oxbow Dam on the Snake River, and a right turn leads to the town of Halfway and I-84. Toward Oxbow Dam, the road passes through successive lava flows of the Imnaha Basalt until reaching the dam, where accreted rocks of the Wallowa terrane are exposed.

Turning toward Halfway, OR 86 quickly ascends through the Imnaha Basalt into the Grande Ronde Basalt and then into a wide valley with an awesome view of the Wallowa Mountains to the north. This valley has a noticeably abrupt southwestern edge where a normal fault has dropped it down relative to a ridge. Notice that the ridge, along the path of the road, contains numerous smaller steps in its surface, the topographic expression of several other smaller faults. Several roadcuts on the way up this ridge confirm that it consists of Grande Ronde Basalt.

Just west of Richland, the road passes by ash deposits of the John Day Formation. These are the easternmost exposures of the John Day deposits, thus outlining the eastern extent of the basin. Just beyond, the road enters Powder River Canyon, eroded into an intrusive complex of the accreted Wallowa terrane. These rocks consist mostly of Triassic diorite and gabbro, likely the foundation on which the Wallowa terrane, an island arc, was built. These rocks are resistant to erosion and form a narrow craggy canyon that continues for about 10 miles (16 km), for as long as the basement rocks are exposed. Between mileposts 31 and 30, a large pull-out on the north side of the road affords a good view of the Hole-in-the-Wall Landslide. After a period of heavy rains in September 1984, a side of the canyon broke loose and some 10 million cubic yards (7.7 million m^3) of rock slid into the river channel. It dammed the river and buried part of the highway, parts of which are still visible near the canyon bottom. Near milepost 23, the road crosses into Grande Ronde Basalt and then into easily eroded Tertiary floodplain and ash-rich lake deposits. There, the canyon opens into a broad floodplain with good views north to the Wallowa Mountains.

The Oregon Trail Interpretive Center, perched some 300 feet (90 m) above the road between mileposts 7 and 6, offers remnants of the Oregon Trail and a

Accreted Mesozoic greenstones and turbidites on Wallowa Mountain Road. Note the steeply dipping sedimentary rocks on the left side of the outcrop. The close-up shows a graded bed in the sedimentary rocks, with the coarsest particles indicating the bottom of an individual graded bed. In these rocks, the bed tops are to the right.

fabulous view of the surrounding area. To the west, you can see Elkhorn Ridge, made of accreted rocks of the Baker terrane. To the north you can see the fault-bounded eastern edge of Baker Valley. Bedrock exposures of Wanapum Basalt line the walkway to the front door. Among other things, the center offers interpretive trails and programs, some wonderful exhibits, and a bookstore.

West of the turnoff to the Oregon Trail Interpretive Center, you can see more scattered exposures of the basement, as well as overlying basalt flows of the Wanapum Basalt and Tertiary sediments. Between mileposts 5 and 4, a large pull-out for the Oregon Trail Memorial allows inspection of a faulted contact between the basalt and steeply dipping sedimentary rocks.

Hells Canyon

Hells Canyon is deep. Depending on where it's measured, Hells Canyon is either the deepest canyon in North America or one of the deepest. From the top of the Seven Devils Mountains, which border it on the Idaho side, the land drops nearly 8,000 feet (2,400 m) to the Snake River. From Hat Point on the Oregon side, it drops less, but still more than 1 mile (1.6 km). This imposing canyon forms a natural boundary between western Idaho and northeastern Oregon for about 100 miles (160 km), with no road crossing it downstream of Hells Canyon Dam.

View of Hells Canyon from Hat Point, looking northward. Columbia River Basalt Group occupies the left foreground. Mesozoic accreted terrane rock lies beneath it in the river canyon and in the sunlit area of the middle ground on the Idaho side.

For rivers to cut deeply and narrowly into bedrock such as at Hells Canyon, the bedrock must be rising from regional uplift. It is difficult to specify when the uplift started in the Hells Canyon area, because there doesn't seem to be a clear record. However, the lava flows of the Columbia River Basalt Group have been uplifted, and the active faults along the edge of the nearby Wallowa Mountains attest to ongoing uplift.

Much of the present course of the Snake River became established through a process called stream capture. When headward erosion by a high-gradient stream cuts through a drainage divide high in a watershed, it can capture a lower-gradient stream and divert it. In Hells Canyon, a tributary of the Salmon River appears to have captured the Snake River in the vicinity of the Oxbow, a large, fault-controlled meander in the Snake River, between about 2.5 and 2 million years ago.

Evidence for this stream capture comes from a variety of places. Upriver from the Oxbow, tributaries join the Snake River so that the acute angle between them points upstream, as if the river flow had been reversed. Downriver from the Oxbow, the tributaries point downstream as would be expected. In addition, much of the Snake River's path in southern Idaho winds across a broad, low relief surface that contains remnants of lakebed deposits that are between 6 and 2.5 million years old. These sediments were deposited in Lake Idaho, a large lake that covered the region from eastern Oregon to Twin Falls and flooded tributary valleys as it grew. One of these valleys was Indian Creek, which today joins the Snake River at the Oxbow.

Before Lake Idaho existed, Indian Creek flowed southward into the ancestral Snake River drainage and away from the ridge that divided the Snake River and Salmon River watersheds. Southward erosion by tributaries of the Salmon River breached this divide and captured part of the drainage immediately west of Indian Creek, which increased erosion rates and further lowered the divide. At the same time, the water level of Lake Idaho was rising, backing up Indian Creek as far as the Oxbow. There, lake water exploited northeast-trending faults in the rock, eroding westward into the next channel. Once water overtopped that divide, it quickly cut it down even farther, creating a gaping hole in the lake basin. As the lake flowed into the Salmon River watershed, it deepened the canyon and established a new course for the Snake River that has been there ever since.

Hells Canyon offers glimpses into Oregon's history of accreted terranes, as well as the eruptions of Columbia River Basalt Group. Just about every canyon overlook reveals thousands of feet of the Imnaha Basalt and Grande Ronde Basalt resting on top the accreted terranes. The Imnaha Basalt is the oldest unit of the Columbia River Basalt Group in

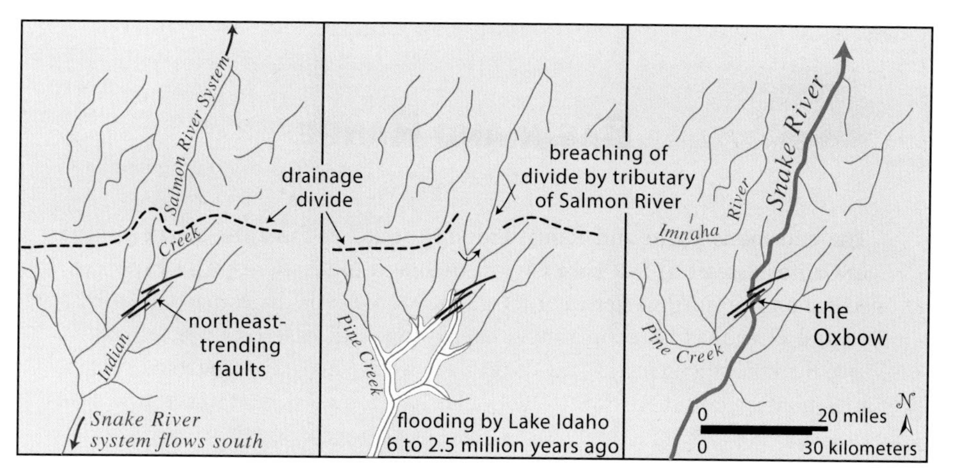

Prior to 6 million years ago (left) a drainage divide separated the southward-flowing tributaries of the Snake River from the northward-flowing tributaries of the Salmon River. As Lake Idaho flooded tributaries of the Snake River (middle), a tributary of the Salmon River breached the divide, and Lake Idaho overflowed to the north. Erosion along northeast-trending faults created the Oxbow. Today (right) the Snake flows north.

northeastern Oregon and has an irregular base where it flowed over uneven terrain. In some places it fills ancient valleys that were eroded into the preexisting rock. By contrast, the base of the Grande Ronde is much more even, because its eruptions followed relatively soon after the Imnaha's.

The accreted terranes in the Hells Canyon area consist mostly of parts of the Wallowa terrane but include some of the overlying Izee terrane in the northern reaches of the canyon. The Wallowa terrane formed as an island arc somewhere in the Pacific, and the Izee terrane was a sequence of marine basin sediments deposited over the top of the subsiding arc. The Wallowa terrane consists mostly of a variety of volcanic rocks, now metamorphosed, as well as sedimentary rocks that accumulated near the islands, including the conspicuous, light gray, well-bedded Martin Bridge Limestone of Triassic age. On the Idaho side of the canyon, the prominent Seven Devils Mountains are part of the Wallowa terrane, uplifted by faulting relative to the Columbia River Basalt Group at lower elevations.

Basin and Range

The enormous Basin and Range Province stretches from the Sierra Nevada in California east to Salt Lake City in Utah and south into much of southern Arizona. The northwestern corner of the province includes parts of south-central Oregon. Mountain ranges alternating with desert valleys, or basins, give the region its name. Many of the valleys have no outlet, so what minimal precipitation falls there flows to the valley low point. There, it may form a lake or evaporate to form a dry playa. In this land of sagebrush, there may be more antelope than people.

The bedrock, well-exposed in the uplifted ranges, consists mostly of volcanic rock that ranges in age from the recently erupted Mazama Ash deposits north of Klamath Falls to the approximately 40-million-year-old lahars and andesitic rocks near Lakeview and Summer Lake. These older rocks resemble the Clarno and John Day Formations of the Blue Mountains and may reflect similar depositional environments, albeit in different areas of the state.

The most abundant bedrock units in Oregon's Basin and Range are Miocene-age basalt, rhyolitic ash-flow tuffs, and volcanically derived sedimentary rocks.

Physiographic map of Oregon's Basin and Range Province. –Base image from US Geological Survey, National Elevation Data Set Shaded Relief of Oregon

The shoreline of Lake Abert is encrusted with salt precipitated from evaporating water.

The Miocene basalt ranges in age from the approximately 17-million-year-old Steens Basalt to about 8 million years old. The Steens Basalt erupted as part of the earliest stages of the Yellowstone hot spot and Columbia River Basalt Group. It dominates the geology near Steens Mountain in the eastern part of Oregon's Basin and Range but can be found as far west as Abert Rim near Lake Abert and Diablo Rim east of Summer Lake. The rhyolitic ash-flow tuffs consist mostly of three units—Prater Creek, Devine Canyon, and Rattlesnake—which erupted from locations in the Lava Plateaus but covered much of what became the Basin and Range. Finally, sedimentary rocks that consist mostly of particles eroding from the volcanic rocks sometimes lie between lava or ash-flow tuffs or in the case of much younger deposits, on top of the older flows.

The Oregon Basin and Range also contains dozens, if not hundreds of Pliocene to recently erupted cinder cones, as well as many small, eroded rhyolitic domes and plugs of mostly Pliocene and Miocene age. The cinder cones, especially, become increasingly abundant toward the western side of the region, suggesting some influence of the Cascades arc on that part of the Basin and Range.

Perhaps the most widespread unit of all, however, is not bedrock, but Quaternary-age alluvium, shown in yellow on the geologic road maps. Alluvium consists of the sand, gravel, and everything else that erodes from the bedrock and is transported by water and deposited on the valley floors. The alluvium in the valleys holds most of the groundwater that is pumped for agriculture. Most of the alluvial deposits are less than about 1,000 feet (300 m) deep.

Normal Faulting

The origin of the Basin and Range topography is crustal extension. This stretching drives normal faulting, which uplifts the ranges relative to the basins. What caused the crustal extension in the first place is an altogether different, more difficult question that relates to plate motions and the region's geologic history. Extension can originate at a subduction zone, when a subducting plate gradually steepens, allowing the overriding continental plate to extend over the top. It can originate after the crust becomes overthickened because of earlier thrust faulting. Or extension can begin because large strike-slip faults affect the plate margin and cause rotation. Each of these factors is present in the Basin and Range Province, so each may contribute to the extension.

If we knew when the normal faulting began, we would know when the mountains and valleys first began to form. Most researchers agree that west of Steens Mountain, the faulting began in earnest around 7 million years ago, because rocks younger than this tend to show evidence of progressively less faulting. Rocks older than 7 million years all show the same amount of faulting. At Steens Mountain, however, the faulting likely began earlier than 10 or 11 million years ago.

The faulting imparts an asymmetry to some of the ranges: they rise steeply from one side of a valley and slope gently down to the next. Because the ranges are bound by faults and their underlying rock layers tilt roughly parallel to the slope, the ranges are called tilted fault blocks. Within Oregon, the best example of a tilted fault block is unquestionably the northern part of Steens Mountain.

A variation on this theme is found elsewhere in Oregon's portion of the Basin and Range. Normal faults run along both sides of the ranges, so the ranges hardly tilt at all and the intervening valleys drop. This fault geometry results in a horst and graben structure, where the uplifted ranges are the horsts and the down-dropped valleys are the grabens. In addition, the low degree of tilting in Oregon reflects the low amount of crustal extension. In Nevada, for example, greater amounts of crustal extension caused greater amounts of fault slip, which caused steeper tilting.

In addition to the main north-northeast-trending faults, another group of smaller normal faults trends in a northwestward direction, essentially chopping up the mountain ranges. The uplands of many of these ranges tend to be relatively smooth landscapes except where cut by these northwest-trending faults. Small steep ridges mark where these faults have displaced the surface. Researchers have determined these faults have slipped at the same time as the

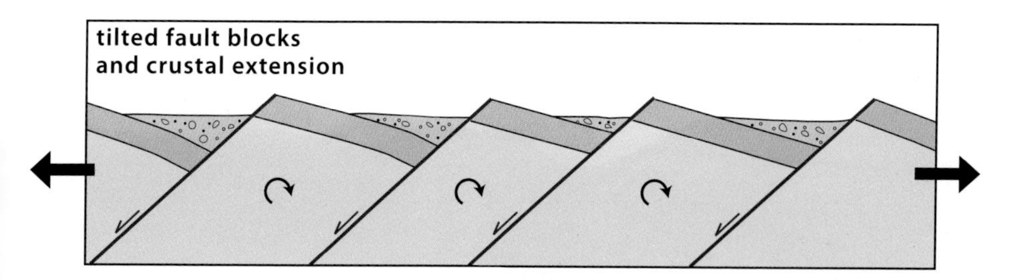

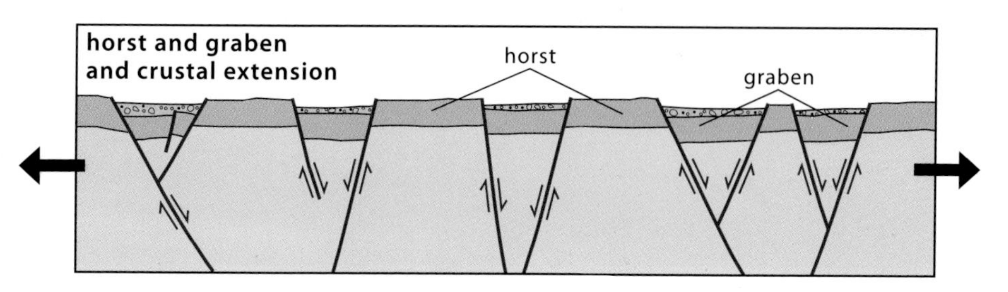

Two styles of Basin and Range faulting. In the top diagram, which represents much of Nevada, normal faults are inclined in the same direction and produce tilted fault blocks. In the lower diagram, normal faults are inclined in both directions, producing uplifted horsts and down-dropped grabens. Southern Oregon's Basin and Range has elements of both styles.

larger range-bounding faults, and that members of both fault groups have been recently active and so could potentially be active in the near future.

Because a fault must be younger than any feature it displaces, we know it's active if it cuts young deposits. In Oregon's Basin and Range, fault scarps, which are steps in the land surface where a fault zone has broken through and offset the land, cut across recently deposited stream gravels, shoreline deposits in valleys, and even ash layers from known volcanic eruptions. In cases where there is no identifiable offset feature, a visible fault scarp still implies recent faulting, because erosion has not had time to erase the scarp.

Recent movement along a normal fault also creates abrupt transitions from the valley floor to the mountain, and straight sections of the range along the fault. The ongoing process of erosion has simply not had enough time to erase the topographic evidence of faulting. In addition, some canyons become narrow and steep at their mouths near the range front because of repeated recent faulting. These canyons are called wineglass canyons because their shapes mimic wineglasses: their steep mouths resemble the stem; the wider, gentler part of the canyon back from the mouth resembles the bowl; and the alluvial fan deposited at the front of the range resembles the base. Faulted mountain ranges also display ridges that suddenly end at the fault as triangular-shaped slopes called triangular facets.

The east side of Steens Mountain showcases features associated with recent faulting. Notice how there is an abrupt transition from basin to range and the range front is straight. Also notice the wineglass canyons and the triangular facets between the canyons, where ridges have been truncated by the fault.

Lakes in the Basins

Most of the down-dropped valleys contain lakes and marshes that are relatively small when compared to the size of the valley. They also contain large flat areas, many of which are covered by dried mud or salt. These dry areas mark former lakebeds, deposited by lakes during the wetter and cooler Pleistocene Epoch. Some of these lakes were huge and filled the valleys. Locations of remnant shorelines, which look like bathtub rings on the surrounding hills, indicate that some lakes exceeded 500 square miles (1,300 km^2) in area. The elevations of these shorelines show depths in excess of 200 feet (60 m). Some lakes even overtopped the low divides separating different basins to form a single body of water. Glacial Lake Chewaucan, for example, filled what is now the Summer Lake basin and the Lake Abert basin. Today, prolonged rainy seasons can cause lower parts of the old lakes to flood, temporarily forming shallow lakes, until evaporation removes the water and leaves behind a thin crust of salt.

The water in the lakes of the Basin and Range stays there until it evaporates, because most of the lake basins are closed systems. The Basin and Range Province in southern Oregon is part of the Great Basin, a vast area of interior drainage that includes nearly all of Nevada, western Utah, and parts of eastern California.

Even though the modern lakes and marshes are much smaller than their Pleistocene counterparts, they still hold a great deal of water in what is otherwise a desert environment. The water originates from snowmelt in the adjacent

mountains. Some of it comes from surface runoff and some from springs that are ultimately fed by snowmelt-derived groundwater. These wetlands provide valuable habitat for a wide variety of mammals and birds. Designated wildlife refuges exist in most of Oregon's Basin and Range valleys and include such well-known areas as Malheur Lake, Summer Lake, Lake Abert, the Warner Lakes, and Upper Klamath Lake.

Wetlands at the southern end of Warner Valley, which is a graben. The mountains on both sides of the valley are long ridges, uplifted relative to the valley floor by normal fault zones along the base of each ridge.

Mud-cracked playa at Summer Lake, north of Lakeview. In the winter months, this surface is flooded by shallow water, but in the summer, the water evaporates.

GUIDES TO THE BASIN AND RANGE

US 97
La Pine—Klamath Falls—California Border
125 miles (201 km)

South of La Pine, US 97 follows the boundary between the Cascades and the Basin and Range Provinces. For the first 60 to 70 miles (100–115 km) south of La Pine, the two provinces overlap and their distinction becomes vague. Although the road crosses recently erupted Cascade lava, which defines the Cascades, it also crosses normal faults, which define the Basin and Range. Approaching Klamath Falls, US 97 passes through what's clearly the Basin and Range. The rocks, although still volcanic, erupted east of the Cascades and the topography becomes unambiguously defined by normal faults.

South of La Pine, US 97 follows the broad floodplain of the Little Deschutes River for some 17 miles (27 km) to Gilchrist. An eroded basaltic cone lies just east of the highway there, lending some topography to the otherwise low-relief landscape. Just south of Gilchrist, the Crescent Cutoff to OR 58 leads past a young cinder cone, lava field, and Odell Butte, an unusually high and symmetrical basaltic cone.

Mazama pumice and ash deposits blanket much of the landscape just south of the intersection of US 97 and OR 58. The thickness of these deposits is about 10 feet (3 m) near the intersection, and they get progressively thicker toward the volcano, some 25 miles (40 km) away. A series of north-trending normal faults breaks the landscape into a series of low, north-trending ridges and valleys, but these features are difficult to see through the car window because of the forest. The road follows several of these valleys, developed on the ash- and pumice-fall deposits, until about 2 miles (3.2 km) south of Chemult.

South of Chemult, US 97 descends a gentle grade and then flattens over Mazama ash- and pumice-fall deposits for about the next 30 miles (48 km). It offers occasional views of Mt. Thielsen, a deeply eroded basaltic volcano. South of the intersection to the north entrance of Crater Lake National Park, you can catch rare glimpses through the trees up the flank of Mt. Mazama, its top removed by the caldera-forming eruption that formed Crater Lake, about 7,700 years ago. Mt. Scott, a satellite cone on Mt. Mazama, forms the most prominent peak.

At milepost 241, US 97 begins its steep descent into the Klamath Basin, a deep, fault-controlled valley. A series of roadcuts, most with ample pull-out space, follows the entire grade. Beginning at the top, the road passes by dacite lavas and then descends through a thick sequence of ash- and tuff-rich sandstone. The material for these deposits was derived from early eruptions of the High Cascades, and the abundant crossbedding in some of the sandstones indicates they were redeposited by rivers. The roadcut about a third of the way down the grade also shows several picturesque normal faults, manifestations

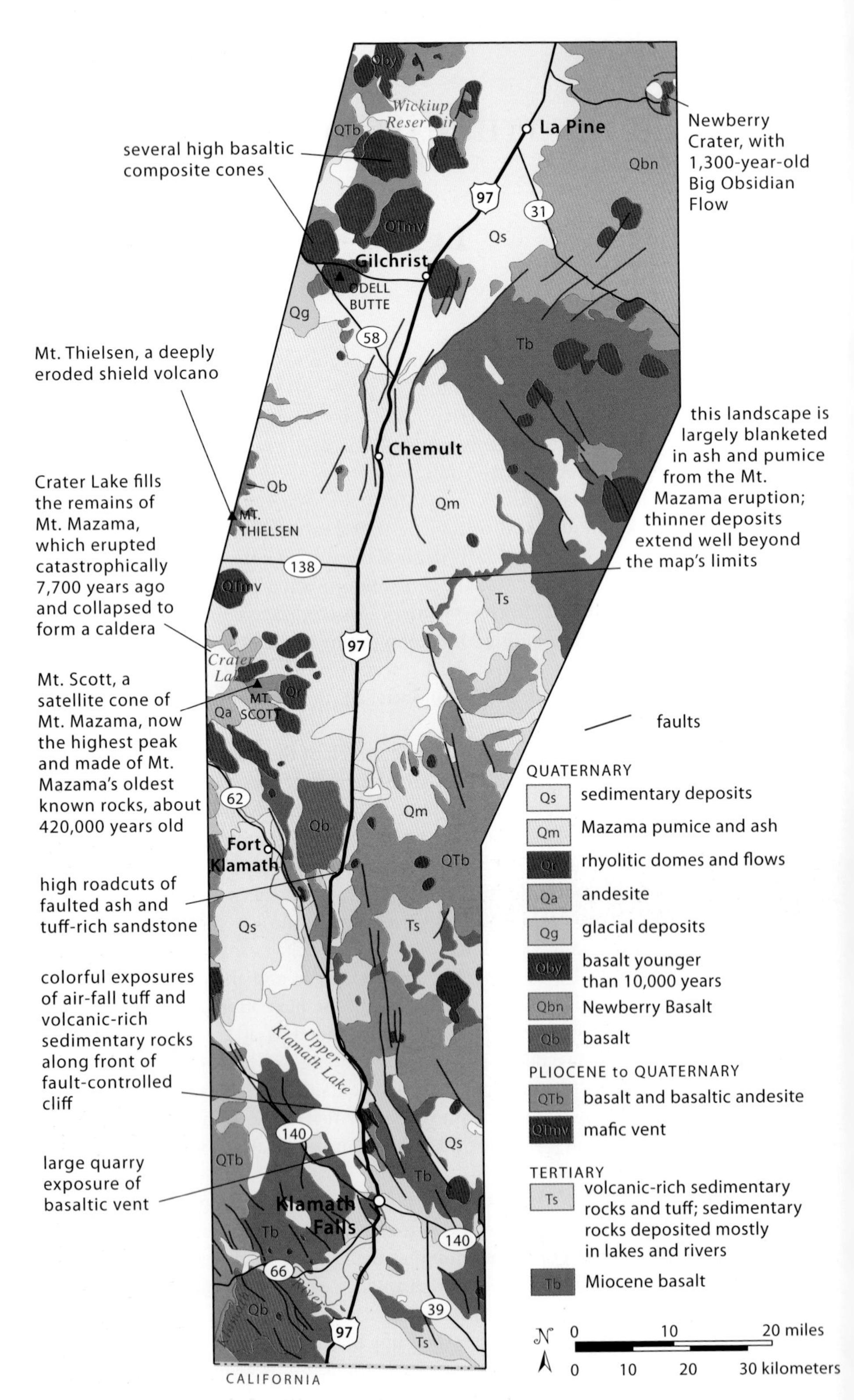

Geology along US 97 between La Pine and the California border.

of recent crustal extension. From the pull-outs, you can see eastward to several shield volcanoes and cinder cones.

At its intersection with OR 62, US 97 meets the bottom of a steep ridge that marks a recently active normal fault, part of the system of faults that has dropped the Klamath Basin on its east side. A similar escarpment lies on the west side of the basin, indicating the basin is a graben, a down-dropped block between faults. The road follows the eastern system of faults to Klamath Falls. An unmarked Forest Service Road (9715), 0.4 mile (640 m) south of milepost 253, leads up the eastern escarpment past rubbly basalt flows to spectacular views in a distance of less than 2 miles (3.2 km).

Between mileposts 265 and 266, US 97 follows the narrow stretch of land between Upper Klamath Lake and the uplifted range front. There, outstanding exposures of colorful, well-bedded air-fall deposits and volcanic-rich sedimentary rocks continue for about 1 mile (1.6 km) along the range front. The highway here is narrow and extremely busy, so in order to view these rocks, pull off into Algoma Road on the north end of the roadcut. At milepost 267, a quarry exposes the discontinuous, alternating layers of cinders and denser basalt typical of basaltic vents in the cliff face.

At the north edge of Klamath Falls, an exposure of diatomite, overlain by a basalt flow, lies next to the road. Diatomite, also called diatomaceous earth, is a bright white rock made of innumerable diatoms, single-celled algae that float

Aerial view looking south-southeastward over the east edge of Klamath Lake. The prominent roadway is US 97. Algoma Road joins US 97 at the bottom of the photo near good exposures of volcanic-rich sedimentary rocks between mileposts 265 and 266. Normal faults of the Basin and Range Province have uplifted each of the long ridges in the photo.

Outcrop of diatomite beneath columnar-jointed basalt along US 97 at the north edge of Klamath Falls.

in freshwater lakes and marine environments and secrete silica shells. Under a microscope, individual shells display an amazing variety of intricate, beautiful patterns. Numerous diatomite deposits exist among the Late Tertiary lake deposits of Oregon. Many of these deposits are mined for a variety of purposes, including filters, abrasives, and insulation.

Upper Klamath Lake fills much of the Klamath Basin graben. With an area of 96 square miles (249 km^2), it is the largest freshwater lake in Oregon. Prior to the twentieth century, it had extensive wetlands, which supported a rich diversity of wildlife. However, most of these areas were drained to create farmland beginning in the early twentieth century. A sizable national wildlife refuge does exist on the west side of the basin. The lake is prone to algal blooms when runoff from the surrounding farmland puts too many nutrients into the lake.

Klamath Falls lies at the southern end of the graben, where the faults on the west and east sides come together. The area is broken by numerous smaller faults, which cause the bedrock to tilt eastward. The highly faulted and fractured bedrock, combined with abundant water and high-temperature rock at shallow depths, make it an ideal place to develop geothermal energy. The city uses geothermal energy from more than six hundred wells to heat government buildings, as well as a hospital and schools, and even sidewalks. Many of the wells provide water that is hotter than 180°F (82°C). Most of these hotter ones lie near the fault zone at the break in slope on the east side of town. The bedrock in these areas exhibits a high degree of alteration because of the hot water. Klamath Falls was named for the waterfalls on the Link River, which connects Upper Klamath Lake with Lake Ewauna, just to the south. Hydropower and irrigation projects have greatly diminished the river's flow and all but eliminated the waterfalls.

South of town, US 97 crosses the Klamath River and passes a roadcut near milepost 276 of east-tilted sedimentary and volcanic rock, faulted against a diatomite deposit. Near milepost 278, US 97 recrosses the river and enters the lower Klamath Basin, which continues several miles into northern California. The basin, once the site of another large lake, has been mostly drained and claimed as farmland. Mt. Shasta rises some 55 miles (89 km) to the southwest. Low hills that punctuate the otherwise flat valley floor consist mostly of faulted basalt.

The Klamath River flows southwest here, through a deep canyon in the Cascade Range. The Klamath is one of only three rivers that begins east of the Cascades and flows directly into the Pacific, joining the Columbia River and the Fraser River in British Columbia for this honor.

US 395

Riley—Lakeview—California Border

126 miles (203 km)

This section of US 395 begins on the High Lava Plains of the Lava Plateaus and crosses into the Basin and Range Province, where it follows two deep fault-block valleys all the way to California. Evidence and remnants of Pleistocene lakes fill parts of the fault block valleys. The entire route travels over volcanic rock that consists mostly of Late Tertiary–age basalt and ash-flow deposits and, south of Lake Abert, Middle Tertiary andesite.

South of Riley, the road crosses a low-relief surface over flat-lying Late Tertiary basalt flows, most of which don't crop out. Some cinder cones, visible as low hills, rise above the surface. A quarry in a cinder cone lies on the east side of the highway about 1.5 miles (2.4 km) south of Riley. A cluster of cinder cones lies near milepost 10, one of which is exposed by a roadcut. The most topographic relief, however, comes from many northwest-trending faults that offset the basalt and create steps in the landscape. These faults are part of the Brothers Fault Zone, a swarm of faults marking the approximate northern edge of the Basin and Range Province in Oregon. Most of them have both normal and minor right-lateral displacements. Several buttes, formed by Tertiary-age mafic vents, break the skyline to the east.

Wagontire Mountain appears as a long ridge northwest of the highway, although it extends some distance back behind the ridge. It consists predominantly of rhyolitic rocks that are as old as 14.7 million years, a contemporary of some of the Columbia River Basalt Group. Near the townsite of Wagontire, the road travels over Rattlesnake Ash-Flow Tuff. About 2 miles (3.2 km) to the south, the ash-flow tuff, as well as some basalt flows, are dropped down to the

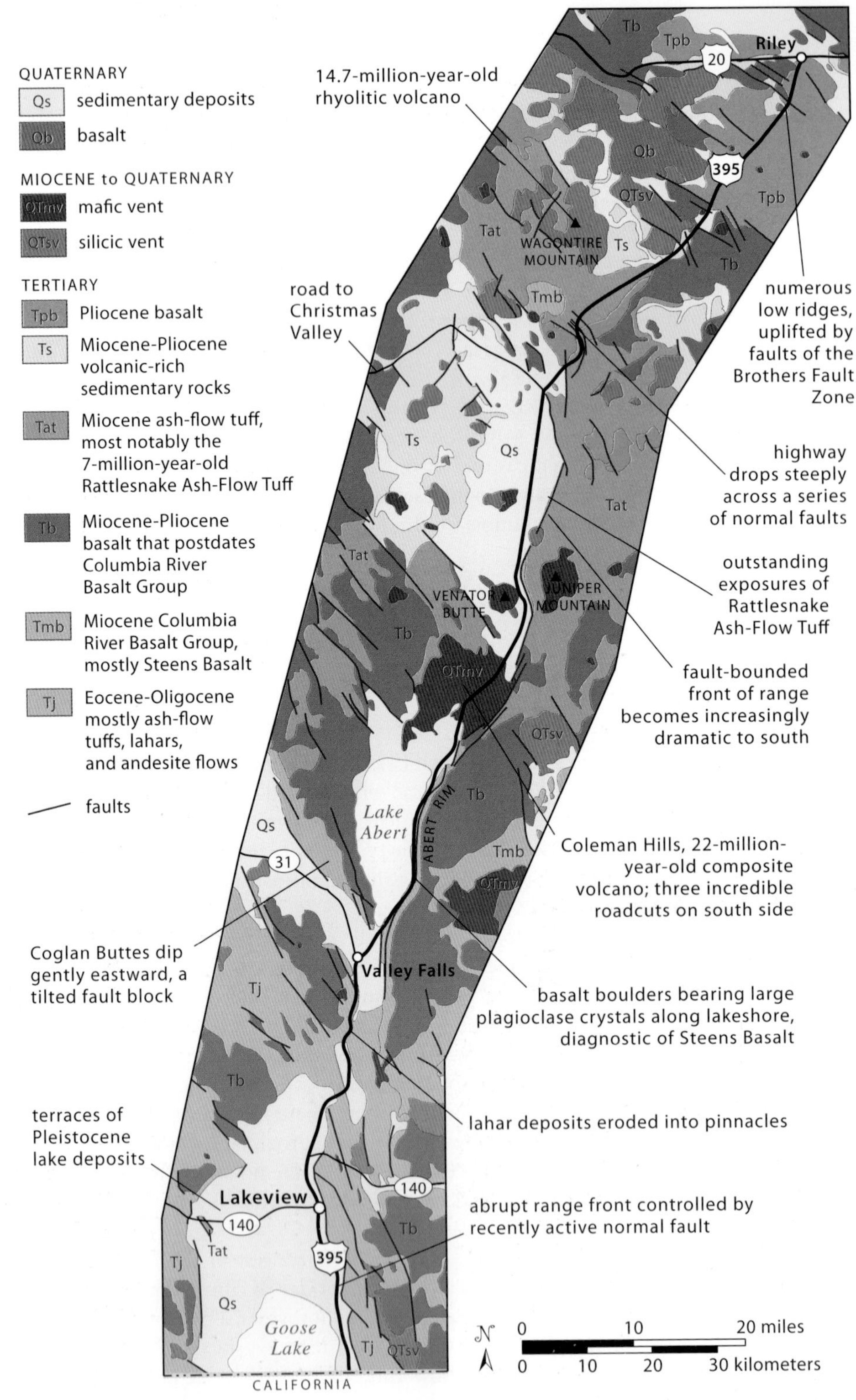

Geology along US 395 between Riley and the California border.

southwest along a series of faults. There the road descends in a series of steps to an open valley, bound by a normal fault on its east side. Good exposures of the tuff lie in the low cliffs just west of the road. Past the turnoff for Christmas Valley, the tuff lies in the cliffs to the east, uplifted by a normal fault.

US 395 follows this valley for the next 15 miles (24 km) or so, crossing a dry lakebed near milepost 40 and sand dunes between mileposts 45 and 47. With every mile, the topographic expression of the fault along the east side becomes more and more distinct, and by milepost 50, you can see a well-defined alluvial fan at the base of the ridge. This fault is the northern extent of the same fault system that uplifts Abert Rim, the 2,200-foot-high (670 m) escarpment some 20 miles (32 km) to the south.

About 3 miles (4.8 km) south of Alkali Lake, near milepost 53, the road passes between Venator Butte to the west, and Juniper Mountain, out of view behind the fault scarp to the east, both of which are small eroded volcanic centers that erupted olivine-rich basalt 7.7 million years ago. The road then turns southwestward and rises some 500 feet (150 m) into the Coleman Hills, an eroded composite volcano that was active about 22 million years ago. Near milepost 63, good views to the north show Venator Butte and Juniper Mountain, as well as the intervening trace of the normal fault.

At milepost 65, US 395 passes the second in a series of three roadcuts that show some of the inner workings of the Coleman Hills, an ancient composite volcano. The roadcuts contain abundant tuff breccias from explosive eruptions, basaltic lavas from quieter times, and some basaltic pillow lavas erupted into a lake. Also present are lahars and numerous basaltic dikes and sills that intruded the tuff and fed later basaltic eruptions. Some of the dikes show finer-grained edges called chilled margins, where the intruding magma cooled quickly against the surrounding host rock. Crosscutting all these rocks is a generation of younger, small normal faults.

The road descends steeply into the Lake Abert basin, which formed the eastern arm of Glacial Lake Chewaucan, the Pleistocene lake that inundated this area, as well as the Summer Lake basin, west of here. The road passes strandlines left by the drying lake from about milepost 69 south. Alkali flats near milepost 70 attest to present-day Lake Abert's fluctuating nature. A brick-red paleosol, developed between a tuff and the overlying basalt flow, crops out on the west side of the road between mileposts 72 and 73.

The road hugs Lake Abert's eastern shoreline for the next 12.5 miles (20.1 km), the cliffs of Abert Rim soaring directly overhead. Because Lake Abert has no outlet, its waters are extremely alkaline, and it precipitates minerals along its shoreline. Some mineral deposits encrust boulders as high as 300 feet (90 m) above the present lake level, more evidence for the high lake levels of Glacial Lake Chewaucan. The high alkalinity of the water prevents the lake from supporting fish, which in turn allows for a thriving population of brine shrimp and small insects, attractive meals for birds. The lake is a critical stopover for migratory birds, as well as one of the few inland habitats for the western snowy plover. The lake is also prone to frequent algal blooms, which can give the water a distinctly green color.

View northward along US 395 toward Alkali Lake. Venator Butte makes up the low hills on the far left of the photo. Juniper Mountain is the highest peak on the right. The low ridges consist mostly of Rattlesnake Ash-Flow Tuff uplifted along normal faults, the northern continuation of the same fault system that uplifts Abert Rim to the south.

Many of the basaltic boulders along the shoreline, eroded from the cliffs of Abert Rim, contain unusually large, elongate crystals of plagioclase. These rocks are part of the Steens Basalt, the oldest member of the Columbia River Basalt Group. Some especially good examples lie scattered around the pull-out at milepost 80.

In contrast with its smooth western shoreline, Lake Abert's eastern shoreline is noticeably irregular, with many small gravel headlands protruding into the lake. Rockfalls off Abert Rim form these talus deposits. You can also see several small landslides on the east side, a larger one between mileposts 81 and 82.

Just north of milepost 86, lakebed deposits from Glacial Lake Chewaucan appear on the east. Wetlands, where the Chewaucan River forms a small delta into the southern end of Lake Abert, appear to the west. Just south of milepost 87, the road enters the Chewaucan River valley, down-dropped as a graben between two normal faults. The view northward shows a beautiful example of a tilted fault-block: the uplifted Coglan Buttes tilt gently eastward toward Lake Abert.

US 395 intersects OR 31 at Valley Falls, at which point mileposts shift to a new set of numbers. Between Valley Falls and Lakeview, the road crosses a low divide that separates the Glacial Lake Chewaucan and Glacial Goose Lake basins. Glacial Goose Lake, however, was not significantly larger than today's lake, because during high water, the lake spills southward into the Pit River, which empties into the Sacramento River. Rocks consist mostly of Oligocene to Early Miocene lava flows, ash-flow tuffs, and lahars. Just north of milepost 125,

View looking south at Lake Abert and Abert Rim, which is uplifted along a recently active normal fault. Boulders in the foreground consist of Steens Basalt, eroded from the cliffs.

Close-up view of Steens Basalt with large crystals of plagioclase.

a lahar has eroded into a striking set of pinnacles on the east side of the road; on the west side, low cuts in the deposit allow an easy close inspection. Ash-flow tuffs are exposed on the highway just uphill from milepost 132.

Lakeview rests on the alluvium-filled northern section of Goose Lake Valley, dropped down along a series of normal faults along its east side. Oligocene andesite flows form much of the bedrock east of the town. The remaining 15 miles (24 km) of this trip to the California border follows the fault-bounded front of the range. Similar to the range front near Lake Abert, this front is abrupt and consists of several short but straight sections, suggestive of several recently active fault strands. Uplifted rocks consist mostly of Oligocene andesite flows and tuff-rich sedimentary rocks and provide a distinct layering to the mountain front. At the border, Goose Lake State Recreation Area provides access to the lakeshore and wetlands.

OR 31
La Pine—Valley Falls
122 miles (196 km)

As it enters the awesome landscape of the Basin and Range, this small highway passes some amazing smaller features of the High Lava Plains. These features include Big Hole, Hole-in-the-Ground, and Fort Rock, all of which formed when basaltic magma encountered groundwater, which flashed to steam. The resulting explosions formed tuff rings, tuff cones, and maar volcanoes. In general, maar volcanoes form when magma encounters fairly deep groundwater,

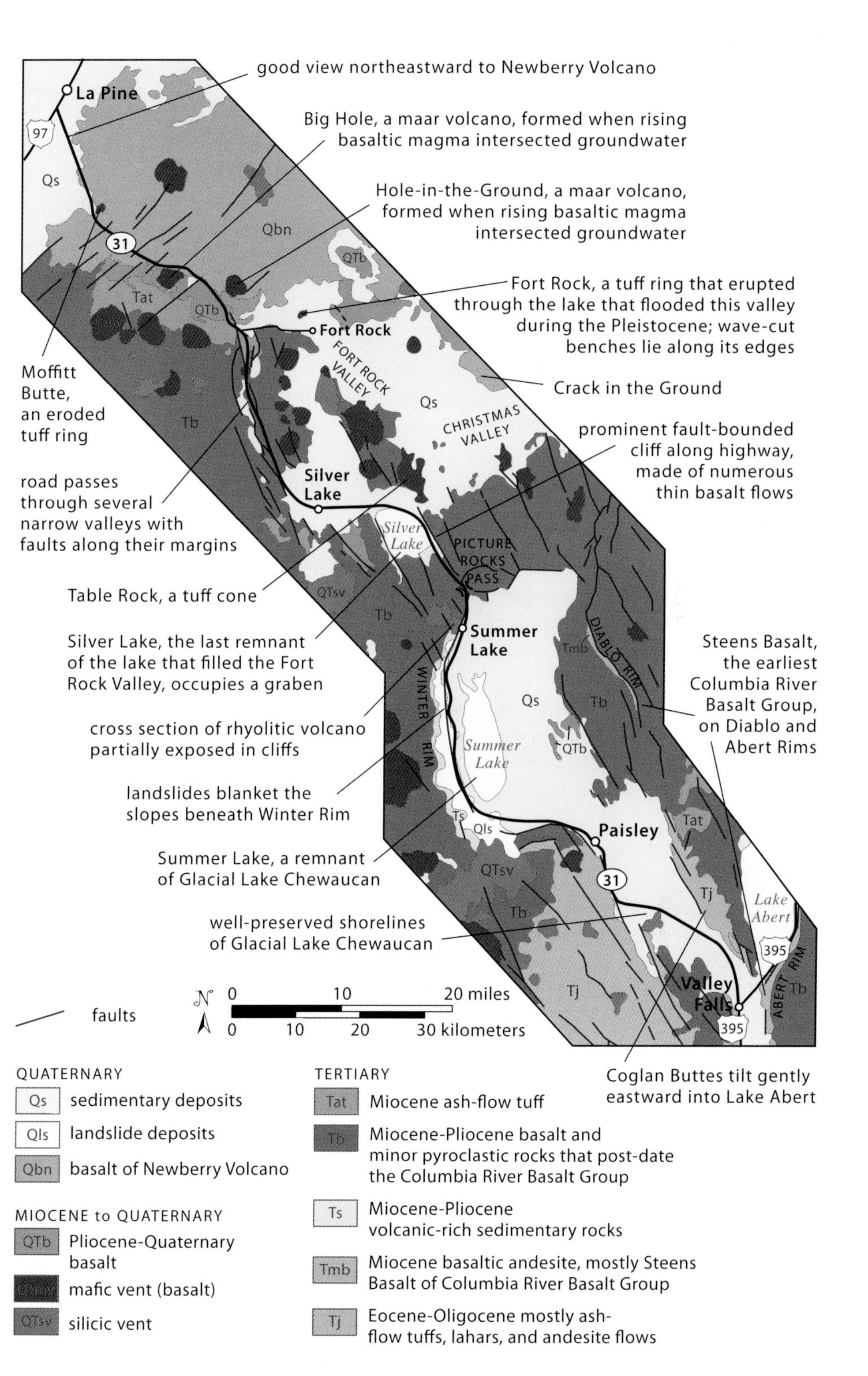

Geology along OR 31 between La Pine and Valley Falls.

tuff rings form when magma encounters shallow groundwater, and tuff cones form when magma encounters a lake. Tuff cones are steeper and higher than tuff rings, whereas maar volcanoes appear as large, circular craters. In the case of tuff rings and cones, which have a much more plentiful water supply, the eruptions tend to pulsate and go on for a long period of time, resulting in a distinctly bedded rock. They formed here because a shallow lake filled the Fort Rock–Christmas Valley basin during Pleistocene time at the time of the eruptions.

For the first 10 miles (16 km) south of La Pine, OR 31 crosses forested land developed on early basalt flows from Newberry Volcano. You can see Newberry from open stretches on the road, the best view within 1 mile (1.6 km) from the junction. Over much of the landscape, the basalt bedrock is covered with pumice and ash erupted by Mt. Mazama during the Crater Lake eruption 7,700 years ago.

Between mileposts 10 and 11, the road bends gently southward at Moffitt Butte, a deeply eroded tuff ring that, although mostly obscured by trees, rises more than 300 feet (100 m) above the road. It is best viewed from its south side. Tuff rings broadly resemble cinder cones, because they are usually basaltic and fairly circular in map view, but tuff rings are made from fragmented shards of altered volcanic glass called hyaloclastite instead of the loose, unconsolidated cinders of cinder cones. The rock of a tuff ring appears fine-grained and brown from the alteration.

Most of the rock along OR 31 until the turnoff to Fort Rock consists of basalt flows from Newberry Volcano. Numerous small ridges in the basalt called pressure ridges can be seen in the relatively light forest. An especially illustrative one, which looks like a gentle anticline, is exposed by a roadcut near milepost 17.

A prominent jog in the road near milepost 19, where it bends left and then right, bypasses Big Hole, a maar volcano that is mostly covered in trees. Forest Service Road 400 leads into the crater. Much more spectacular, however, is Hole-in-the-Ground, accessed by an easy 4-mile (6.4 km) Forest Service road marked by a road sign at milepost 22. Hole-in-the-Ground is between 300 and 400 feet (90–120 m) deep and has a distinct ridge of tuff surrounding it. When rising basaltic magma encountered groundwater some 1,000 feet (300 m) or more beneath the surface, the water flashed to steam and produced a huge explosion. The eruption likely occurred between 18,000 and 13,000 years ago and produced an eruption cloud more than 3 miles (4.8 km) high.

The road to Fort Rock intersects the highway between mileposts 29 and 30. Fort Rock, visible to the east, is a large tuff ring volcano. It consists of well-layered deposits of cemented hyaloclastite, mostly orange from the presence of palagonite, an alteration product that forms along with the fragmentation of the hyaloclastite. To get a sense of the nature of this material, try to imagine hot, wet, clay-rich mud that's full of glass fragments. Within a short period of time, the heat and pressure of the overlying material causes it to solidify into a hard rock. These deposits rise some 200 feet (60 m) above the valley floor, three-quarters of the way around a central crater that is about 0.75 mile (1.2 km) in diameter.

Aerial view of Hole-in-the-Ground crater, a maar volcano. View looking northward to Newberry Volcano on horizon.

Fort Rock's crater floor rests some 40 feet (12 m) above the surrounding valley floor, typical of other tuff rings that erupt through standing water. Moreover, several wave-cut benches along its margins indicate Fort Rock was eroded by the lake it erupted through. This erosion likely explains why the south side of the volcano lacks crater walls. During Pleistocene time, when the eruption occurred and the valley was filled with water, winds came from the south and would naturally drive waves against the south side of the volcano, eroding it.

Between the Fort Rock turnoff and Silver Lake, OR 31 follows a series of normal faults that break the basalt into numerous low, linear ridges. These faults seem to be the northernmost extent of the Basin and Range Province in Oregon, although they are relatively small. The ridges uplifted along the faults are especially noticeable at milepost 32, where the road follows a ridge along its east side, and between mileposts 43 and 44, where an uplifted ridge of basalt lies immediately to the west. Silver Lake, dry during the summer months, marks the last remnant of the large shallow lake that filled the Fort Rock–Christmas Valley basin during Pleistocene time. Some lakebed deposits show up along the road just south of milepost 45.

From Silver Lake, you can take a side trip a few miles northward to Table Rock, a tuff cone, or venture farther northeast to Crack in the Ground, a unique geologic site just north of Christmas Valley. This 2-mile-long (3.2 km) fracture in basalt reaches a maximum depth of 70 feet (21 m). It formed along a recently active normal fault that offsets basalt flows as young as 12,000 years. In some places, however, the same flows entered the crack, which indicates faulting prior to 12,000 years as well. A trail leads through much of the crack.

Aerial view to the north over Fort Rock, a tuff ring that erupted through a lake that filled this valley during glacial times. Newberry Volcano lies in the background, its caldera about 28 miles (45 km) away.

Silver Lake and points south are in true Basin and Range country, marked by normal faults with much larger offsets and more striking effects on the landscape. Silver Lake lies within a graben, down-dropped between normal faults on both its western and eastern sides. Beginning near milepost 58, the road follows the eastern fault, which uplifted the high linear ridge just east of Picture Rock Pass. The ridge consists of numerous thin basaltic lava flows of Tertiary age. At the pass, you can find petroglyphs carved into the basaltic boulders and cliffs.

The Summer Lake basin stretches southward from Picture Rock Pass, although the best views come from 1 mile (1.6 km) or so south of the pass. On its west side, a large normal fault, called the Winter Rim Fault, uplifts Winter Rim, the large ridge along the lake. Summer Lake occupies the western edge of the basin, with the valley rising gently toward the east, accompanied by some minor faulting. Thermal springs exist throughout the basin, some of which are developed on the west side. During Pleistocene time, Summer Lake was part of the much larger Glacial Lake Chewaucan, which covered the valley floor south to Valley Falls, including Lake Abert's basin. All told, Glacial Lake Chewaucan covered some 460 square miles (1,190 km^2) and had a maximum depth of about 300 feet (100 m). Old shorelines are visible in several places along the route. The road descends steeply to the valley floor, passing numerous normal faults along the way.

Near milepost 68, just north of the townsite of Summer Lake, the uplifted cliff face displays light-colored, well-stratified, air-fall tuffs that both underlie and overlie the edge of a basalt flow. The effect looks something like a giant eye peering down from the cliffs! This relationship demonstrates that the tuffs, which formed adjacent to a small rhyolitic volcano, erupted at the same time as the basalt, sometime during Late Miocene time. A second basalt flow lies above the upper tuffs. The dark cliffs below the tuffs consist mostly of lahars.

Summer Lake is dry along its margins, especially during dry periods. Near the deepest part of the basin, year-round water is extremely shallow, occasionally reaching depths of only 5 to 7 feet (1.5–2.1 m), and the water is highly alkaline. Marshlands, especially at the north end of the lake, offer valuable bird habitat.

On the west side of the road, Winter Rim soars some 2,500 feet (762 m) above. Numerous large landslides are causing its retreat for almost the entire length of the lake. These slides appear as hummocky, jumbled topography below the rim, in many cases continuing all the way down to near road level. An especially large slide is visible near milepost 87. These slides repeat a familiar theme in Oregon geology: where weak, tuff-rich sedimentary rocks lie beneath dense, rigid basalt, the sedimentary rocks are prone to failure and can bring down large sections of the mountain. Fortunately, the timescale of these events is such that they seem to happen on the order of thousands of years. Numerous

The cliffs behind the townsite of Summer Lake consist of well-layered air-fall tuffs (white) that partially enclose a basalt flow. Note the basalt flow at the top of the ridge. Dark lahar deposits in the cliffs below the white tuff reflect earlier episodes of volcanism.

smaller, slow-moving slides also exist on the slopes and cause frequent headaches for the scattered homeowners in the valley.

Paisley Caves, the site of the oldest known human artifacts in North America, resides in the hills east of Summer Lake. Archeologists have dated the site to approximately 14,000 years before present, at which time Summer Lake would have held more water. The caves were carved in Pliocene basalt by waves of Glacial Lake Chewaucan. When inhabited, the caves were probably about 1 mile (1.6 km) from water.

Entering Paisley from the north, OR 31 descends an alluvial fan built by the Chewaucan River as it emerged from the fault-bounded mountains to the west. As the fan grew during the latter part of Pleistocene time, it eventually blocked river flow northward into the Summer Lake basin and isolated it as a closed basin. A 3-megawatt geothermal powerplant began construction in November 2013 adjacent to the mountain front west of Paisley.

South of Paisley, the road continues along the faulted range front. In some places, the range front displays triangle-shaped facets, indicative of recent faulting. Near milepost 108, the road passes by an orange-weathered silicic vent to the south, with prominent shorelines from Glacial Lake Chewaucan along its side. Deposits of the lake, a fine-grained silt that is broken into polygon-shaped mud cracks, blanket the flat valley floor.

View northward along the west edge of the Summer Lake basin and Winter Rim. The hummocky landscape along the edge of the escarpment consists of landslide material.

Coglan Buttes border the east side of the valley. They consist of faulted and tilted andesite and ash-rich deposits of Oligocene age, overlain by tilted, Miocene-age basalt flows. Near milepost 113, you can see a couple of dikes cutting the flows of the southern hills. At Valley Falls at the intersection with US 395, the view northward shows the Coglan Buttes forming a beautiful fault block that tilts eastward into the basin of Lake Abert.

Shorelines of Glacial Lake Chewaucan appear as distinct benches south of Paisley. Behind them are rhyolitic rocks of a silicic vent.

OR 140
Klamath Falls—Lakeview
95 miles (153 km)

From Klamath Falls, OR 140 skirts around and between faulted ridges to Bly Mountain Pass, then descends into the Sprague River valley, which it follows upstream to its headwaters at Quartz Mountain Pass. From there, the road descends east into the Goose Lake Valley and Lakeview. Tertiary volcanic and sedimentary rocks along the way are covered with scattered ponderosa pine and sagebrush. Much of the route follows the OC&E Woods Line State Trail, which utilizes the bed of a former railroad used to haul timber.

East of Klamath Falls, OR 140 initially travels over floodplain deposits. A high ridge of Miocene and Pliocene basalt flows, uplifted relative to the valley along a normal fault, rises immediately north of the road. At Olene, the road cuts through a gap in this ridge, which provides a natural, although poorly exposed, cross section of what's in store for the next 30 miles (48 km) or so.

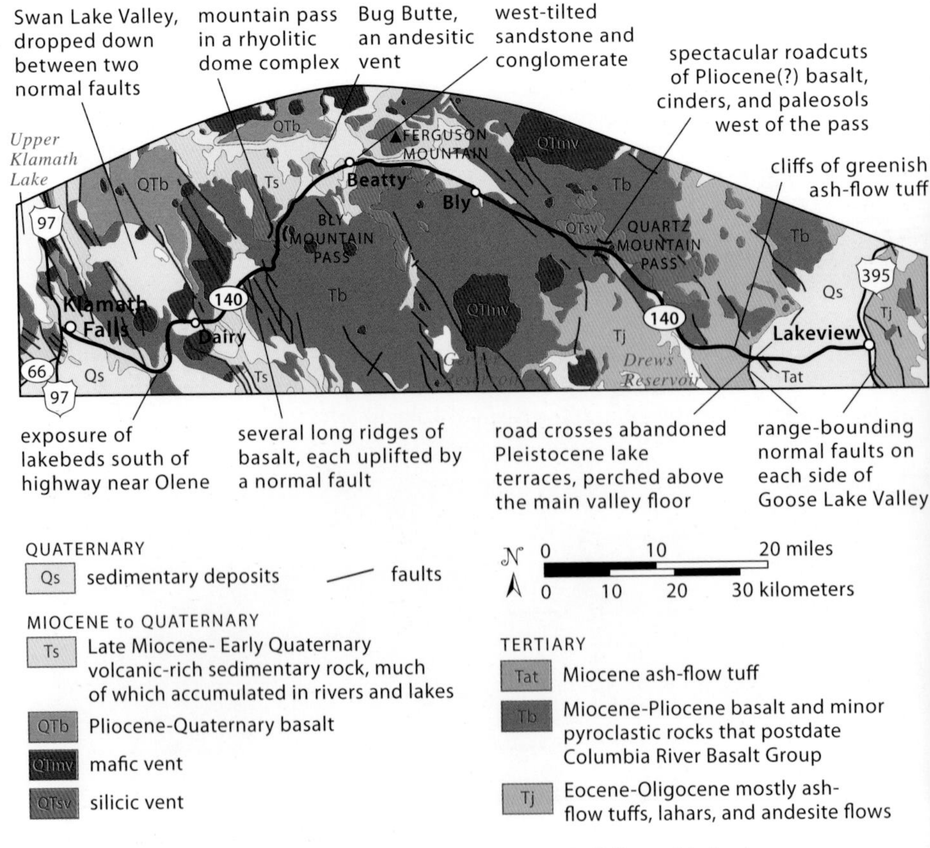

Geology along OR 140 between Klamath Falls and Lakeview.

Near road level are Pliocene lake deposits—the best exposure showing up along a small road on the south side of the gap. Above the lake deposits lie basalt flows. The next several valleys are also floored by the poorly exposed lakebeds and surrounded by ridges and hills made of the basalt. This area must have been inundated by a large lake prior to the faulting that created most of today's topography.

At milepost 15, the road turns abruptly eastward and descends a basalt flow as it crosses a normal fault into a valley floored by the lake and river deposits. This valley is the southern end of Swan Lake Valley, a graben down-dropped between two normal faults. Similar to its western side, the eastern side is marked by an uplifted basaltic ridge. Two large hills, volcanic vents that erupted basalt, lie on either side of the road near milepost 16.

East of Dairy, OR 140 crosses a valley dropped down by normal faults and floored by the Pliocene lake deposits. Numerous low ridges of basalt, especially on the eastern side, punctuate the otherwise flat valley floor. Each ridge follows a small normal fault. The road climbs through the younger basalt toward Bly Mountain Pass, with good views southward to the faulted ridges. Near the pass, the road crosses through a rhyolitic dome complex. Along the highway you can see numerous exposures of crystal-rich rhyolite. By milepost 34, the road is back into basalt. Bug Butte, a small andesitic vent, lies just northwest of the road near milepost 36.

Just east of Beatty, the road passes a long roadcut of west-tilted sandstone and conglomerate. Lahars lie at the base of the sequence. It's not clear how these tilted rocks fit into the overall picture, because 0.25 mile (400 m) down the road to the north, the rocks are flat lying, with Pliocene lake deposits overlain by

Faulted, volcanic-rich sedimentary rocks just east of Beatty. These rocks become more steeply dipping farther to the east (left) and eventually overlie lahar deposits.

rhyolitic rocks. The rhyolites are part of a silicic vent complex, the high points of which are Ferguson and Medicine Mountains, north and south of the road, respectively. Some poor exposures of these rhyolites exist along the road for the next couple of miles.

Near Bly, at milepost 54, the character of the bedrock along this route changes. East of here, the road gradually passes into older rock without the Pliocene lake deposits on the valley floors. The lake in which those deposits accumulated apparently did not extend this far to the east.

While ascending the west side of Quartz Mountain Pass, the road passes pyroclastic flows just west of milepost 64. These deposits originated from a large silicic vent complex that includes Quartz Mountain, immediately north of the road. Just east of milepost 64, some spectacular roadcuts of basalt flows and cinders extend well over 1 mile (1.6 km). At the east side of the roadcuts, a basalt flow fills a small channel cut into the cinders. A paleosol underlies the basalt. Numerous other exposures of basalt, although not quite as interesting, lie along the road east of the pass. Beyond them lies Drews Reservoir, dammed in 1911 for irrigation.

On the way toward Drews Gap Summit, the road passes occasional outcrops of greenish ash-flow tuffs, then drops steeply along Antelope Creek into Goose Lake Valley. The green color comes from the mineral celadonite, which forms through weathering of the rock. Exposures of these ash-flow tuffs lie on both sides of the road near the canyon bottom, at milepost 84. The rocks consist of a hodgepodge of variously colored rock fragments suspended in an ashy matrix. These rocks are likely Oligocene or Eocene in age and possibly related to the John Day and Clarno Formations that are so well-known in the Blue Mountains to the north. Some greenish shale and sandstone lie at the bottom of these tuffs at the mouth of the canyon.

Part of the long roadcut that exposes basalt flows and cinders. In this photo, the basalt occupies a former channel. Notice the red paleosol at its base.

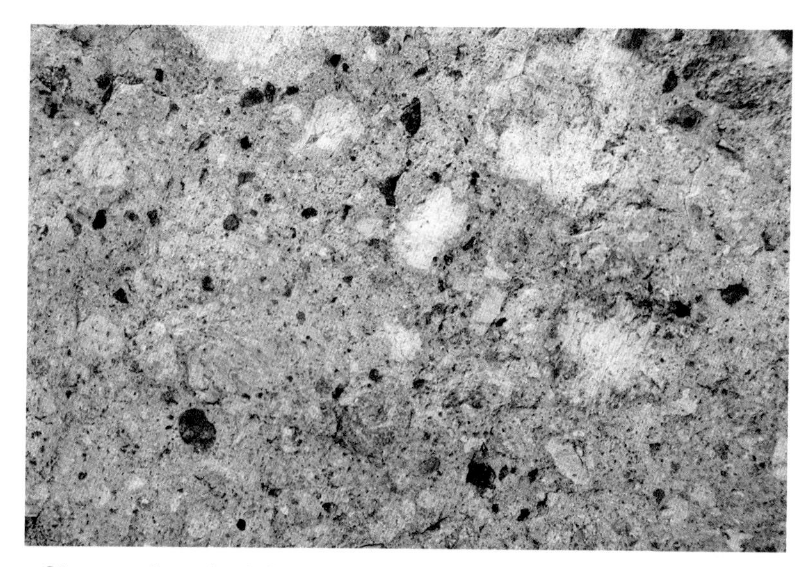

Close-up view of ash-flow tuff in a canyon near milepost 84. Photo is 6 inches (15 cm) across. Rock fragments are various volcanic rocks, and the matrix consists of ash. The green color comes from the mineral celadonite.

The road abruptly leaves the canyon and crosses onto terraces of lakebed deposits. A glance back shows that the range front is remarkably straight, because normal faulting has dropped the valley relative to the mountains. The terraces formed during the ice age when Goose Lake was at its highest level. The terraces are faintly visible about 5 miles (8 km) to the south. A double roadcut of ash-flow tuff, locally uplifted by faults on either side, is exposed just east of milepost 86, and a small roadcut of the lake deposits is exposed just east of milepost 89. The road drops down from the terraces between mileposts 91 and 92. While approaching Lakeview, notice the abrupt edge of the eastern side of the valley, also caused by a normal fault, as well as the high terrace that follows the front of the range.

OR 205

Burns—Fields—Nevada Border

133 miles (214 km)

This highway plunges straight through Oregon's Basin and Range Province. Beginning at Burns in the High Lava Plains, it heads southward directly to Steens Mountain, Oregon's largest fault-block mountain range, and then crosses Steens Mountain to reach the next valley to the east. In addition to Steens Mountain, this road provides access to two other geologically stimulating destinations: Diamond Craters, a recently erupted cinder cone and lava

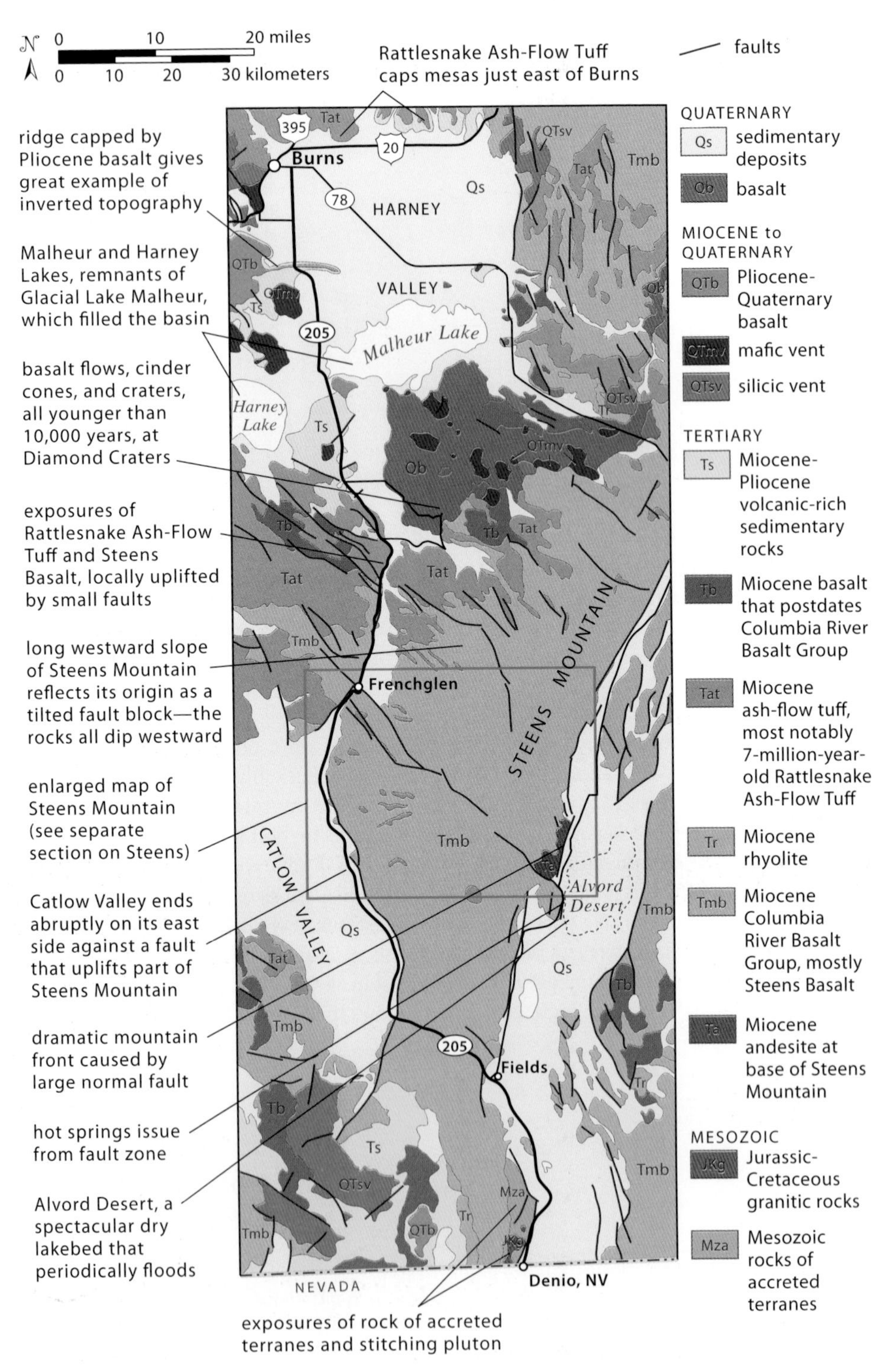

Geology along OR 205 between Burns and Nevada.

field, and Alvord Desert, a playa at the foot of the Steens's imposing eastern face.

For about the first 20 miles (32 km) south of Burns, OR 205 crosses the flat Harney Valley, once inundated by Glacial Lake Malheur. With an area of 920 square miles (2,380 km^2), Glacial Lake Malheur was Oregon's largest Pleistocene lake. Its remnants, Harney and Malheur Lakes, provide important wetland habitat for migratory birds.

Halfway across the basin near milepost 9, OR 205 climbs and then descends more than 200 feet (60 m) over a narrow ridge that protrudes nearly 10 miles (16 km) into the basin. This ridge, which consists of Pliocene basalt on top of Late Miocene volcanic-rich sediment, is one of Oregon's finest examples of inverted topography. The basalt originated as a single flow that poured down a long valley cut into the sedimentary rock. Being far more resistant to erosion, however, the basalt flow remained while everything around it eroded away. It now sits perched over the valley like an extraordinarily long finger. The best exposures of both rock types lie on the south side of the ridge. Be sure to note the paleosol at the base of the lava flow near the top of the ridge and the beautiful crossbedding in the sedimentary rock near the bottom.

More exposures of the volcanic-rich sedimentary rock exist near milepost 23, where the road rises south out of Harney Valley; an especially good roadcut lies on the west side of the road near milepost 26. A few miles farther south, several buttes on both sides of the road mark volcanic vents that erupted beneath Glacial Lake Malheur. The large butte east of the road between mileposts 31 and 32, however, is a basalt flow overlying the volcanic-rich sedimentary rock.

Pliocene basalt resting on top volcanic-rich sedimentary rock. The basalt, which originally flowed down a narrow valley, now forms a long ridge stretching eastward into Harney Valley, a dramatic example of inverted topography.

Beginning at about milepost 35, the road enters the Brothers Fault Zone, a swarm of closely spaced, northwest-trending normal faults that run all the way to Newberry Volcano south of Bend. The faults punctuate the landscape as straight-edged cliffs of various heights. In some places, the road follows the base of these cliffs; in other places, it cuts across them.

About 1.5 miles (2.4 km) south of the Diamond Craters turnoff, OR 205 turns sharply southwestward and follows a canyon cut into the Rattlesnake Ash-Flow Tuff. This tuff, which blankets much of southeastern Oregon, formed during a cataclysmic eruption about 7 million years ago. Its vent was probably

Diamond Craters

Diamond Craters, with fresh basalt, cinders, craters, and cones, is at once remote and accessible. Lying some 60 miles (100 km) south of Burns in vast open terrain along the High Lava Plains–Basin and Range transition, it's an easy side trip, reachable by a 10-mile (16 km) drive from OR 205. The volcanic field formed within the last 10,000 years. The name Diamond Craters derives from its proximity to a local ranch that once used a diamond shape to brand its cattle.

The volcanic field originated from a small, somewhat circular shield volcano, some 6 miles (10 km) in diameter. Early eruptions produced pahoehoe flows that spread radially outward from a central vent, which was obscured by later eruptions. These early flows reach thicknesses of 75 to 100 feet (23–30 m) in the central area but thin rapidly to only about 1 foot (0.3 m) on the margins. The rocks are porphyritic olivine basalts, because they contain larger crystals of olivine surrounded by a much finer-grained matrix.

Small-scale, yet explosive eruptions followed the initial nonexplosive phase. Doming of the earlier flows preceded these eruptions, as magma intruded into shallow levels. In some cases, the magma interacted with groundwater, caused steam explosions, and formed craters called maars. More than twenty craters decorate the surface of the volcano, at least twelve of which are in the central crater complex. Some of these craters are parts of cinder cones, but others appear as funnel-shaped holes in the ground. Many of the craters are actually several craters nested into one larger feature.

A prominent feature of the volcanic field is a graben centered over one of the domed areas. The graben, a down-dropped block of crust, formed because of collapse into a recently emptied magma chamber, much in the way a more circular caldera would form. Late-stage lava flows, which lie along the road that runs along the south side of the area, seem to reflect a return to quiet conditions following the explosive activity, possibly because

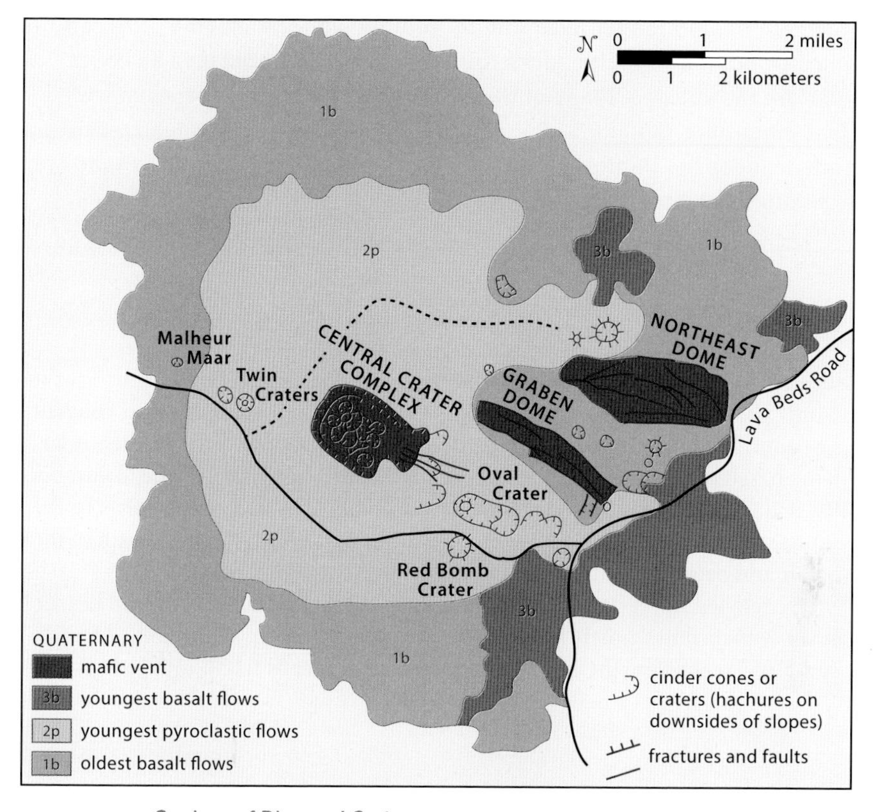

Geology of Diamond Craters. –Modified from Peterson and Groh, 1964

Two volcanic bombs, the one on the right being about 8 inches (20 cm) long. Their streamlined shapes result from cooling as the ejected lava is flying through the air.

groundwater was no longer available to interact with the magma. You can find volcanic bombs—basaltic rocks formed and shaped from lava as it was ejected from the volcano. Many of these bombs are cored by pieces of wall rock that must have fallen into the magma before it erupted violently. This wall rock includes the Danforth Formation, a Pliocene-aged deposit of sandstone, siltstone, and tuff that underlies the basalt.

---

located near Burns. Near milepost 45, the road makes several tight turns as it winds past some fault-bounded ridges in the tuff and passes a good roadcut just south of milepost 46. South of there, the road passes through the underlying basalt of Steens Mountain, which marked the beginning of the Columbia River Basalt Group eruptions about 16.7 million years ago. Good views to the south show Steens Mountain's gentle western slope, which follows the tilt of the basalt and overlying Rattlesnake Ash-Flow Tuff.

Frenchglen sits at the south end of a valley carved by the Donner und Blitzen River, which flows north from its headwaters on Steens Mountain. Directly south of Frenchglen, the road ascends a ridge of Steens Basalt uplifted relative

to the valley by a large normal fault. From there it passes southward into Catlow Valley, bordered on the east by Steens Mountain.

The northern part of Steens Mountain differs from the southern part. To the north, Steens Mountain is a tilted fault block: the western slope is gradual and not faulted, but the east side is uplifted along a fault. South of Frenchglen, however, Steens Mountain is a horst, because both the west and east sides rise along faults. Even so, the range is somewhat asymmetrical, as reflected in the road gradient beginning near the south end of Catlow Valley, where it turns east and cuts through the mountain. Over the first 7 miles (11 km), it ascends 1,000 feet (300 m) to a pass, and then, in just 5 miles (8 km), it drops 1,400 feet (430 m) into Alvord Valley.

Catlow and Alvord Valleys hosted large lakes during glacial times and are floored by fine sand and clay deposited in the lakes. The valleys display faint "bathtub rings," marking former shorelines, along their edges. In Catlow Valley, the most prominent of these former shorelines lies low on the mountain front between mileposts 73 and 76. Today, Catlow Valley does not contain a lake, but Alvord Valley contains several remnants. The most notable of these remnants are Alvord Lake and Alvord Desert, both of which periodically flood. Alvord Desert, a vast flat surface broken everywhere by innumerable mud cracks, is especially well worth the side trip, about 18 miles (29 km) north beneath the soaring east face of Steens Mountain.

The abrupt transition from the west edge of Steens Mountain down to the east side of Catlow Valley is suggestive of recent faulting. Note the wineglass-shaped canyons along the mountain front. The low bench along the mountain front is likely a shoreline of the lake that filled Catlow Valley during glacial times.

South of Fields, the highway follows the faulted southern section of Steens Mountain, called the Pueblo Mountains. Some small lakes and wetlands exist in the valley, left over from Glacial Lake Alvord. The mountains consist predominantly of Steens Basalt, but younger volcanic-rich sediments form the hills along the road. About 9 miles (14 km) south of Fields, however, the nature of the mountains changes to a more rugged character, reflecting bedrock made of accreted metamorphosed volcanic rocks of Triassic and Jurassic age. Near the Nevada border, these rocks are intruded by granitic rocks of probable Cretaceous age.

View of Steens Mountain from a temporarily flooded Alvord Desert. At an elevation of 9,730 feet (2,966 m), Steens rises more than 1 mile (1.6 km) above the Alvord Desert, which lies at 4,060 feet (1,237 m).

Steens Mountain

Perhaps more than anywhere else in Oregon, Steens Mountain exemplifies the Basin and Range Province. The mountain is a tilted fault block of the 16.7-million-year-old Steens Basalt. It has a comparatively gentle western slope, a long ridgeline that serves as its summit area, and a steep eastern face that descends over 1 mile (1.6 km) to a recently active normal fault and the Alvord Valley. Over a period of some 10 million years, this fault has uplifted the mountain range and tilted it some 10 degrees westward. Steens Mountain is therefore more of an uplifted range rather than a single mountain, being the eroded crest of an uplifted block. The Steens Mountain Loop, a 4-wheel-drive, high-clearance road that climbs to the crest of the ridge on the gentle western slope, begins and ends on OR 205.

Superimposed on the tilted fault-block geometry is a fault along the east side of the range that separates it from the down-dropped Catlow Valley,

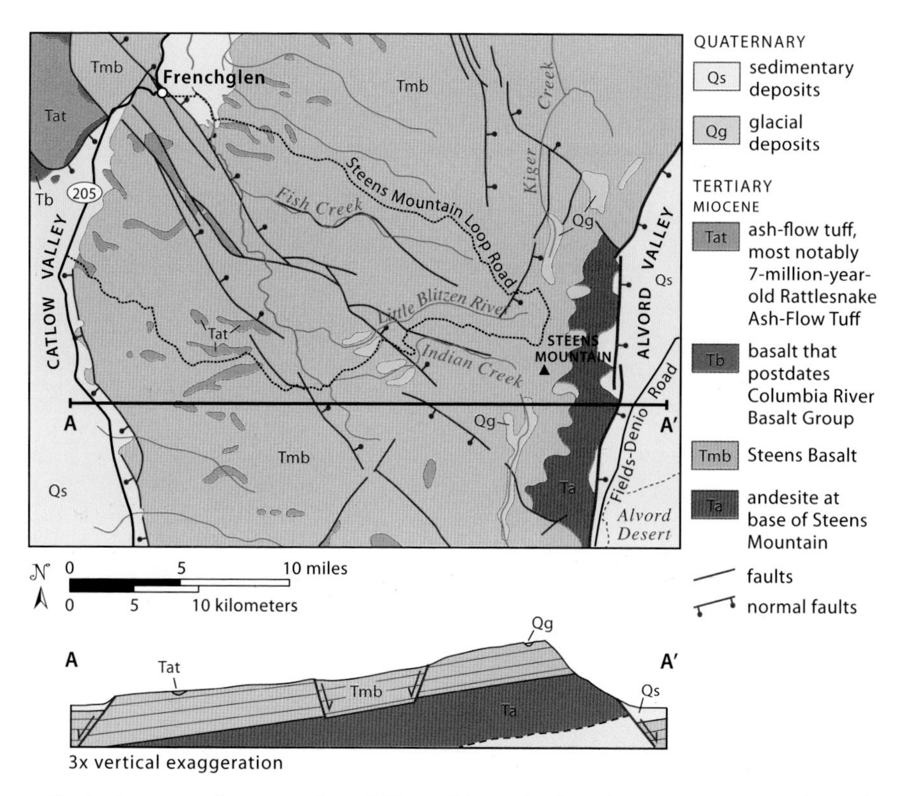

Geologic map and cross section of Steens Mountain. Note that about 3,000 feet (900 m) of andesitic flows and minor sedimentary rocks lie beneath the Steens Basalt.

as well as a series of faults that break it up into smaller fragments. These faults are the eastern end of the Brothers Fault Zone, a collection of normal faults that run northwestward from here to just south of Bend. The highest part of Steens Mountain, which reaches 9,730 feet (2,966 m) in elevation, lies in its central part, uplifted by these faults relative to its southern and northern parts. Several of the larger canyons of Steens were eroded along these faults.

The canyons of Steens Mountain beautifully display many of the lava flows of the Steens Basalt, which attain a cumulative thickness of over 3,000 feet (900 m) in some places. Some of the flows contain unusually large crystals of plagioclase, which formed in the magma chamber prior to their eruption. The basalt flows extend well beyond Steens Mountain over an area of some 19,000 square miles (50,000 km^2) of southern Oregon, northeastern California, and northwestern Nevada. Its eruptions marked the first flood basalts of the Columbia River Basalt Group, which inundated most of eastern Oregon and Washington and led to the Yellowstone hot spot eruptions.

View southeastward from the crest of Steens Mountain, showing Alvord Desert at the bottom of the mountain and layers of Steens Basalt in the upper slopes of the mountain.

Overlying the basalt are scattered exposures of the 9.3-million-year-old Devine Canyon Tuff and 7-million-year-old Rattlesnake Ash-Flow Tuff, products of two large rhyolitic eruptions in the vicinity of Burns. These remnants are all that remains after erosion otherwise stripped everything down to the more resistant basalt. In some places, the Devine Canyon Tuff occupies ancient stream channels carved into the Steens Basalt, indicating that some uplift of the region had already occurred by the time the tuff erupted. Beneath the basalt, some 3,000 feet (900 m) of older andesitic rock can be found on the range's eastern side. Interbedded with these volcanic rocks are lakebed deposits that contain fossils indicative of a warm, but more temperate climate than the near-desert conditions of today.

At elevations above about 6,000 feet (1,800 m), Steens Mountain supported glaciers during the Pleistocene Epoch. The most striking evidence for these glaciers comes in the form of the U-shaped valleys visible from the loop road. At their highest reaches, these valleys end in glacial cirques, while at their lower reaches, they display V-shaped profiles, characteristic of river erosion. The transition from U-shaped to V-shaped profiles occurs between 5,800 and 6,500 feet (1,770–1,980 m) on Steens. Numerous deposits of glacial till, which exist as moraines in some of the valleys and cirques, also indicate glacial activity.

Glossary

aa. A Hawaiian term for a lava flow that has a rough, jagged surface.

accreted. Something that has been added onto, such as an accreted terrane that has been added to the North American continent through plate motions.

air fall. An accumulation of material that fell onto the ground from the atmosphere, such as volcanic ash or pumice.

alluvial fan. A gently sloping, fan-shaped accumulation of sediments deposited by a stream where it flows out of a narrow valley onto a wider, flatter area.

alluvium. Water-transported sedimentary material.

amphibolite. A type of rock formed by regional metamorphism of basalt under medium to high pressures and temperatures. It consists largely of the minerals amphibole and plagioclase.

andesite. A medium- to dark-colored volcanic rock that is between basalt and rhyolite in silica content. It was named after the Andes Mountains in South America, where it is common.

anticline. A fold with the oldest rock in the core; most anticlines have limbs that dip away from the core.

ash. Tiny particles of volcanic glass blown into the air during volcanic eruptions.

ash fall. Accumulation of volcanic ash particles that have fallen out of the atmosphere.

ash-flow tuff. The rock formed from the consolidation and compaction of an ash-flow deposit.

asthenosphere. The zone of somewhat malleable rock beneath the lithosphere. It is the zone over which the lithospheric plates move.

back-arc basin. An ocean basin formed by crustal extension on the opposite side of a volcanic island arc from the compressional environment of the oceanic trench of a subduction zone.

basalt. A dark-colored volcanic rock that contains less than 52 percent silica. It is unquestionably the most abundant rock of the Oregon landscape.

basement. The deepest crustal rocks of a given area. They are typically igneous or metamorphic rock, but in the case of some accreted terranes of Oregon, may be sedimentary in some places.

bedding. The layering as seen in a sedimentary rock. A single layer is called a bed. When different rock types are interlayered with each other, they are described as **interbedded**.

bedrock. Rock that remains in its place of origin and has not been moved by erosional processes.

biotite. Dark mica, a platy mineral. It is a minor but common mineral in igneous and metamorphic rocks.

block-and-ash flow. A pyroclastic flow that consists of ash and large blocks of rock.

blueschist. A metamorphic rock with the blue-colored minerals glaucophane and lawsonite, which form from basaltic rocks under high-pressure and low-temperature conditions. Blueschists most commonly form in subduction zones.

breccia. A rock consisting of angular fragments.

caldera. A steep-walled, subcircular depression in a volcano, at least 1 mile (1.6 km) across, that formed by collapse into an emptied or partially emptied magma chamber below.

chert. An extremely fine-grained sedimentary rock made of silica.

cinder cone. A steep-sided, cone-shaped accumulation of cinders that surrounds a basaltic vent.

clast. A grain or fragment of a rock. Clastic rock is sedimentary rock composed of broken fragments, such as sand grains, derived from preexisting rocks.

clay-size. A sedimentary particle with a grain size less than 0.004 millimeter in diameter.

coal. An organic-rich, dark-colored to black rock that burns. Coal forms from the compaction and long-term, low-temperature heating of plant material. Coal typically is founds in layers called seams.

coarse-grained. A term used to describe a rock with large particles or crystals, typically visible to the naked eye, about 1 millimeter in diameter or larger.

coastal plain. A low, gently sloping region on the margin of an ocean.

colonnade. The part of a basaltic lava flow that shows near-parallel, typically steep fractures that break the rock into regular-shaped columns.

columnar jointing. The fracturing in a lava flow that causes the flow to break into columns. This is the prevalent style of fracturing in the colonnade.

composite volcano. Synonymous with stratovolcano but used to explicitly describe a volcano that consists of several parts, typically a series of domes and vents.

concretion. A subspherical to irregular-shaped body of sedimentary rock that is more resistant to weathering than the rest of the rock. This resistance usually occurs because it is unusually well cemented.

conglomerate. A sedimentary rock composed of particles that exceed 2 millimeters in diameter.

continental shelf. The gently inclined part of the continental landmass between the shoreline and the more steeply inclined continental slope.

crossbedding. Layering in a sedimentary rock that forms at an angle to horizontal.

crust. The uppermost layer of Earth. Continental crust consists mainly of an igneous and/or metamorphic basement overlain by sedimentary and volcanic rock. Oceanic crust consists of basalt.

dacite. A volcanic rock that is intermediate in silica content between andesite and rhyolite.

delta. A nearly flat accumulation of clay, sand, and gravel deposited in a lake or ocean at the mouth of a river.

diatomite. A sedimentary rock that consists mostly of diatoms, single-celled freshwater or marine algae that consist of silica.

dike. A tabular intrusive body that cuts across layering in the host rock.

diorite. An intrusive igneous rock that is between gabbro and granite in silica content. It is the intrusive equivalent of andesite.

drainage basin. The land area drained by a stream and all of its tributaries.

drainage divide. A line of high elevations, typically a ridge, that separates one drainage basin from another.

entablature. The part of a basaltic lava flow that shows numerous closely spaced, typically steep fractures that break the rock into thin, irregular columns.

erosion. Movement or transport of weathered material by water, ice, wind, or gravity.

erratic. A block of rock transported by glacial ice and deposited at some distance from the bedrock outcrop it came from.

escarpment/scarp. A sudden steep rise in the landscape.

fault. A fracture or zone of fractures in Earth's crust along which blocks of rock on either side have shifted. A **normal fault** forms under extensional stresses, and one side drops relative to the other side. A **reverse fault** forms under compressional stresses, and one side is pushed up and over the other side. In a **strike-slip fault**, rock on one side moves sideways relative to rock on the other side.

fault scarp. An abrupt cliff or steep section in an otherwise even landscape, formed by offset along a fault.

feldspar. The most abundant rock-forming mineral group, making up 60 percent of Earth's crust and including calcium, sodium, or potassium with aluminum silicate. Includes plagioclase feldspars and alkali feldspars.

fine-grained. A term used to describe a rock with small particles or crystals, typically not visible to the naked eye, less than about 1 millimeter in diameter.

fissure. An open crack

floodplain. The portion of a river valley adjacent to the river that is built of sediments deposited when the river overflows its banks during flooding.

foliation. Layering in rock caused by metamorphism or deformation.

fore-arc basin. A sedimentary basin located between a magmatic arc and a subduction zone.

formation. A body of sedimentary, igneous, or metamorphic rock that can be recognized over a large area. It is the basic stratigraphic unit in geologic mapping. A formation may be part of a larger group and may be broken into members.

fossils. Remains, imprints, or traces of plants or animals preserved in rock.

gabbro. Dark-colored intrusive igneous rock that consists of less than 52 percent silica. When magma of the same composition erupts at the surface, it forms basalt.

glacial outwash. Sediment deposited by the large quantities of meltwater emerging from the terminus of a glacier.

glacial till. An unsorted mixture of silt, sand, and gravel left by a melting glacier.

glacier. A large and long-lasting mass of ice on land that flows downhill in response to gravity.

gneiss. A regional metamorphic rock that has a banded appearance caused by the segregation of minerals at high temperatures.

graben. A crustal block that is down-dropped between two inwardly dipping normal faults.

graded beds. Sedimentary layers in which the grain size is coarsest at the bottom and becomes finer toward the top.

granite. A light-colored, coarse-grained igneous rock with a silica content that exceeds 66 percent. It is the intrusive equivalent of rhyolite.

greenstone. Volcanic rocks, typically basalt, that have been metamorphosed and have grown green metamorphic minerals.

groundwater. Subsurface water contained in fractures and pores of rock and soil.

group. Two or more formations that occur together.

horst. A crustal block that is uplifted between two outwardly dipping normal faults.

igneous rock. Rock that solidified from the cooling of molten magma.

impermeable. Having a texture that does not permit water to move through. Clay is often considered a relatively impermeable sediment.

intrusive igneous rocks. Rocks that cool from magma beneath the surface of Earth. The body of rock is called an **intrusion**.

island arc. An offshore volcanic arc or linear chain of volcanoes formed along a convergent plate margin.

lahar. A volcanic mudflow deposit.

laterites. Soils that consist predominantly of aluminum and iron through deep, tropical weathering.

lava. Molten rock erupted on the surface of Earth.

limestone. A sedimentary rock composed of calcium carbonate formed by precipitation in warm water, usually aided by biological activity.

lithification. Compaction and cementation of sediment into sedimentary rock.

lithosphere. The outer rigid shell of Earth that is broken into the tectonic plates. On average, continental lithosphere is about 100 miles (160 km) thick and old oceanic lithosphere is about 60 miles (100 km) thick.

loess. Silt and dust picked up from glacial streambeds and redeposited by the wind.

mafic. Said of an igneous rock with the approximate silica content of basalt, typically rich in dark-colored, iron- and magnesium-rich minerals.

magma. Molten rock within Earth.

magmatic arc. The zone of magma production that stretches in an arc-like fashion above and parallel to a subduction zone.

magnetite. A black, strongly magnetic iron mineral that is an important ore of iron and occurs commonly as a heavy mineral in sands.

mantle. The part of Earth between the interior core and the outer crust.

marble. Metamorphosed limestone.

marine. Pertaining to the sea.

marine sandstone. Sandstone deposited in a shallow or deep ocean environment.

mélange. A mixture of rocks that may not have formed together. Mélanges can form at a variety of scales and most frequently indicate subduction zone settings.

metamorphic rock. Rock derived from preexisting rock that has changed mineralogically, texturally, or both in response to changes in temperature and/or pressure, usually deep within Earth.

metamorphism. Recrystallization of an existing rock. Metamorphism typically occurs at high temperatures and often high pressures.

mica. A family of silicate minerals, including biotite and muscovite, that breaks easily into thin flakes. Micas are common in many kinds of igneous and metamorphic rocks.

moraine. A mound or ridge of an unsorted mixture of silt, sand, and gravel (glacial till) left by a melting glacier.

mountain building event. An event in which mountains rise. During these events, rocks are typically folded, faulted, and/or metamorphosed. Intrusive and extrusive igneous activity often accompanies it. Also known as orogeny.

mouth. The end of a stream or river where it discharges into a larger body of water, such as a larger river, a bay, or the ocean.

mudflow. A mixture of water, mud, and assorted particles that flows downhill in response to gravity.

mudstone. A sedimentary rock composed of mud.

normal fault. A fault in which rocks on the hanging wall side move down relative to rocks on the footwall side in response to extensional stresses.

obsidian. Volcanic glass, typically high in silica and dark gray to black. Impurities may give rise to brown or red colors in obsidian.

olivine. An iron and magnesium silicate mineral that typically forms glassy green crystals. A common mineral in gabbro, basalt, and peridotite.

ophiolite. The sequence of rocks that makes up the oceanic lithosphere.

orogeny. A mountain building event.

outwash. Sediment deposited by the large quantities of meltwater emerging from the terminus of a glacier.

pahoehoe. Hawaiian term for a basaltic lava flow that has a smooth, undulating, or ropy surface. These surface features are due to the movement of fluid lava under a congealing surface crust.

paleosol. A fossil soil.

peat. A deposit of semicoalified plant remains in a bog.

pebble. A rounded rock particle 0.16 to 2.5 inches (4–64 mm) in diameter.

peperite. A rock made from the mixing of magma with wet sediments, most commonly by intrusion. Typically, the igneous portion is highly fragmented and glassy, while the sedimentary portion shows evidence of baking.

peridotite. Low-silica igneous rock that makes up Earth's mantle.

phyllite. A metamorphic rock that has been heated more than slate but not as much as schist; typically starts as a fine-grained sedimentary rock.

pillow basalt. Basalt that takes on a bulbous, pillow shape due to the lava's interaction with water, from either erupting underwater,or flowing into it.

plagioclase. A feldspar mineral rich in sodium and calcium. One of the most common rock-forming minerals in igneous and metamorphic rocks.

plate tectonics. The theory that Earth's lithosphere is broken into large fragments or plates that move slowly over the somewhat malleable asthenosphere, with intense geological activity at plate boundaries.

Pleistocene. The last 2 million years of geologic time, during which periods of extensive continental glaciation alternated with warmer interglacial periods of glacial retreat.

pluton. A body of intrusive igneous rock. A **stitching pluton** intrudes across the boundaries of two or more terranes.

pumice. A pyroclastic rock that consists of volcanic glass with a frothy texture because of an abundance of air holes.

pyrite. A brass- or gold-colored mineral composed of iron sulfide.

pyroclastic. Describing fragmental volcanic particles, broken during explosive eruptions.

pyroxene. An iron and magnesium-bearing silicate mineral; abundant in basaltic rock.

quartz. A mineral composed entirely of silica; one of the most common rock-forming minerals.

radiolarian chert. A variety of chert (a sedimentary rock formed from microcrystalline silica) formed by the accumulation of radiolaria: tiny, single-celled, silica-producing marine organisms.

radiometric dating. The calculation of age based on the rate of time it takes for radioactive elements to decay.

rhyolite. A typically light-colored volcanic rock with more than 66 percent silica. It is the volcanic equivalent of granite. Some of the rhyolites in Oregon are dark-colored to black because they contain small, disseminated particles of black obsidian.

rift zone. On volcanoes, a zone of fissuring along which lava has erupted.

sand. Weathered mineral grains, most commonly quartz, between 0.06 and 2 millimeters in diameter.

sandstone. A sedimentary rock made primarily of sand.

schist. A metamorphic rock that is strongly layered due to an abundance of visible, platy minerals.

sea stack. A tall outcropping of bedrock on a beach or offshore, left as a remnant of the coastline as erosion causes the coastline to retreat.

sedimentary rock. A rock formed from the compaction and cementation of sediment.

serpentinite. A rock made of minerals of the serpentine group, which are formed by low-grade metamorphism of iron- and magnesium-rich rocks, usually of the oceanic lithosphere.

shale. A thinly layered rock made of sedimentary particles less than 0.004 millimeter in diameter.

shearing. The action and deformation caused by two bodies of rock sliding past each other. The sheared part of a rock is called a **shear zone**.

shield volcano. A gently sloped volcano typically made of basalt. In profile, shield volcanoes resemble shields.

silica. The compound **silicon dioxide**. The most common mineral made entirely of silica is quartz.

sill. An igneous intrusion that parallels the planar structure or bedding of the host rock.

silt. Sedimentary particles larger than clay but smaller than sand (between 0.004 and 0.06 millimeter in diameter).

siltstone. A sedimentary rock made primarily of silt.

slate. Slightly metamorphosed shale or mudstone that breaks easily along parallel surfaces.

spit. A long, narrow, fingerlike ridge of sand extending into the water from the shore.

stratification/stratified. Sequentially layered.

stratovolcano. A steep-sided volcano, typically made of andesite, although basaltic andesite and dacite make up several stratovolcanoes of the Cascades.

strike-slip fault. A fault showing sideways movement or offset of the adjacent rock.

subduction zone. A long, narrow zone where an oceanic plate descends beneath another plate at a convergent boundary.

suspended sediment. Sediment in a stream that remains lifted or transported due to the turbulence and energy of the water. When the currents slow, suspended sediment falls to the bottom.

syncline. A fold with the youngest rock in the core; most synclines have limbs that dip toward the core.

talus. An accumulation of rock fragments derived from and resting at the base of a cliff.

tectonics. The study of regional-scale deformation of Earth's crust.

tephra. A general term for the material ejected from a volcano.

terrace. An erosional remnant of a former floodplain or coastline, standing above the present river or coast.

terrane. A fault-bounded crustal fragment with a geologic history that differs from adjacent fragments.

thrust fault. A fault dipping less than 45 degrees, in which the rock above the fault has moved upward and over the rock below the fault. Thrust faults typically form by horizontal compression.

till. An unlayered and unsorted mixture of clay, silt, sand, gravel, and boulders deposited directly by a glacier.

tilted fault block. A mountain or mountain range that is uplifted along one side by a normal fault and shows a consistent tilt in its bedding away from the fault.

transform fault. A special type of strike-slip fault along a midocean ridge or plate boundary.

tuff. A volcanic rock made mostly of consolidated pyroclastic material, chiefly ash and pumice, derived from ash falls or pyroclastic flows. A **welded tuff** is distinctly harder because the heat of its particles caused them to weld together. An **ash-flow tuff** is derived from an ash flow.

turbidite. Sands and muds that settle on the seafloor from clouds of sandy, muddy water that flow as submarine density currents. They form alternating layers of sandstone and shale. Some of the sandstone layers grade upward from coarse to fine particles.

ultramafic. An igneous rock with a silica content below 50 percent. Ultramafic rocks typically contain a great deal of iron and magnesium.

unconformity. A depositional contact across which the rock record is missing. As a result, the rocks beneath the unconformity are typically significantly older than the ones above it.

vein. A deposit of minerals that fills a fracture in rock.

vent (silicic and mafic). The actual place where volcanic materials erupt. Vents are either eruptive localities on large volcanoes or mark much smaller volcanoes, shown on geologic maps as silicic or mafic vents. Silicic vents are usually marked by silica-rich rocks, such as dacite and rhyolite. Mafic vents are marked by silica-poor rocks, such as basaltic cinders, basalt, and basaltic andesite.

volcanic arc. A chain of island volcanoes that formed above an ocean-floor subduction zone.

weathering. The physical disintegration and chemical decomposition of rock at Earth's surface.

Further Reading and References

Oregon contains such a wealth of amazing geology that it is impossible to do more than scratch its surface in a book such as this. However, there is a trove of detailed literature for those who want to dig deeper. The following books, articles, and maps reflect only a sampling of them—but they are the ones I found most helpful in my research. The technical readings and maps are organized first by those pertaining to the regional geology, including the Missoula Floods and the Columbia River Basalt Group, followed by individual chapters.

Oregon has a multitude of geologic maps, published by the US Geological Survey and the Oregon Department of Geology and Mineral Industries (DOGAMI). DOGAMI maintains an easily navigable website that highlights all its available maps at http://www.oregongeology.org/sub/publications/GMS/gms.htm.

Nontechnical Reading

Allen, J. E., Burns, M., and S. Burns. 2009. *Cataclysms on the Columbia: the Great Missoula Floods*, 2nd ed. Ooligan Press.

Atwater, B. F., Satoko, M., Kenji, S., Yoshinobu, T., Kazue, U., and D. K. Yamaguchi. 2005. *The Orphan Tsunami of 1700: Japanese Clues to a Parent Earthquake in North America.* US Geological Survey Professional Paper 1707.

Baldwin, E. M. 1976. *Geology of Oregon,* 2nd ed. Kendall-Hunt Publishing. Out-of-print but worth looking for in used bookstores.

Bishop, E. M. 2006. *In Search of Ancient Oregon: A Geological and Natural History.* Timber Press.

Bishop, E. M., and J. E. Allen. 2004. *Hiking Oregon's Geology*, 2nd ed. Mountaineers Books.

Freed, M. 1979. Silver Falls State Park. *Oregon Geology* 41 (1): 3–10.

Henderson, B. 2014. *The Next Tsunami: Living on a Restless Coast.* Oregon State University Press.

Jennings, A., and T. Jennings. 2003. *Estuary Management in the Pacific Northwest.* Oregon State University.

Komar, P. D. 1998. *The Pacific Northwest Coast: Living with the Shores of Oregon and Washington.* Duke University Press.

Komar, P. D. 2010. Oregon. In *Encyclopedia of the World's Coastal Landforms,* ed. E. C. F. Bird, p. 33–41. Springer Science.

Loy, W. G., Allan, S., Buckley, A. R., and J. E. Meacham. 2001. *Atlas of Oregon,* 2nd ed. University of Oregon Press.

Norman, D. K., and J. M. Roloff. 2004. A self-guided tour of the geology of the Columbia River Gorge: Portland Airport to Skamania Lodge, Stevenson, Washington. Washington Division of Geology and Earth Resources Open File Report 2004-7.

Orr, E. L., and W. N. Orr. 2009. *Oregon Fossils,* 2nd. ed. Oregon State University Press.

Orr, E. L., and W. N. Orr. 2012. *Oregon Geology.* 6th ed. Oregon University Press.

Orr, W. N., and E. L. Orr. 2002. *Geology of the Pacific Northwest,* 2nd. ed. Waveland Press.

Vallier, T. 1998. *Islands and Rapids: A Geologic Story of Hells Canyon.* Confluence Press.

Regional Geology, Missoula Floods, and Columbia River Basalt Group

Benito, G., and J. E. O'Connor. 2003. Number and size of last-glacial Missoula Floods in the Columbia River valley between the Pasco Basin, Washington, and Portland, Oregon. *Geological Society of America Bulletin* 115 (5): 624–38.

Burchfiel, B. C., Cowan, D. S., and G. A. Davis. 1992. Tectonic overview of the Cordilleran orogeny in the western United States. In *The Cordilleran Orogeny, Conterminous US,* The Geology of North America Volume G-3, Geological Society of America, eds. B. C. Burchfiel, P. W. Lipman, and M. L. Zoback, p. 407–79.

Camp, V. E. 1995. Mid-Miocene propagation of the Yellowstone mantle plume head beneath the Columbia River basalt source region. *Geology* 23 (5): 435–38.

Camp, V. E., Ross, M. E., and W. E. Hanson. 2003. Genesis of flood basalts and Basin and Range volcanic rocks from Steens Mountain to the Malheur River Gorge, Oregon. *Geological Society of America Bulletin* 115 (1): 105–28.

Christiansen, R. L., and R. S. Yeats, with contributions by Graham, S. A., Niem, W. A., Niem, A. R., and P. D. Snavely. 1992. Post-Laramide geology of the US Cordilleran region. In *The Cordilleran Orogeny, Conterminous US,* The Geology of North America Volume G-3, Geological Society of America, eds. B. C. Burchfiel, P. W. Lipman, and M. L. Zoback, p. 261–406.

Ferns, M. L., and D. F. Huber. 1984. *Mineral Resources Map of Oregon.* Oregon Department of Geology and Mineral Industries, map GMS-36, 1:500,000.

Hooper, P. R., Camp, V. E., Reidel, S. P., and M. E. Ross. 2007. The origin of the Columbia River flood basalt province: Plume vs. nonplume models. In *Plates, Plumes, and Planetary Processes,* Geological Society of America Special Paper 430, eds. G. R. Foulger and D. M. Jurdy, 635–68.

Madin, I. P. 2009. *Oregon: A Geologic History.* Oregon Department of Geology and Mineral Industries Interpretive Series Map 28.

Minervini, J. M., O'Connor, J. E., and R. E. Wells. 2003. *Maps Showing Inundation Depths, Ice-Rafted Erratics, and Sedimentary Facies of Late Pleistocene Missoula Floods in the Willamette Valley, Oregon.* US Geological Survey Open-File Report 03-408.

O'Connor, J. E., and S. F. Burns. 2009. Cataclysms and controversy: Aspects of the geomorphology of the Columbia River Gorge. In *Volcanoes to Vineyards: Geologic Field Trips through the Dynamic Landscape of the Pacific Northwest,* Geological Society of America Field Guide 15, eds. J. E. O'Connor, R. J. Dorsey, and I. P. Madin, p. 237–51.

O'Connor, J. E., and Waitt, R. B. 1995. Beyond the Channeled Scabland: A field trip to the Missoula Flood features in the Columbia, Yakima, and Walla Walla valleys of Washington and Oregon—Part 2: Field trip, day one. *Oregon Geology* 57 (4): 75–86.

Reidel, S. P., Camp, V. E., Tolan, T. L., and B. S Martin. 2013. The Columbia River flood basalt province: Stratigraphy, areal extent, volume, and physical volcanology. In *The Columbia River Basalt Province,* Geological Society of America Special Paper 497, eds. S. P. Reidel, V. E. Camp, M. E. Ross, J. A. Wolff, B. S. Martin, T. L. Tolan, and R. E. Wells, p. 1–44.

Thelin, G. P., and R. J.Pike. 1991. *Landforms of the Conterminous United States: A Digital Shaded-Relief Portrayal.* US Geological Survey Miscellaneous Investigations Map, I-2206, 1:3,500,00.

Tolan, T. L, Martin, B. S., Reidel, S. P., Anderson, J. L., Lindsey, K. A., and W. Burt. 2009. An introduction to the stratigraphy, structural geology, and hydrogeology of the Columbia River flood basalt province: A primer for the GSA Columbia River Basalt Group field trips. In *Volcanoes to Vineyards: Geologic Field Trips through the Dynamic Landscape of the Pacific Northwest*, Geological Society of America Field Guide 15, eds. J. E. O'Connor, R. J. Dorsey, and I. P. Madin, p. 599–643.

Tolan, T. L., Reidel, S. P., Beeson, M. H., Anderson, J. L., Fecht, K. R., and D. A. Swanson. 1989. Revisions to the estimates of the areal extent and volume of the Columbia River Basalt Group. In *Volcanism and Tectonism in the Columbia River Flood Basalt Province.* Geological Society of America Special Paper 239, eds. S. P. Reidel and P. R. Hooper, p. 1–20.

Waitt, R. B., Denlinger, R. P., and J. E. O'Connor. 2009. Many monstrous Missoula Floods down Channeled Scabland and Columbia Valley. In *Volcanoes to Vineyards: Geologic Field Trips through the Dynamic Landscape of the Pacific Northwest*, Geological Society of America Field Guide 15, eds. J. E. O'Connor, R. J. Dorsey, and I. P. Madin, p. 775–844.

Walker, G. W., and N. S. MacLeod. 1991. *Geologic Map of Oregon.* US Geological Survey, 1:500,000.

Wells, R. E., Engebretson, D. C., Snavely, P. D., Jr., and R. S. Coe. 1984. Cenozoic plate motions and the volcano-tectonic evolution of western Oregon and Washington. *Tectonics* 3: 275–94.

Wells, R. E., and P. L. Heller. 1988. The relative contribution of accretion, shear, and extension to Cenozoic tectonic rotations in the Pacific Northwest. *Geological Society of America Bulletin* 100: 324–38.

Wells, R. E., Niem, A. R., Evarts, R. C., and J. T. Hagstrum. 2009. The Columbia River Basalt Group: From the gorge to the sea. In *Volcanoes to Vineyards: Geologic Field Trips through the Dynamic Landscape of the Pacific Northwest*, Geological Society of America Field Guide 15, eds. J. E. O'Connor, R. J. Dorsey, and I. P. Madin, p. 737–74.

Wells, R. E., Simpson, R. W., Bentley, R. D., Beeson, M. H., Mangan, M. T., and T. L. Wright. 1989. Correlation of Miocene flows of the Columbia River Basalt Group from the central Columbia River Plateau to the coast of Oregon and Washington. In *Volcanism and Tectonism in the Columbia River Flood Basalt Province*, Geological Society of America Special Paper 239, eds. P. Hooper and S. Reidel, p. 113–30.

Coast Range and Willamette Valley

Armentrout, J. M. 1980. Field trip road log for the Cenozoic stratigraphy of Coos Bay and Cape Blanco, southwestern Oregon. In *Geologic Field Trips in Western Oregon and Southwestern Washington*, Oregon Department of Geology and Mineral Resources Bulletin 101, eds. K. F. Oles, J. G. Johnson, A. R. Niem, and W. A. Niem, p. 177–216.

Atwater, B. F., and D. K. Yamaguchi. 1991. Sudden, probably coseismic submergence of Holocene trees and grass in coastal Washington State. *Geology* 19: 706–9.

Baitis, K., and M. E. James. 2005. Willamette Valley clay linked to thick blankets of Mount Mazama airfall. *Current Archaeological Happenings in Oregon.* 30: 14–19.

Baldwin, E. M., and J. D. Beaulieu. 1973. *Geology and Mineral Resources of Coos County, Oregon.* Oregon Department of Geology and Mineral Resources Bulletin 80.

Beeson, M. H., Perttu, R., and J. Perttu. 1979. The origin of the Miocene basalts of coastal Oregon and Washington: An alternative hypothesis. *Oregon Geology* 41 (10): 159–65.

Beeson, M. H., Tolan, T. L., and I. P. Madin. 1991. *Geologic Map of Portland Quadrangle, Multnomah and Washington Cos., Oregon, and Clark Co., Washington.* Oregon Department of Geology and Mineral Industries map GMS-75, 1:24,000.

Blakely, R. J., Wells, R. E., Yelin, T. S., Madin, I. P., and M. H. Beeson. 1995. Tectonic setting of the Portland-Vancouver area, Oregon and Washington: Constraints from low-altitude aeromagnetic data. *Geological Society of America Bulletin* 107: 1051–62.

Burns, W. J., Hofmeister, R. J., and Y. Wang. 2008. *Geologic Hazards, Earthquake and Landslide Hazard Maps, and Future Earthquake Damage Estimates for Six Counties in the Mid/Southern Willamette Valley including Yamhill, Marion, Polk, Benton, Linn, and Lane Counties, and the City of Albany, Oregon.* Oregon Department of Geology and Mineral Industries Interpretive Map Series, IMS-24.

Chan, M. A., and R. H. Dott, Jr. 1983. Shelf and deep-sea sedimentation in Eocene forearc basin, western Oregon: fan or non-fan? *American Association of Petroleum Geologists Bulletin* 67 (11): 2100–16.

Clemens, K. E, and P. D. Komar. 1988. Oregon beach-sand compositions produced by mixing of sediments from multiple sources under a transgressing sea. *Journal of Sedimentary Petrology* 56: 15–22.

Davis, A. S., Snavely, P. D., Jr., Gray, L. B., and D. L. Minasian. 1995. *Petrology of Late Eocene Lavas Erupted in the Fore-arc of Central Oregon.* US Geological Survey Open File Report 95-40.

Evarts, R. C., Conrey, R. M., Fleck, R. J., and J. T. Hagstrum. 2009. The Boring volcanic field of the Portland-Vancouver area, Oregon and Washington: Tectonically anomalous forearc volcanism in an urban setting. In *Volcanoes to Vineyards: Geologic Field Trips through the Dynamic Landscape of the Pacific Northwest,* Geological Society of America Field Guide 15, eds. J. E. O'Connor, R. J. Dorsey, and I. P. Madin, p. 253–70.

Evarts, R. C., O'Connor, J. E., Wells, R. E., and I. P. Madin. 2009. The Portland Basin: A (big) river runs through it. *GSA Today* 19 (9): 4–10.

Hart, R. 1997. Episodically buried forests in the Oregon surf zone. *Oregon Geology* 59 (6): 131–43.

Heller, P. L., Peterman, Z. E., O'Neil, J. R., and M. Shafiqullah. 1985. Isotopic provenance of sandstones from the Eocene Tyee Formation, Oregon Coast Range. *Geological Society of America Bulletin* 96: 770–80.

Kelsey, H. M. 1990. Late Quaternary deformation of marine terraces on the Cascadia subduction zone near Cape Blanco, Oregon. *Tectonics* 9 (5): 983–1014.

Komar, P. D. 1992. Ocean processes and hazards along the Oregon coast. *Oregon Geology* 54 (1): 3–19.

Komar, P. D., McManus, J., and M. Styllas. 2004. Sediment accumulation in Tillamook Bay, Oregon: Natural processes versus human impacts. *Journal of Geology* 112: 455–69.

Liberty, L. M., Hemphill-Haley, M. A., and I. P. Madin. 2003. The Portland Hills Fault: Unconvering a hidden fault in Portland, Oregon, using high-resolution geophysical methods. *Tectonophysics* 368: 89–103.

Madin, I. P. 2009. Portland, Oregon, geology by tram, train, and foot. *Oregon Geology* 69 (1): 1–20.

Madin, I. P., and R. B. Murray. 2006. *Preliminary Geologic Map of the Eugene East and Eugene West 7.5 Minute Quadrangles, Lane County, Oregon.* Oregon Department of Geology and Mineral Industries Open-File Report O-03-11.

Mardock, C. L. 1994. A geologic overview of Yaquina Head, Oregon. *Oregon Geology* 56 (2): 27–33.

McClaughry, J. D., Wiley, T. J., Ferns, M. L., and I. P. Madin. 2010. *Digital Geologic Map of the Southern Willamette Valley, Benton, Lane, Linn, Marion, and Polk Counties, Oregon.* Oregon Department of Geology and Mineral Industries Open File Report O-10-03.

Muhs, D. R., Kelsey, H. M., Miller, G. H., Kennedy, G. L., Whelan, J. F., and G. W. McInelly. 1990. Age estimates and uplift rates for Late Pleistocene marine terraces: Southern Oregon portion of the Cascadia forearc. *Journal of Geophysical Research* 95: 6685–98.

Niem, A. R. 1975. Geology of Hug Point State Park, northern Oregon coast. *Ore Bin* 37 (2): 17–36.

O'Connor, J. E., Sarna-Wojcicki, Wozniak, K. C., Polette, D. J., and R. J. Fleck. 2001. *Origin, Extent, and Thickness of Quaternary Geologic Units in the Willamette Valley, Oregon.* US Geologic Survey Professional Paper 1620, 1:250,000.

Perttu, R. K., and G. T. Benson. 1980. Deposition and deformation of the Eocene Umpqua Group, Sutherlin area, southwestern Oregon. *Oregon Geology* 42 (8): 135–40.

Peterson, C. D. 1997. Coseismic paleoliquefaction evidence in the central Cascadia margin, USA. *Oregon Geology* 59 (3): 51–74.

Peterson, C. D., and M. E. Darienzo. 1988. Coastal neotectonic field trip guide for Netarts Bay, Oregon. *Oregon Geology* 50 (9/10): 99–106.

Retallack, G. J., Orr, W. N., Prothero, D. R., Duncan, R. A., Kester, P. R., and C. P. Ambers. 2004. Eocene-Oligocene extinction and paleoclimatic change near Eugene, Oregon. *Geological Society of America Bulletin* 116 (7/8): 817–39.

Snavely, P. D., Jr. 1987. Depoe Bay, Oregon. In *Geological Society of America Centennial Field Guide, Cordilleran Section,* Geological Society of America, ed. M. Hill, p. 307–10.

Snavely, P. D., Jr., Wells, R. E., and D. Minasian. 1993. *The Cenozoic Geology of the Oregon and Washington Coast Range and Road Log for the Northwest Petroleum Association 9th Annual Field Trip.* US Geological Survey Open-File Report 93-189.

Snyder, S. L., Felger, T. J., Blakely, R. J., and R. E. Wells. 1993. *Aeromagnetic Map of the Portland-Vancouver Metropolitan Area, Oregon and Washington.* US Geological Survey Open-File Report 93-211, 1:100,000.

Thomas, G. C., Crosson, R. S., Carver, D. L., and T. S. Yelin. 1996. The 25 March 1993 Scotts Mills, Oregon, earthquake and aftershock sequence: Spatial distribution, focal mechanisms, and the Mount Angel Fault. *Bulletin of the Seismological Society of America* 86: 925–35.

Wells, R. E., Jayko, A. S., Niem, A. R., Black, G., Wiley, T., Baldwin, E, Molenaar, K. M., Wheeler, K. L., Duross, C. B., and R. W. Givler. 2000. *Geologic Map and Database of the Roseburg 30 x 60 Minute Quadrangles, Douglas and Coos Counties, Oregon.* US Geological Survey Open File Report 00-376.

Wells, R. E., Niem, A. R., MacLeod, N. S., Snavely, P. D., Jr., and W. A. Niem. 1983. *Preliminary Geologic Map of the West Half of the Vancouver (WA-OR) 1° x 2° Quadrangle, Oregon.* US Geological Survey Open File Report 83-591, 1:250,000.

Wells, R. E., Snavely, P. D., Jr., Macleod, N. S., Kelly, M. M., Parker, M. J., Fenton, J., and T. Felger. 1995. *Geologic Map of the Tillamook Highlands, Northwest Oregon Coast Range: A Digital Database.* US Geological Survey Open-File Report 95-670.

Wong, I. G., Hemphill-Haley, M. A., Liberty, L. M., and I. P. Madin. 2001. The Portland Hills Fault: An earthquake generator or just another old fault? *Oregon Geology* 63 (2): 39–50.

Cascade Range

Bacon, C. R. 1989. Mount Mazama and Crater Lake caldera, Oregon. In *South Cascades Arc Volcanism, California and Southern Oregon.* International Geological Congress field trip T312, eds. L. J. P. Muffler, C. R. Bacon, R. L. Christiansen, M. A. Clynne, J. M. Donnelly-Nolan, C. D. Miller, D. R. Sherrod, and J. G. Smith, p. 26–37.

Bacon, C. R. 2008. *Geologic Map of Mount Mazama and Crater Lake Caldera, Oregon.* US Geological Survey Scientific Investigations Map I-2832, 1:24,000.

Bacon, C. R. 1987. Mount Mazama and Crater Lake caldera, Oregon. In *Geological Society of America Centennial Field Guide, Cordilleran Section,* Geological Society of America, ed. M. Hill, p. 301–6.

Beeson, M. H., and T. L. Tolan. 1987. Columbia River Gorge: The geologic evolution of the Columbia River in northwestern Oregon and southwestern Washington. In *Geological Society of America Centennial Field Guide, Cordilleran Section,* Geological Society of America, ed. M. Hill, p. 321–26.

Cashman, K. V., Deligne, N. I., Gannett, M. W., Grant, G. E., and A. Jefferson. 2009. Fire and water: Volcanology, geomorphology, and hydrogeology of the Cascade Range, central Oregon. In *Volcanoes to Vineyards: Geologic Field Trips through the Dynamic Landscape of the Pacific Northwest,* Geological Society of America Field Guide 15, eds. J. E. O'Connor, R. J. Dorsey, and I. P. Madin, p. 539–82.

Cameron, K. A., and P. Pringle. 1986. Post-glacial lahars of the Sandy River Basin, Mount Hood, Oregon. *Northwest Science* 60 (4): 225–37.

Hildreth, W. 2007. *Quaternary Magmatism in the Cascades: Geologic Perspectives.* US Geological Survey Professional Paper 1744.

Peck, D. L., Griggs, A. B., Schlicker, H. G., Wells, F. G., and Dole, H. M. 1964. *Geology of the Central and Northern Parts of the Western Cascade Range in Oregon.* US Geological Survey Professional Paper 449, 1:250,000.

Scott, W. E., Gardner, C. A., Sherrod, D. R., Tilling, R. I., Lanphere, M. A., and R. M. Conrey. 1997. *A Geologic History of Mount Hood Volcano, Oregon: A Field Trip Guidebook.* US Geological Survey Open File Report 97–263.

Sherrod, D. R. 1991. *Geologic Map of a Part of the Cascade Range between Latitudes 43 degrees–44 degrees, Central Oregon.* US Geological Survey Geologic Investigations Series Map I-1291, 1:25,000.

Sherrod, D. R., and W. E. Scott. 1995. *Preliminary Geologic Map of the Mount Hood 30 x 60-Minute Quadrangle, Northern Cascade Range, Oregon.* US Geologic Survey Open-File Report 95-219.

Sherrod, D. R., and J. G. Smith. 2000. *Geologic Map of Upper Eocene to Holocene Volcanic and Related Rocks of the Cascade Range, Oregon.* US Geological Survey Geologic Investigations Series Map I-2569.

Sherrod, D. R., Taylor, E. M., Ferns, M. L., Scott, W. E., Conrey, R. M., and G. A. Smith. 2004. *Geologic Map of the Bend 30 x 60-Minute Quadrangle, Central Oregon.* US Geologic Survey Geologic Investigations Series Map I-2683.

Smith, G. A. 1989. Western Cascades, southern Oregon and northern California. In *South Cascades Arc Volcanism, California and Southern Oregon: International Geological Congress Field Trip T312,* eds, L. J. P. Muffler, C. R. Bacon, R. L. Christiansen, M. A. Clynne, J. M. Donnelly-Nolan, C. D. Miller, D. R. Sherrod, and J. G. Smith, p. 37–42.

Smith, G. A., and G. R. Priest. 1983. A field guide to the central Oregon Cascades. *Oregon Geology* 45 (11): 119–26.

Snavely, P. D., Jr., Wells, R. E., and D. Minasian. 1993. *The Cenozoic Geology of the Oregon and Washington Coast Range and Road Log for the Northwest Petroleum Association 9th Annual Field Trip: Cenozoic Geology of Coastal Northwest Oregon.* US Geological Survey Open-File Report 93–189.

Taylor, E. M. 1987. Late High Cascade volcanism from summit of McKenzie Pass, Oregon: Pleistocene composite cones on platform of shield volcanoes: Holocene eruptive centers and lava fields. In *Geological Society of America Centennial Field Guide, Cordilleran Section,* Geological Society of America, ed. M. Hill, p. 311–12.

Taylor, E. M. 1990. Volcanic history and tectonic development of the central High Cascades range, Oregon. *Journal of Geophysical Research* 95 (B12): 19,611–22.

Tolan, T. L., and M. H. Beeson. 1984. Intracanyon flows of the Columbia River Basalt Group in the lower Columbia River Gorge and their relationship to the Troutdale Formation. *Geological Society of America Bulletin* 95: 463–77.

Tolan, T. L, Beeson, M. H., and B. F. Vogt. 1984. Exploring the Neogene history of the Columbia River: Discussion and geologic field trip guide to the Columbia River Gorge. Part I. Discussion. *Oregon Geology* 46 (8): 87–97.

Tolan, T. L, Beeson, M. H., and B. F. Vogt. 1984. Exploring the Neogene history of the Columbia River: Discussion and geologic field trip guide to the Columbia River Gorge. Part II. Road log and comments. *Oregon Geology* 46 (9): 103–12.

Wang, Y., Hofmeister, R. J., McConnell, V. S., Burns, S. F., Pringle, P. T., and G. L. Peterson. 2002. Columbia River Gorge landslides. In *Field Guide to Geologic Processes in Cascadia, Oregon.* Department of Geology and Mineral Industries Special Paper 36, ed. G. W. Moore, p. 273–88.

Klamath Mountains

Aalto, K. R. 2006. The Klamath peneplain: A review of J. S. Diller's classic erosion surface. In *Geological Studies in the Klamath Mountains Province, California and Oregon: A Volume in Honor of William P. Irwin,* Geological Society of America Special Paper 410, eds. A. W. Snoke and C. G. Barnes, p. 451–63.

Bestland, E. A. 1987. Volcanic stratigraphy of the Oligocene Colestin Formation in the Siskiyou Pass area of southern Oregon. *Oregon Geology* 49 (7): 79–86.

Blake, M. C., Jr., Engebretson, D. C., Jayko, A. S., and D. L. Jones. 1985. Tectonostratigraphic terranes in southwest Oregon. In *Tectonostratigraphic Terranes of the Circum-Pacific Region,* Circum-Pacific Council for Energy and Mineral Resources Earth Science Series 1, ed. D. G. Howell, p. 159–71.

Bourgeois, J., and R. H. Dott, Jr. 1985. Stratigraphy and sedimentology of Upper Cretaceous rocks in coastal southwest Oregon: Evidence for wrench-fault tectonics in a postulated accretionary terrane. *Geological Society of America Bulletin* 96: 1007–19.

Dott, R. H., Jr. 1971. *Geology of the Southwestern Oregon Coast West of the 124th Meridian.* Oregon Department of Geology and Mineral Industries Bulletin 69. Maps 1:250,000 and 1:62,500.

Harper, G. D. 1984. The Josephine Ophiolite, northwestern California. *Geological Society of America Bulletin* 95: 1009–26.

Harper, G. D., and J. E. Wright. 1984. Middle to Late Jurassic tectonic evolution of the Klamath Mountains, California-Oregon. *Tectonics* 3 (7): 759–72.

Hunter, R. E., and H. E. Clifton. 1987. Shelf and deep-marine deposits of Late Cretaceous age, Cape Sebastian area, southwest Oregon. In *Geological Society of America Centennial Field Guide, Cordilleran Section,* Geological Society of America, ed. M. Hill, p. 295–300.

Irwin, W. P. 1994. *Geologic Map of the Klamath Mountains, California and Oregon.* US Geological Survey Miscellaneous Investigations Map I-2148, 1:500,000.

Irwin, W. P. 1997. *Preliminary Map of Selected Post-Nevadan Geologic Features of the Klamath Mountains and Adjacent Areas, California and Oregon.* US Geological Survey Open File Report 97-465. 1:500,000 map and 19-page report.

Irwin, W. P., and J. L. Wooden. 1999. *Plutons and Accretionary Episodes of the Klamath Mountains, California and Oregon.* US Geological Survey Open File Report 99-374.

KellerLynn, K. 2011. *Oregon Caves National Monument: Geologic Resources Inventory Report.* Natural Resource Report NPS/NRSS/GRD/NRR—2011/457. National Park Service.

MacDonald, J. H., Jr., Harper, G. D., and B. Zhu. 2006. Petrology, geochemistry, and provenance of the Galice Formation, Klamath Mountains, Oregon and California. In *Geological Studies in the Klamath Mountains Province, California and Oregon: A Volume in Honor of William P. Irwin,* Geological Society of America Special Paper 410, eds. A. W. Snoke and C. G. Barnes, p. 77–101.

Snoke, A. W., and C. G. Barnes. 2006. The development of tectonic concepts for the Klamath Mountains province, California and Oregon. In *Geological Studies in the Klamath Mountains Province, California and Oregon: A Volume in Honor of William P. Irwin,* Geological Society of America Special Paper 410, eds. A. W. Snoke and C. G. Barnes, p. 1–29.

Yule, D., Wiley, T., Kays, M. A., and R. Murray. 2009. Late Triassic to Late Jurassic petrotectonic history of the Oregon Klamath Mountains. In *Volcanoes to Vineyards: Geologic Field Trips through the Dynamic Landscape of the Pacific Northwest,* Geological Society of America Field Guide 15, eds. J. E. O'Connor, R. J. Dorsey, and I. P. Madin, p. 165–85.

Lava Plateaus

Beachly, M. W., Hooft, E. E. E., Toomey, D. R., and G. P. Waite. 2012. Upper crustal structure of Newberry Volcano from P-wave tomography and finite difference waveform modeling. *Journal of Geophysical Research* 117: B10311.

Bishop, E. M. 1990. Field trip guide to Cove Palisades State Park and the Deschutes Basin. *Oregon Geology* 52 (1): 13–16.

Bishop, E. M., and G. A. Smith. 1990. A field guide to the geology of Cove Palisades State Park and the Deschutes Basin in central Oregon. *Oregon Geology* 52 (1): 3–12.

Brand, B. D., and G. Heiken. 2009. Tuff cones, tuff rings, and maars of the Fort Rock–Christmas Valley basin, Oregon: Exploring the vast array of pyroclastic features that record violent hydrovolcanism at Fort Rock and the Table Rock Complex. In *Volcanoes to Vineyards: Geologic Field Trips through the Dynamic Landscape of the Pacific Northwest,* Geological Society of America Field Guide 15, eds. J. E. O'Connor, R. J. Dorsey, and I. P. Madin, p. 521–38.

Cummings, M. L., Evans, J. G., Ferns, M. L., and K. R. Lees. 2000. Stratigraphic and structural evolution of the middle Miocene synvolcanic Oregon-Idaho graben. *Geological Society of America Bulletin* 112: 668–82.

Donnelly-Nolan, J. M., Stovall, W. K., Ramsey, D. W., Ewert, J. W., and R. A. Jensen. 2011. *Newberry Volcano: Central Oregon's Sleeping Giant.* US Geological Survey Fact Sheet 2011-3145.

Ferns, M. L., Brooks, H. C., Evans, J. G., and M. L. Cummings. 1993. *Geologic Map of Vale 30 x 60 Minute Quadrangle, Malheur County, Oregon.* Oregon Department of Geology and Mineral Industries Map GMS-77. 1:100,000.

Ferns, M. L., Evans, J. G., and M. L. Cummings. 1993. *Geologic Map of Mahogany Mountain 30 x 60 Minute Quadrangle, Malheur County, Oregon*. Oregon Department of Geology and Mineral Industries Map GMS-78, 1:100,000.

Greene, R. C., Walker, G. W., and R. E. Corcoran. 1972. *Geologic Map of the Burns Quadrangle, Oregon*. US Geological Survey Map I-680, 1:250,000.

Jensen, R. A., Donnelly-Nolan, J. M., and D. Mckay. 2009. A field guide to Newberry Volcano, Oregon. In *Volcanoes to Vineyards: Geologic Field Trips through the Dynamic Landscape of the Pacific Northwest*, Geological Society of America Field Guide 15, eds. J. E. O'Connor, R. J. Dorsey, and I. P. Madin, p. 53–79.

Johnson, K. E., and E. V. Ciancanelli. 1984. Geothermal exploration at Glass Buttes, Oregon. *Oregon Geology* 46 (2): 15–20.

Jordan, B. T. 2005. Age-progressive volcanism of the Oregon High Lava Plains: Overview and evaluation of tectonic models. In *Plates, Plumes, and Planetary Processes*, Geological Society of America Special Paper 430, eds. G. R. Foulger and D. M. Jurdy, p. 503–15.

Jordan, B. T., Streck, M. J., and A. L. Grunder. 2002. Bimodal volcanism and tectonism of the High Lava Plains, Oregon. In *Field Guide to Geologic Processes in Cascadia, Oregon*. Department of Geology and Mineral Industries Special Paper 36, ed. G. W. Moore, p. 23–46.

MacLeod, N. S., Sherrod, D. R., Chitwood, L. A., and R. A. Jensen. 1995. Geologic map of Newberry volcano, Deschutes, Klamath, and Lake Counties, Oregon. US Geological Survey Miscellaneous Investigations Series Map I–2455, 1:24,000.

McClaughry, J. D., Ferns, M. L., Streck, M. J., Patridge, K. A., and C. L. Gordon. 2009. Paleogene calderas of central and eastern Oregon: Eruptive sources of widespread tuffs in the John Day and Clarno Formations. In *Volcanoes to Vineyards: Geologic Field Trips through the Dynamic Landscape of the Pacific Northwest*, Geological Society of America Field Guide 15, eds. J. E. O'Connor, R. J. Dorsey, and I. P. Madin, p. 407–34.

Nash, B. P., and M. E. Perkins. 2012. Neogene fallout tuffs from the Yellowstone Hotspot in the Columbia Plateau Region, Oregon, Washington, and Idaho, USA. *PLoS ONE* 7(10): e44205.

Peterson, N. V., and E. A. Groh. 1964. Diamond Craters, Oregon. *Ore Bin* 26: 17–34.

Smith, G. 1991. A field guide to depositional processes and facies geometry of Neogene continental volcaniclastic rocks, Deschutes Basin, central Oregon. *Oregon Geology* 53 (1): 3–20.

Smith, G. 1998. Geology along US Highways 197 and 97 between The Dalles and Sunriver, Oregon. *Oregon Geology* 60 (1): 3–17.

Streck, M., and Ferns, M. 2004. The Rattlesnake Tuff and other Miocene silicic volcanism in eastern Oregon. In *Geological Field Trips in Southern Idaho, Eastern Oregon, and Northern Nevada*, Rocky Mountain and Cordilleran sections of the Geological Society of America annual meeting guidebook, eds. K. M. Haller and S. H. Wood, p. 2–17.

Taggart, R. E., and A. T. Cross. 1980. Vegetation change in the Miocene Succor Creek Flora of Oregon and Idaho: A case study in paleosuccession. In *Biostratigraphy of Fossil Plants*, eds. D. L. Dilcher and T. N. Taylor, p. 185–210.

Taggart, R. E., and A. T. Cross. 1990. Plant successions and interruptions in Miocene volcanic deposits, Pacific Northwest. In *Volcanism and Fossil Biotas*, Geological Society of America Special Paper 244, eds. M. G. Lockley and A. Rice, p. 57–68.

Taylor, E. M., and G. A. Smith. 1987. Record of High Cascade volcanism at Cove Palisades, Oregon: Deschutes Formation volcanic and sedimentary rocks. In *Geological Society of America Centennial Field Guide, Cordilleran Section*, Geological Society of America, ed. M. Hill, p. 313–15.

Waters, A. C. 1968. *Reconnaissance Geologic Map of the Dufur Quadrangle, Hood River, Sherman, and Wasco Counties, Oregon.* US Geological Survey Geologic Investigations Series Map I-556, 1:25,000.

Wood, S. H., and D. M. Clemens. 2002. Geologic and tectonic history of the western Snake River Plain, Idaho and Oregon. In *Tectonic and Magmatic Evolution of the Snake River Plain Volcanic Province*, Idaho Geological Survey Bulletin 30, eds. B. Bonnichsen, C. M White, and M. McCurry, p. 69–103.

Blue Mountains

Blome, C. D. 1992. Field guide to the geology and paleontology of pre-Tertiary volcanic arc and mélange rocks, Grindstone, Izee, and Baker terranes, east-central Oregon. *Oregon Geology* 54 (6): 123–41.

Blome, C. D., and M. K. Nestell. 1991. Evolution of a Permo-Triassic sedimentary mélange, Grindstone terrane, east-central Oregon. *Geological Society of America Bulletin* 103: 1280–96.

Brooks, H. C., McIntyre, J. R., and G. W. Walker. 1976. *Geology of Oregon Part of Baker 1° x 2° Quadrangle.* Oregon Department of Geology and Mineral Industries Map, 1:250,000.

Dillhoff, R. M., Dillhoff, T. A., Dunn, R. E., Myers, J. A., and C. A. E. Stromberg. 2009. Cenozoic paleobotany of the John Day Basin, central Oregon. In *Volcanoes to Vineyards: Geologic Field Trips through the Dynamic Landscape of the Pacific Northwest*, Geological Society of America Field Guide 15, eds. J. E. O'Connor, R. J. Dorsey, and I. P. Madin, p. 135–64.

Dorsey, R. J., and T. A. LaMaskin. 2007. Stratigraphic record of Triassic-Jurassic collisional tectonics in the Blue Mountains Province, northeastern Oregon. *American Journal of Science* 307: 1167–93.

Dorsey, R. J., and T. A. LaMaskin. 2008. Mesozoic collision and accretion of oceanic terranes in the Blue Mountains Province of northeastern Oregon: New insights from the stratigraphic record. *Arizona Geological Society Digest* 22: 325–32.

Fremd, T. 2010. *Guidebook: John Day Basin Field Conference.* Society of Vertebrate Paleontologists Field Symposium,

Fremd, T., Bestland, E. A., and G. J. Retallack. 1994. *John Day Basin Paleontology Field Trip Guide and Road Log.* Society of Vertebrate Paleontology 1994 Annual Meeting Guidebook.

Hales, T. C., Abt, D. L., Humphreys, E. D., and J. J. Roering. 2005. A lithospheric instability origin for Columbia River flood basalts and Wallowa Mountains uplift in northeast Oregon. *Nature* 438: 842–45.

Hooper, P. R., Houseman, M. D., Beane, J. E., Caffrey, G. M., Engh, K. R., Scrivner, J. V., and A. J. Watkinson. 1995. Geology of the northern part of the Ironside Mountain inlier, northeastern Oregon. In *Geology of the Blue Mountains Region of Oregon, Idaho, and Washington: Petrology and Tectonic Evolution of Pre-Tertiary Rocks of the Blue Mountains Region*, US Geological Survey Professional Paper 1432, eds. T. L. Vallier and H. C. Brooks, p. 415–57.

Humphreys, E., and B. Schmandt. 2011. Looking for mantle plumes. *Physics Today* August: 34–39.

Kays, M. A., Stimac, J. P., and P. M. Goebel. 2006. Permian-Jurassic growth and amalgamation of the Wallowa composite terrane, northeastern Oregon. In *Geological Studies in the Klamath Mountains Province, California and Oregon: A Volume in Honor of William P. Irwin*, Geological Society of America Special Paper 410, eds. A. W. Snoke and C. G. Barnes, p. 465–94.

Kleinhans, L. C., Balcells-Baldwin, E. A, and R. E. Jones. 1984. A paleogeographic reinterpretation of some middle Cretaceous units, north-central Oregon: Evidence for a submarine turbidite system. In *Geology of the Upper Cretaceous Hornbrook Formation, Oregon and California*, Pacific Section Society of Economic Paleontologists and Mineralogists 42, ed. T. H. Nilsen, p. 239–57.

LaMaskin, T. A., Stanley, G. D., Jr., Caruthers, A., and M. Rosenblatt. 2011. Detrital record of Upper Triassic reef facies in the Blue Mountains Province, northeastern Oregon: Implications for early Mesozoic paleogeography of western North American Cordilleran terranes. *PALAIOS* 26: 779–89.

Mullen, E. D. 1983. Paleozoic and Triassic terranes of the Blue Mountains, northeast Oregon: Discussion and field trip guide. *Oregon Geology* 45 (7/8): 75–82.

Noblett, J. B. 1981. Subduction-related origin of the volcanic rocks of the Eocene Clarno Formation near Cherry Creek, Oregon. *Oregon Geology* 43 (7): 91–99.

Oles, K. F., and H. E. Enlows. 1971. *Bedrock Geology of the Mitchell Quadrangle, Wheeler County, Oregon*. Oregon Department of Geology and Mineral Industries Bulletin 72, 1:48,000.

Retallack, G. J. 1991. A field guide to mid-Tertiary paleosols and paleoclimatic changes in the high desert of central Oregon—Part 1. *Oregon Geology* 53 (3): 51–74.

Retallack, G. J. 1991. A field guide to mid-Tertiary paleosols and paleoclimatic changes in the high desert of central Oregon—Part 2. *Oregon Geology* 53 (4): 75–80.

Retallack, G. J. 2008. Cenozoic cooling and grassland expansion in Oregon and Washington. *PaleoBios* 28 (3): 89–113.

Robinson, P. T. 1987. John Day Fossil Beds National Monument, Oregon: Painted Hills Unit. In *Geological Society of America Centennial Field Guide, Cordilleran Section*, ed. M. Hill, p. 317–20.

Robinson, P. T., Brem, G. F., and E. H. McKee. 1984. The John Day Formation: A distal record of early Cascade volcanism. *Geology* 12: 229–32.

Schwartz, J. J., Snoke, A. W., Cordey, F., Johnson, K., Frost, C. D., Barnes, C. G., LaMaskin, T. A., and J. L. Wooden. 2011. Late Jurassic magmatism, metamorphism, and deformation in the Blue Mountains Province, northeast Oregon. *Geological Society of America Bulletin* 123 (9/10): 2083–111.

Taubeneck, W. H. 1987. The Wallowa Mountains, northeast Oregon. In *Geological Society of America Centennial Field Guide, Cordilleran Section*, ed. M. Hill, p. 327–32.

Vallier, T. L. 1995. Petrology of pre-Tertiary igneous rocks in the Blue Mountains region of Oregon, Idaho, and Washington: Implications for the geologic evolution of a complex island arc. In *Geology of the Blue Mountains Region of Oregon, Idaho, and Washington: Petrology, and Tectonic Evolution of Pre-Tertiary Rocks of the Blue Mountains Region*. US Geological Survey Professional Paper 1438, eds. T. L. Vallier and H. C. Brooks, p. 125–209.

Wheeler, G. 1982. Problems in the regional stratigraphy of the Strawberry Volcanics. *Oregon Geology* 44 (1): 3–7.

Wood, S. H., and Clemens, D. M. 2002. Geologic and tectonic history of the western Snake River Plain, Idaho and Oregon. In *Tectonic and Magmatic Evolution of the*

Snake River Plain Volcanic Province. Idaho Geological Survey Bulletin 30, eds. B. Bonnichsen, C. M. White, and M. McCurry, p. 69–103.

Basin and Range

Badger, T. C., and R. J. Watters. 2009. Landslides along the Winter Rim Fault, Summer Lake, Oregon. In *Volcanoes to Vineyards: Geologic Field Trips through the Dynamic Landscape of the Pacific Northwest*, Geological Society of America Field Guide 15, eds. J. E. O'Connor, R. J. Dorsey, and I. P. Madin, p. 203–20.

Evans, J. G., T. Geisler. 2010. *Geologic Field Trip Guide to Steens Mountain Loop Road, Harney County, Oregon*. US Geological Survey Bulletin 2183, 1:100,000.

Licciardi, J. M. 2001. Chronology of latest Pleistocene lake-level fluctuations in the pluvial Lake Chewaucan basin, Oregon, USA. *Journal of Quaternary Science* 16 (6): 545–53.

Meigs, A., Scarberry, K., Grunder, A., Carlson, R., Ford, M. T., Fouch, M., Grove, T., Hart, W. K., Iademarco, M., Jordan, B., Milliard, J., Streck, M. J., Trench, D., and R. Weldon. 2009. Geological and geophysical perspectives on the magmatic and tectonic development, High Lava Plains and northwest Basin and Range. In *Volcanoes to Vineyards: Geologic Field Trips through the Dynamic Landscape of the Pacific Northwest*, Geological Society of America Field Guide 15, eds. J. E. O'Connor, R. J. Dorsey, and I. P. Madin, p. 435–70.

Minor, S. A., Plouff, D., Esparza, L. E., and T. J. Peters. 1987. *Geologic Map of the High Steens and Little Blitzen Gorge Wilderness Study Areas, Harney County, Oregon*. US Geological Survey Miscellaneous Field Studies Map MF-1976, 1:24,000.

Peterson, N. V., and J. R. McIntyre. 1970. *The Reconnaissance Geology and Mineral Resources of Eastern Klamath County and Western Lake County, Oregon*. Oregon Department of Geology and Mineral Industries Bulletin 66, 1:250,000.

Pezzopane, S. K., and R. J. Weldon. 1993. Tectonic role of active faulting in central Oregon. *Tectonics* 12: 1140–69.

Scarberry, K. C., Meigs, A. J., and A. L. Grunder. 2010. Faulting in a propagating continental rift: Insight from the Late Miocene structural development of the Abert Rim Fault, southern Oregon, USA. *Tectonophysics* 488: 71–86.

Snyder, C. T., Hardman, G., and F. F. Zdenek. 1964. *Pleistocene Lakes in the Great Basin*. US Geological Survey Miscellaneous Geologic Investigations Map I-416.

Wiley, T. J., Sherrod, D. R., Keefer, D. K., Qamar, A., Schuster, R. L., Dewey, J. W., Mabey, M. A., Black, G. L., and R. E. Wells. 1993. Klamath Falls earthquakes, September 20, 1993—including the strongest quake ever measured in Oregon. *Oregon Geology* 55: 127–34.

Index

Page numbers in bold indicate a photo.

–KRISTINA WALOWSKI

Marli B. Miller is a senior instructor and researcher in geology at the University of Oregon. She completed a BA in geology at Colorado College in 1982 and an MS and PhD in structural geology at the University of Washington in 1987 and 1992, respectively. She teaches a variety of courses, including geology of national parks, structural geology, field geology, and geophotography. In addition to numerous technical papers, she is the author of *Geology of Death Valley National Park*, with coauthor Lauren A. Wright, and the photographer for *What's So Great About Granite?*, written by Jennifer Carey, the editor of this Roadside Geology book! Marli has two daughters, Lindsay and Megan.